Ice Mechanics

Risks to Offshore Structures

Cover photograph The cover shows Dome Petroleum's Single Steel Drilling Caisson (SSDC) surrounded by broken sea-ice at the Uviluk P-66 site, Canadian Beaufort Sea, in April 1983 (photograph by the author, courtesy of BP Petroleum Development Ltd).

Ice Mechanics
Risks to Offshore Structures

T. J. O. Sanderson

BP Petroleum Development Ltd, London

Graham & Trotman
A member of the Kluwer Academic Publishers Group
LONDON/DORDRECHT/BOSTON

First published in 1988 by

Graham and Trotman Limited
Sterling House
66 Wilton Road
London
SW1V 1DE
UK

Graham and Trotman Inc.
Kluwer Academic Publishers Group
101 Philip Drive
Assinippi Park
Norwell, MA 02061
USA

British Library Cataloguing in Publication Data

Sanderson, T. J. O.
Ice Mechanics.
1. Cold regions. Offshore structures.
Structural engineering
I. Title
627′.98′0911

ISBN 0 86010 785 X

LCCCN 88–3195

Typeset and printed in Great Britain by
Henry Ling, The Dorset Press

Contents

Foreword ix

Acknowledgements xiii

Units, Symbols and Conversion Factors xiv

Chapter 1 Introduction **1**

Background 1
Ice mechanics 2
Structure of this book 2
Topics which are not treated 4

Chapter 2 Types of Ice **5**

Introduction 5
The crystal structure of ice 5
Sea ice 7
  Development of an ice cover 7
  First-year ice 9
  Second-year ice 16
  Multi-year ice 18
Polynyas 21
Icebergs 21
  Sources 21
  Types of iceberg 24
  Ice islands 27

Chapter 3 Worldwide Occurrence of Ice **30**

Introduction 30
Arctic ice conditions 32
  The Arctic regions: general 32
  Central Arctic Ocean 35
  Beaufort Sea 39
  Chukchi Sea 47
  Bering Sea 50
  Canadian Arctic Islands 50
  Labrador, Newfoundland and West Greenland 54
  Greenland Sea and Svalbard 55
  Soviet Union 61

Antarctica 65
General 65
First-year ice 66
Multi-year ice 66
Icebergs 66
Long-term perspective 67

Chapter 4 Mechanical Properties of Ice: Laboratory Studies 70

Introduction: the concepts of strength and failure 70
Continuum behaviour 72
Pure ice: general 72
Microscopic deformation mechanisms 76
Pure ice: secondary creep 77
Sea ice: secondary creep 80
General constitutive equations for secondary creep 85
Fracture behaviour 88
Nucleation versus propagation 88
Tensile fracture 88
Compressive fracture 92
Multiaxial behaviour 96
Indentation tests 97
Test geometry 97
Laboratory indentation tests 98
Medium-scale *in situ* experiments 101

Chapter 5 Full-scale Measurements 104

Introduction 104
Measurement techniques 105
Interfacial method 105
Structural response 106
In situ ice measurements 107
Floe deceleration 110
Measurements on narrow structures 110
Cook Inlet, Alaska 110
Lighthouses 112
Bridge piers 113
Measurements on wide structures 113
Artificial and natural islands 113
Introduction 113
Adgo P-25 117
Netserk South B-44 117
Netserk North F-40 117
Kannerk G-42 and Arnak L-30 119
Issungnak O-61 120
Tarsiut N-44 120
Adams Island, NWT 121
Kadluk caisson-retained island 121
Amerk O-90 122
Hans Island 123
Molikpaq 130

Icebreakers and floating structures 137
General 137
Global loads 137
Local loads 138
A pressure-area curve for ice 139
Full-thickness indentation 139
Aspect ratio or area? 143
Triaxial indentation: icebreakers and impact hammers 144

Chapter 6 Theoretical Analysis of Ice Failure 145

Introduction 145
Continuum behaviour 145
Elastic buckling 145
Creep buckling 148
Indentation analysis 151
Transition to fracture 161
Imperfect contact: local fracture 161
Loading-rate conditions 162
Fracture analysis 164
Crushing: the effects of flaw-size 164
Radial cracking 167
Spalling 169
Flexural failure 170
Non-simultaneous failure 176
Korzhavin's equation 186
Creep indentation with perfect contact 187
Creep indentation with partial contact 187
Fast fracture 187

Chapter 7 Calculation of Design Ice Loads 190

Introduction 190
Design load philosophy: risk analysis 190
Types of ice loading 194
Types of structure 196
Surface slope 196
Structural response 196
Rubble pile formation 196
Multi-legged structures 198
Driving forces 198
Wind and current forces 199
Thermal expansion 201
Pack-ice action 204
Static ice loading 205
General 205
Design load recommendation 210
Dynamic impact: multi-year floes 215
Introduction 215
Penetration distance 216
Impact probability 218
Design load recommendation 222

Dynamic impact: icebergs 226
Winter blocking 232
Local loads 232
  Consequences 233
  Local "events" 233
Epilogue 234

Bibliography 236

Index 247

Foreword

Most engineering works are carried out in the temperate regions of our planet. Therefore to the average practising engineer, the topic addressed by this book may seem highly specialized and obscure. Certainly, until recently, even in Northern countries, where the subject has some relevance, only a few universities offered either ice mechanics or ice engineering as part of their curriculum. Yet as modern man's engineering activities have progressed into higher latitudes there has been a growing need to develop an understanding of how to deal with ice: ice in the ground; ice on lakes and rivers; and most challenging of all, ice in the marine environment which covers such a large part of the Arctic region.

Methods of operating in ice-covered regions have always challenged man's ingenuity. Long before Europeans ventured into the Arctic, the native peoples of the North developed a lifestyle compatible with living on the fringes of the frozen sea. They adapted their methods of hunting and transportation to make use of the ice. In fact, their collective intuitive knowledge of ice, gathered over many centuries, is as impressive in its own way as that gained recently by modern scientists.

For several centuries the issue of operating ships in ice has been one of concern for Northern nations. Wooden ships propelled by sail were not able to navigate through ice to any significant extent. But starting from the advent of steam and steel, ship technology has advanced to the point where ice-breaking vessels can now ply the northern seas year-round.

Only recently has there been a need to build offshore structures in the Arctic regions. This need has been generated by the possibility that the vast continental shelves of the ice-covered oceans might contain large deposits of oil and gas. Furthermore, northern nations have recognized that development of offshore oil and gas in the Arctic might be their only hope of replacing fast depleting onshore reserves, and avoidance of too heavy a reliance on imported oil. Similarly, as this book demonstrates, the multi-national oil companies have shown a willingness to develop the technology for oil and gas exploration and production from these harsh regions. The key issues are (and always will be) whether these resources can be developed safely, economically and with negligible risk to the environment. Several factors influence these issues, including the price of oil, the size of the deposits and last but not least, the technology. Paramount among the technological issues is the problem of designing offshore platforms to withstand ice, which is, of course, the topic of this book.

When engineers in the late 1960s began to look at the problem of drilling offshore in the Arctic and building platforms for production, there were few precedents. Granted, there were structures which had been built to resist ice forces, but these existed only in the subarctic regions, and it was not clear how to transfer this limited experience to the Arctic where the ice was thicker and stronger.

So began a sustained and systematic set of research activities which commenced in about the late 1960s and gathered momentum as the imperative to move offshore in the Arctic became apparent. Scientists and engineers of various disciplines dedicated themselves to the task, and demonstrated an exemplary spirit of cooperation and a high level of innovation. They have been supported and sponsored by diverse organizations including the large oil companies, government organizations, research institutes and the universities. The presence of ice recognizes no political boundaries and perhaps this has driven the spirit of international cooperation which has been exhibited. The reader of this book will recognize that it draws on the research and experience of the worldwide technical community, and this is one of the book's great strengths.

The issue of predicting ice forces on structures essentially reduces to what at first sight appears to be a relatively simple problem, that of understanding how ice deforms and fails under stress. This apparent simplicity turns out to be a paradox, first because of the complex nature of ice, and second because of the many ways it can fail against structures of various shapes. As Dr Sanderson so ably describes, the deformation and strength properties of ice are affected by two major characteristics. Ice is a solid close to its melting point, and therefore it exhibits creep and its strength is temperature dependent. Also ice is a material much more brittle than glass; therefore cracks propagate through it with ease. During the past two decades our understanding of these aspects of ice mechanics has advanced considerably. The response of laboratory samples of ice to stress is now well quantified. Where we have always had difficulty is in moving from the knowledge of small-scale ice properties to prediction of how ice behaves on a large scale. As this book clearly emphasizes, it is fair to say that even now, despite extensive research, we do not yet have generally accepted and reliable methods for predicting ice loads on structures.

Yet there has been remarkable progress in designing and operating structures in the offshore Arctic, mostly for exploration drilling. The first offshore structure was built in 1972 (a dredged island in 3 m of water in the Canadian Beaufort Sea). By 1987 the deeper-water locations were being drilled using bottom-founded caisson structures in water depths to 32 m (again in the Canadian Beaufort Sea). Between these two milestones approximately 35 structures have been successfully placed and drilled from in the Canadian and US Arctic offshore regions.

It has perhaps been fortunate that the use of offshore structures in the Arctic has progressed from shallow to deep locations. In the shallow locations the ice is landfast for much of the winter and extreme thicknesses do not occur. Consequently the earlier locations, which could be designed conservatively, provided an outdoor laboratory where ice interaction could be studied and ice loads measured. This knowledge and experience was then used to test theories and to enable extrapolation of design criteria to deeper locations with less conservatism but without excessive risk. As this book well describes, approaches to predicting ice loads have used various tools including theories of material behaviour, model testing and empirical methods based on real data. Looking back on all the work done and all the experience gained, there is no doubt that without the insights gained from real-world observations and measurements, little progress would have been made. I am very pleased therefore, that this book devotes a large portion of its content to full-scale measurements and their interpretation.

There are other texts available on ice mechanics, some of which cover the topic of ice forces. However, this book provides the technical community with something it has been missing to date; a reference text which is focused entirely on the issue of ice loads on marine structures. Furthermore, in contrast to previous publications, this book places heavy emphasis on full-scale measurements. Also it provides adequate caution on the use of theoretical models without qualification. Yet the comprehensive review of theoretical approaches provides direction on those offering the most promise for further development and research. In addition, the review of ice types and the worldwide occurrence of ice provides the reader with a single reference text which can be used to conduct a preliminary assessment of ice action on structures for almost any part of the globe.

I am particularly pleased to write this Foreword because the book covers one of my favourite topics, and one in which I have been involved for almost twenty years, as both a researcher and engineer. I believe the book ably captures the exciting progress in the field of ice mechanics and the technology of Arctic offshore structures which I have witnessed during this period. Clearly, advances in ice mechanics have been made by individuals and organizations from all over the world. However, because of our geographical location, we in Canada have had to address this topic with perhaps more urgency than others. Certainly to date, more offshore wells have been drilled in Canada's ice-covered water than anywhere else, and despite the recent slide in oil prices there are at least two major offshore oil discoveries which are considered commercial. It is likely that these will be developed during the next few years using fixed offshore platforms designed to resist ice.

In the future I am confident that more offshore hydrocarbons will be discovered in ice-covered regions, not only in Canada but in other parts of the world which are now receiving the attention of the explorationists. The engineering challenge will be to design and build safe, and cost-effective structures for a wide range of water depths, sea floor properties and ice conditions. Those involved in this exciting work will surely find this book an invaluable reference text, and I highly recommend it.

K. R. Croasdale,
Calgary, Alberta
February, 1988

Acknowledgements

I am grateful to The British Petroleum Company plc for permission to publish this book, and for the use of library and information services.

I am also greatly indebted to the many individuals within the BP Group – BP Petroleum Development Ltd, BP Research Centre Sunbury and Standard Oil – who have over the years provided constant stimulus and lively discussion as new ideas were developed and tested. I have also greatly enjoyed their company and unfailing good humour on many Arctic field programmes. In particular, I thank Dr D. J. Goodman, Mr M. W. Howard, Dr S. D. Hallam, Dr R. Duckworth, Mr A. J. Child, Dr P. H. Westermann, Dr N. Riley, Dr N. K. Gibbs, Mr J. T. Baldwin and Mr J. T. Walden.

I thank The British Petroleum Company plc for permission to reproduce the photographs used in Figures 2.5, 2.8, 2.15, 2.17, 2.19, 2.23(a) and (b), 2.27, 5.9, 6.4 and 6.20, and for permission to use information on sea-ice distribution prepared for the Company by the Climatic Research Unit of the University of East Anglia. I thank Dome Petroleum for permission to reproduce the photographs used in Figures 5.16, 5.20 and 6.30; and I thank Gulf Canada Resources Inc. for permission to reproduce the photograph used in Figure 5.31(a).

I am most grateful to Mr J. T. Baldwin, Mr K. R. Croasdale, Dr R. Duckworth and Dr S. D. Hallam for reading and reviewing the typescript and making many helpful suggestions.

The views expressed in this book remain, however, the views of the author, and do not necessarily represent those of The British Petroleum Company plc.

Units, symbols and conversion factors

The units and symbols used in this book are metric throughout, and generally conform also to the SI (Système International) and British Standard 3763:1964 "The International System (SI) Units".

SYMBOLS

m	metre
cm	centimetre
km	kilometre
kg	kilogramme
s	second
N	newton
MN	meganewton
Pa	pascal
kPa	kilopascal
MPa	megapascal
°C	degrees celsius (centigrade)
K	degrees kelvin

PREFIXES

m	milli-	= 1/1000 = 10^{-3}
c	centi-	= 1/100 = 10^{-2}
k	kilo-	= 1000 = 10^{3}
M	mega-	= 1 000 000 = 10^{6}
G	giga-	= 1 000 000 000 = 10^{9}

CONVERSION FACTORS

The following conversion factors may be useful for translating into traditional engineering units. This, however, is to be discouraged.

Length

1 cm = 0.394 inches
1 m = 3.281 feet
1 km = 0.621 miles (statute)
1 km = 0.540 nautical miles

1 inch = 2.54 cm
1 foot = 0.3048 m
1 mile = 1.609 km
1 nautical mile = 1.852 km

Mass

1 kg = 2.2046 pounds mass (lb)
1 tonne = 1000 kg = 0.984 tons

1 lb = 0.4536 kg
1 ton = 1.016 tonnes

Force

1 MN = 102.0 tonnes force
1 MN = 100.4 tons force
1 MN = 225 kips

1 tonne force = 9.8 kN
1 ton force = 9.96 kN
1 kip = 4.44 kN

Pressure

$1\,MPa = 1\,MN\,m^{-2} = 145\,psi$ $1\,psi = 6.90\,kPa$
$1\,MPa = 10\,bar$ $1\,bar = 100\,kPa$

Velocity

$1\,m\,s^{-1} = 1.944\,knots$ $1\,knot = 0.5144\,m\,s^{-1}$

Density

$1\,Mg\,m^{-3} = 1\,g\,cm^{-3}$

Chapter 1

Introduction

1.1 BACKGROUND

Ice engineering is a young subject. Mariners have for centuries been fascinated by the hazards and obstacles presented to polar navigation, and engineers have for a long time been challenged by the problems of constructing bridges across ice-covered waters; but it is only since about 1960 that a concerted effort has been made to achieve a scientific understanding of ice interaction with man-made structures.

The effort gained great impetus – and the funding to sustain the impetus – during the 1970s and early 1980s, when oil companies initiated a major operation to explore the offshore Arctic frontier areas for oil and gas. Exploration has centred around the Arctic coasts of North America: the Beaufort Sea, the Bering Sea, the Queen Elizabeth Islands, the Labrador coast and the Grand Banks of Newfoundland. The exploration effort was fanfared as the grand solution to the Western world's oil supply problems: by tapping the vast expected resources of the North, we could survive the threat of OPEC monopoly and secure our own oil supply to take us into the 21st century. The rush to explore for offshore Arctic oil was, of course, encouraged greatly by the dramatic leaps in oil price during 1973 and 1979, and colossal sums of money were spent in acquiring acreage for exploration and devising means to explore for it in the hostile frozen oceans. Most importantly, from our point of view here, a large amount of money and expertise was channelled into Arctic ice research in order to achieve an understanding of ice mechanics and to be able to quantify the ice hazards and risks associated with offshore Arctic exploration for hydrocarbons.

In the event, exploration success has been modest. Significant reserves of oil have been discovered in the Beaufort Sea, but not the giant fields which had been hoped for; the Queen Elizabeth Islands were found to be predominantly gas-prone, the gas proving problematical to export and market; and offshore Newfoundland turned out not to be as exciting as originally thought. Added to this disappointing success rate, the collapse of oil prices over the 1985–6 period has left the Arctic frontiers looking distinctly less attractive to the oil explorationist than they did in the mid-1970s.

The research, however, has been done, and much of it which was carried out under closely guarded confidentiality in the 1970s and early 1980s has now emerged into the public domain. This book is an attempt to summarise the most important aspects of this research and ensure that it runs no risk of being lost as a result of capricious changes in the energy sector outlook. The research is fascinating both to the pure scientist and to the practising engineer, and there is no doubt in my own mind that it will continue to find important applications in spite of an apparent downturn in interest at the time of writing. The information presented here is my own view of the state-of-the-art in late 1987. It includes data which were in, or due to come into, the public domain up to late 1987.

The central question addressed by this book is the ice design load question. That is: what ice load must a structure be designed to withstand if it is to be placed in Arctic waters? The answer is complex, and involves an integrated examination of several subjects, in particular: ice conditions, ice mechanics and risk analysis. The statistics of ice conditions are relatively straightforward; there are still gaps in the data-set, but these are fast being filled. Risk analysis, itself quite a young science, is also quite straightforward and has become especially easy to apply since Monte Carlo computing facilities became readily accessible. Ice mechanics, however, remains a difficult and incomplete science.

1.2 ICE MECHANICS

Our understanding of ice design loads is limited by our understanding of ice mechanics. Ice is an extraordinary material, and its behaviour under load has led to great difficulties in quantifying the maximum loads it might exert on a structure. It is a creeping, ductile material at low stresses, yet an extremely brittle material at high stresses, and its behaviour is dependent on rate of deformation and deformation history. These mechanical properties have been extensively and thoroughly studied in the laboratory, leading to an apparently secure and fairly complete understanding of the mechanical properties of ice.

This security is, however, beguiling. The "strength" of ice appears, as we shall see, to be strongly scale-dependent: average stresses at failure measured under small-scale laboratory conditions tend to lie in the region of 10 MPa, while measured average stresses during full-scale interaction with a structure generally lie in the region of 1 MPa. This behaviour is illustrated by the data plotted in Fig. 1.1, which shows peak measured indentation pressure (MPa) against gross contact area (m^2), as ice is indented by a wide variety of different-sized indentors. None of the data are in any way corrected or normalised. This plot, which is generally referred to as a "pressure–area" curve, covers a vast range of scales, from the smallest laboratory test on a scale of several millimetres to descriptions of behaviour at the scale of many kilometres. It illustrates the problems involved in applying data acquired in the laboratory to interaction conditions which are different in scale by several orders of magnitude. The pressure–area curve concept is central to the treatment given in this book.

There are many potential explanations for the striking scale-dependence shown by the pressure–area curve, and I shall hope to shed some light on the subject. It is my firm opinion that the answer lies with the phenomena of imperfect contact and non-simultaneous failure, and I try in this book to clarify our understanding of these processes. There may be wide disagreement over the appropriate explanation, and I hope that argument and debate will continue to flourish until the truth emerges. There is, however, little doubt about the data. The data speak for themselves: ice at large scale appears to fail at a lower average stress than ice at small scale. Until new data appear to contradict this statement, it is a statement about which there can be no disagreement.

Add to this interesting state of affairs another curious feature of ice engineering. In analysing the forces exerted by ice acting on a structure we are at every turn contradicting the practising engineer's natural instincts. When we build a bridge we are interested in how strong a material is, and how weak it might be: that is, what is the maximum load that the material (stone, concrete, steel) can be relied upon to sustain without risk of failure? In analysing ice hazards the reverse applies. We must look at the ice and examine how weak it is, and how strong it might be. That is, what is the minimum load that it can be relied upon not to sustain? Can we rely on it failing? In looking at the mechanical properties of ice we are continually looking for ways in which it can be encouraged to fail, and we are only satisfied when we feel sure that it will fail. This switch of viewpoint of looking at ice as a material colours our whole approach to the subject.

1.3 STRUCTURE OF THIS BOOK

This book may either be read straight through, as an introduction to the ice design load question, or it may be used as a reference.

It begins, in Chapter 2, with an account of what ice is: its genesis, structure and the forms it takes in nature. Chapter 3 then gives a summary

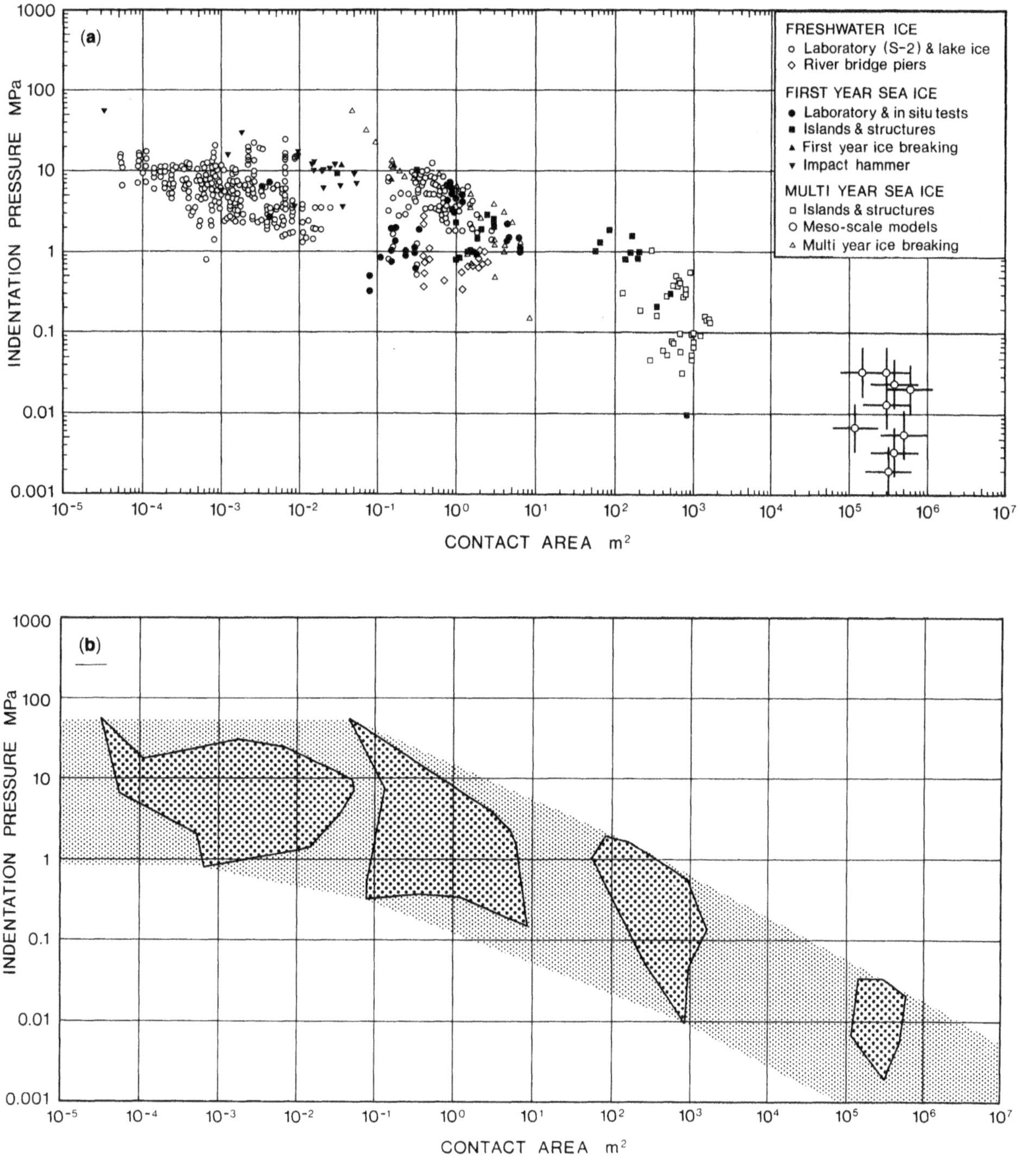

Fig. 1.1 A pressure–area curve for ice: the average stress required to fail a large area of ice is much lower than that required to fail a small area of ice. These data are all for indentation of ice, including laboratory experiments, ice-breakers and full-scale interactions with islands, and are discussed in detail in Chapters 4 and 5. No corrections have been applied. (a) The full data set; (b) highlighted trend, showing the apparent decrease of failure pressure with increasing area.

account of the geographical distribution of ice worldwide, including maps and statistics on ice distribution, thicknesses, floe sizes and drift rates. In Chapters 4 and 5 we then look at the mechanical properties of ice: at small scale in the laboratory and at large scale in the field. I have tried, in Chapter 5, to summarise all the important data available from large-scale Arctic field measurement programmes, and I have also included an introduction to the subject of field measurement techniques. In many ways, Chapter 5 proves to be the most important chapter to the book, as it is the principal source of fundamental data for calculating design loads.

Chapter 6 examines the theoretical "tools-of-the-trade" available for analysis of ice interaction with structures. These tools do not supply all the answers. It is a central thesis of this book that the theory of ice mechanics is not yet sufficiently far advanced to provide a thoroughly satisfactory theoretical analysis of ice interaction with structures. Everyone working in the field realises this. It is not a science like electromagnetism or Newtonian mechanics, which are governed by robust immutable laws. However, there are now enough data available (and described in Chapter 5), to enable quite precise statistical analysis of ice interaction processes and allow quantitative empirically based guidance on the ice design load question.

Because there is a gulf between the data available and the theoretical ability we have to explain it, Chapter 6 contains some untied ends. It contains theoretical analyses and developments which are not used in Chapter 7, where I aim to draw together all the available threads and provide a rationale for design load specification. This is perhaps unfortunate, but an account of the theoretical analysis available is necessary for two reasons: for completeness, in order to provide a reference for future developments; but, just as importantly, to provide a theoretical understanding of the basic principles which underpin the largely empirical analysis which is ultimately proposed.

The quantitative answers are given in Chapter 7. The answers are given on the basis of environmental data, driving force assessment, risk analysis and an empirical rationalisation of large-scale ice mechanics. I am in no way ashamed of the highly empirical basis of the methods used: after all, the data come first, and a theoretical analysis has to conform with them – and not *vice versa*.

As to the use of the book as a reference: I would be the last person to encourage the engineer-in-a-hurry, but the impatient reader may very well begin the book at Chapter 7 and refer back to other chapters as necessary. Chapter 7 contains some fairly bold assumptions and some fairly bald statements, but they are in general supported by the chapters which have gone before.

1.4 TOPICS WHICH ARE NOT TREATED

Finally, there are a number of areas and topics which are not treated in this book. This is principally for either or both of two reasons: that they are, in my judgement, not relevant to the theme of the book; or, that I simply do not know enough about them to contribute usefully. Among the topics I have chosen not to treat in any detail are:

- scale model ice tank experiments
- design and performance of floating structures and ice breakers
- vertical (uplift) forces on structures
- ice pile-up and ride-up problems
- seabed scouring processes

If there are others, they are ones I have simply forgotten.

Chapter 2

Types of Ice

2.1 INTRODUCTION

Floating ice in polar regions comes in many different forms and shows distinct regional variations in character, but it is essentially of two different origins:

(a) *frozen sea water*, which gives rise to diverse forms of sea ice;
(b) *land-based glacier ice*, which gives rise to icebergs.

This chapter gives an account of the growth and formation of these various forms of marine ice cover, and in Chapter 3 we shall look at their regional distribution worldwide.

First, however, it is worth looking briefly at the microscopic structure and crystallography of ice.

2.2 THE CRYSTAL STRUCTURE OF ICE

There is only one microscopic type of ice which need concern us here: ice Ih (h for hexagonal), which is the stable form at normal temperatures and pressures. At least eight other atomic configurations of ice exist at high pressures (greater than 200 MPa) or low temperatures (below about −110°C) but they are of concern only under laboratory conditions. Even at the most extreme depths beneath the Antarctic ice cap, which reaches a thickness of 4000 m in places, the hydrostatic pressure (35 MPa) is insufficient to allow other forms of ice to exist stably.

Ice forms, of course, from liquid water, and it preserves some of the geometrical features of the water molecule. Liquid water, H_2O, consists of an oxygen atom bonded by electrons to two hydrogen atoms, forming an angle of 104° 31′ (Fig. 2.1). The non-linearity of the molecule is due to the presence of two other lone-pair electron orbitals, which form an approximately tetrahedral system with the two bonding orbitals.

The atomic structure of the ice Ih crystal builds on this tetrahedral geometry of the water molecule: the solid crystal is a repeating tetrahedral coordination of oxygen atoms, in which each oxygen atom is bonded by hydrogen to four other oxygen atoms (Fig. 2.2). Each oxygen atom has two hydrogen atoms closely associated with it, at a distance of about 0.95 Å ($1\ \text{Å} = 10^{-10}$ m); and two less closely associated with it, at a distance of about 1.76 Å. The position of hydrogen atoms on the bonds as shown in Fig. 2.2 is only one possible permutation. There are many other ways of allocating the hydrogen atoms in such a way that each oxygen atom has two hydrogen atoms closely associated with it. The angle between bonds is about 109° 30′, and the individual units of oxygen

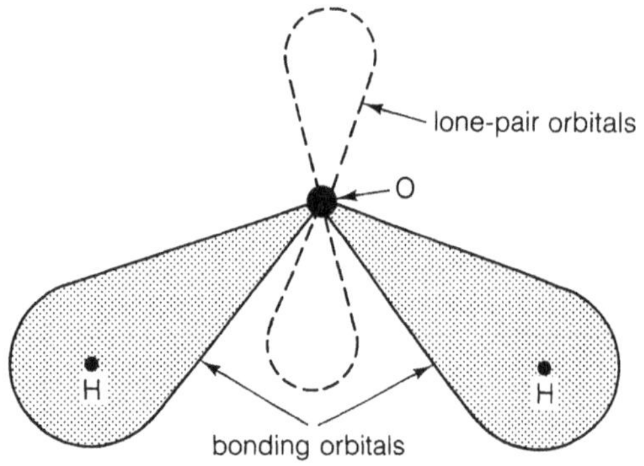

Fig. 2.1 The molecular structure of a liquid water molecule. The central oxygen atom is bonded by electron orbitals to two hydrogen atoms. The electrons are shown as shaded clouds. In addition to these two bonding orbitals, two lone-pair electron orbitals exist; the four electron orbitals repel each other and form an approximately tetrahedral system (after Hobbs, 1974).

and hydrogen in ice in fact differ only slightly from the molecular structure of liquid water.

The resulting crystal geometry is shown from two other viewpoints in Figs 2.2(b) and (c). This illustrates that oxygen atoms are well bonded in layers of hexagonal symmetry, but rather less firmly attached from layer to layer: each oxygen atom has three bonds within a layer but only one bond across to the next layer. This defines a "basal plane", parallel to the layer structure, along which deformation (gliding and cleavage) preferentially takes place. The direction perpendicular to the basal plane is referred to as the crystal "*c*-axis" and is important when describing the various macroscopic crystalline forms of sea ice and glacier ice. Figure 2.2 shows projections of the three-dimensional lattice along the most significant crystal directions: looking down the *c*-axis (Fig. 2.2(b)), in which the hexagonal plan structure of the basal plane layers is clearly visible; and looking along the layers, perpendicular to the *c*-axis (Fig. 2.2(c)), showing the relatively sparse bonding from

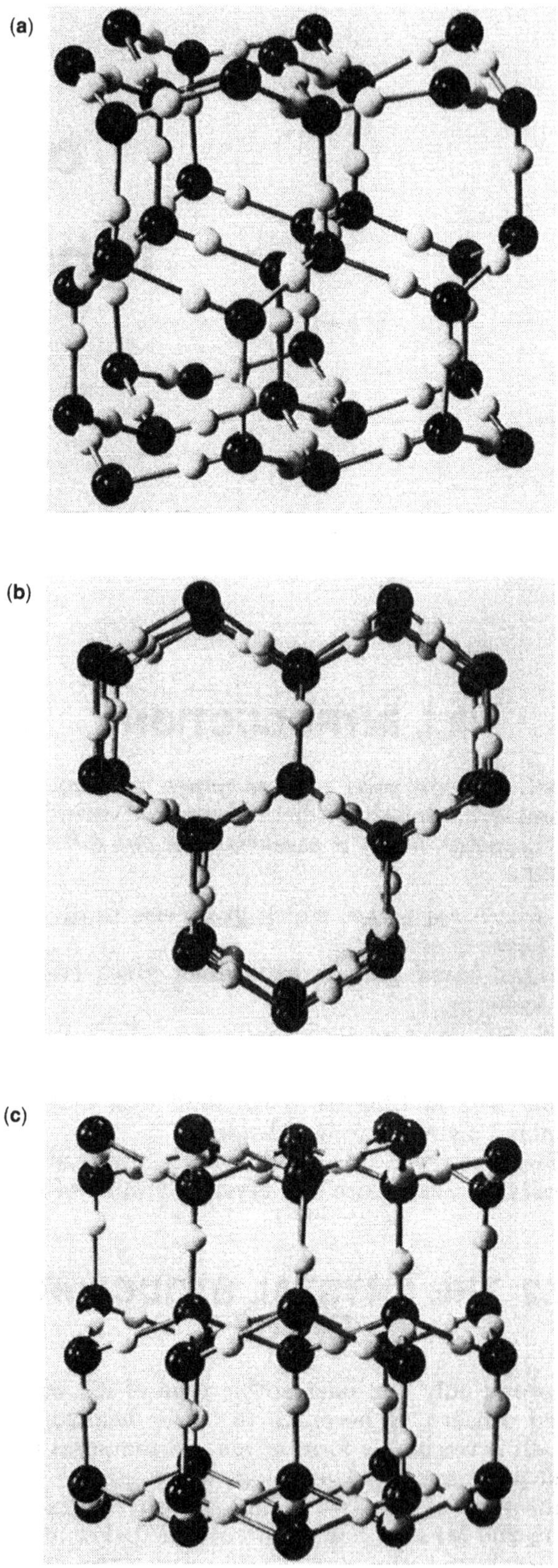

Fig. 2.2 (a) General view of the crystal structure of ice Ih, showing the tetrahedral constituent unit. Oxygen atoms are shown as black, hydrogen atoms as white. Each bond is associated with a hydrogen atom, and each oxygen atom has two hydrogen atoms particularly closely associated with it. One possible arrangement of hydrogen atoms is schematically illustrated. (b) View of crystal lattice looking along the *c*-axis. The hexagonal structure within a layer is clearly visible. (c) View of crystal lattice looking along basal plane layers, perpendicular to the *c*-axis.

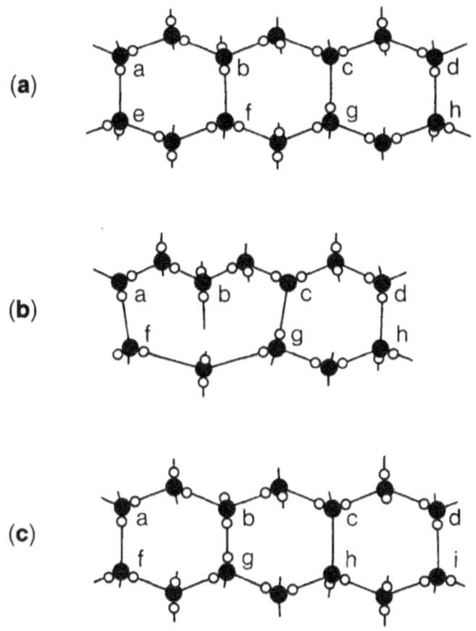

Fig. 2.3 The process of dislocation movement through an ice crystal (after Glen, 1968). (a) Perfect "ideal" crystal of ice Ih, with one hydrogen atom at each bond site. (b) Passage of a dislocation system through the crystal, creating defects. (c) Defects of two types, as the dislocation continues moving through the crystal: type D, with two hydrogen atoms at a bond site (b–g in the figure); type L, with no hydrogen atom at a bond site (c–h in the figure).

layer to layer. A single hydrogen atom is associated with each bond.

An ideal perfect crystal of ice in this form would be extraordinarily difficult to deform permanently, since all possible bonds on the H_2O molecular unit have been made. In practice, however, numerous point defects exist in ice, where a bonding is formed either by two hydrogen atoms or not at all (Fig. 2.3). This feature allows easy travel of dislocations through the ice crystal by a process of switching of hydrogen atoms between bond-sites, and is responsible for the well known creeping properties of ice. Figures 2.3(b) and (c) illustrate this process: as a defect travels through the crystal, the net effect is a permanent deformation of the crystal.

Ice crystals are generally of extremely high purity, irrespective of the composition of the water from which they form. This is because very few chemical substances have the correct size and valence to substitute effectively into the ice lattice. There are some possibilities, including some fluoride- and ammonia-based compounds (for instance, F^-, HF, NH_4^+, NH_3, NH_4OH, NH_4F) and some hydro-halogen acids (Weeks and Mellor, 1984) but, in general, impurities are very efficiently rejected as the crystal lattice is formed. Impurities seldom account for more than 0.02 mole per cent of the ice crystal. This is an important point to recognise when dealing with sea ice: although formed from a strong saline solution (sea water), the solid ice itself contains a negligible amount of salt. Any gross saltiness of the ice is due only to trapped solution or trapped salt crystals – but not to any incorporation of salts into the ice crystal structure. This means that the mechanical properties of sea ice are not substantially different from those of pure ice, provided that appropriate allowance is made for the presence of fluid pockets trapped within the matrix of pure ice.

2.3 SEA ICE

2.3.1 Development of an ice cover

Sea water freezes at approximately −1.9°C. The exact freezing point is not invariable since it depends on the seawater salinity: standard sea water (known as Copenhagen water) has a salinity of 35 parts per thousand (ppt) by weight, and freezes at −1.91°C, while water of salinity 30 ppt freezes at −1.63°C. Arctic Ocean surface water is generally of salinity in the range 30–34 ppt during winter months. During summer, however, much of the surface water is relatively warm and fresh, as a result of river run-off and melting of sea ice. Typical summer salinities are in the range 25–30 ppt, and some regions, particularly the Laptev Sea near Severnaya Zemlya, have summer surface salinities as low as 15–20 ppt (US Navy Hydrographic Office, 1958).

Let us now look at what happens as air temperature drops and cools the surface layers of sea water. Unlike fresh water, which has an anomalous maximum density at +4°C, well before it freezes, sea water becomes progressively more dense as its temperature drops towards freezing point. This is true for all normal sea water with salinity higher than 24.7 ppt. The consequences of this are important, since it is one reason why sea ice forms less readily than lake ice. As air temperature over a fresh water lake drops below 4°C and approaches 0°C, a stable layer of cold, light, fresh water forms on the lake surface and does not mix convectively with the underlying denser water at 4°C. This means that once the air temperature drops below 0°C the surface layer of water will very readily freeze and form a stably growing ice cover. Pure

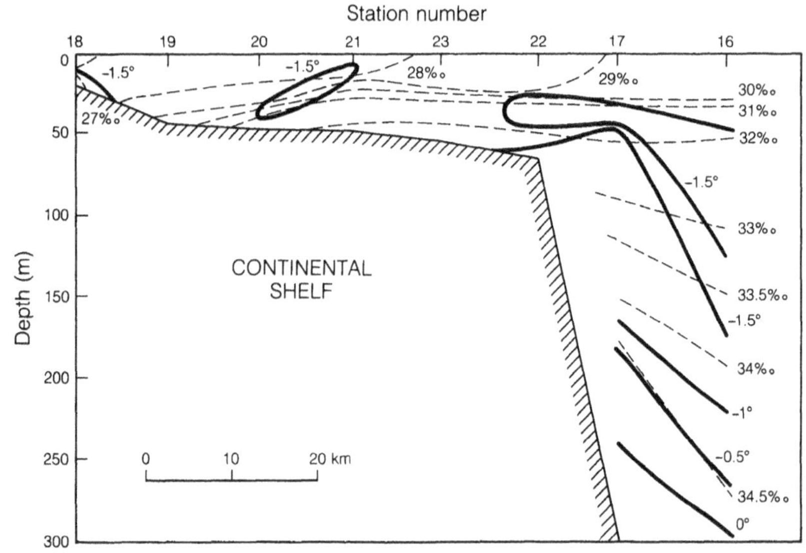

Fig. 2.4 A temperature and salinity section taken in autumn through the Beaufort Sea continental shelf and upper slope region near longitude 142°W (after Aagaard, 1984). The upper layers are cold and of low salinity while the lower layers are relatively warm and of high salinity.

ice, of course, floats on water, having a density of 0.917 Mg m^{-3}.

Sea water, in contrast, undergoes a complex mixing process as the surface is cooled. As the upper layers of sea water cool down, they densify and sink convectively until they reach stable equilibrium with lower denser layers (these lower layers may be denser on account of lower temperature or higher salinity or a combination of the two). The sinking water is replaced by warmer, or fresher, water.

If the oceans were of quite uniform salinity, this would mean that the entire ocean would have to be mixed in this way until it reached a uniform temperature of −1.9°C throughout the entire water column. Only then would it start to freeze. In practice, lower layers of the ocean are generally of higher salinity (35 ppt rather than 30–34 ppt) and are denser than the cooling surface water on account of their higher salt content (Fig. 2.4). Thus only a shallow surface layer in the upper 10–20 m of the ocean has to be cooled to its freezing point, and it does not mix appreciably with the underlying high-salinity waters.

Once the surface waters are thoroughly cooled and air temperature remains below about −2°C, small crystals of *frazil ice*, up to a few centimetres across, begin to form and float to the surface. These crystals, especially if they are nucleated at the surface by snowflakes, often have predominantly vertical *c*-axes. In the very early stages of formation the crystals are only loosely linked together and form a thin layer of slush at the surface. They give the surface of the sea a leaden tint and a smooth viscous appearance, and this is often known as *grease ice*. Subsequent early stages of development depend very much on the degree of calmness or agitation of the sea surface, but ice crystals soon coalesce to form an *ice rind* of solid surface ice up to 5 cm thick. Under even moderate wave action this rapidly breaks into thin plates of fragile ice which jostle and abrade each other and form irregular rounded discs, between 30 cm and 3 m thick. This is known as *pancake ice*.

The next stage of development involves the coalescence of pancake ice or fragments of ice rind into a stable, solid surface layer some 5–30 cm thick, referred to as *young ice*. Young ice is greenish-blue in appearance and is generally moist at the surface. A pattern of rings is often visible, showing the position of individual constituent pancakes as they coalesced. This layer generally consists of crystals about 1 mm in diameter, frozen together more or less randomly, and is known as *granular ice* (or frazil ice, after its crystal origin). A horizontal section of such ice, photographed through crossed polarizing filters, is shown in Fig. 2.5(a).

The crystals themselves are of pure ice, without appreciable salt content, since salt is expelled during the freezing process. However, brine (and indeed air and gas) becomes incorporated into the ice structure in the form of fluid or gas pockets trapped within the solid matrix. On account of this,

(a)

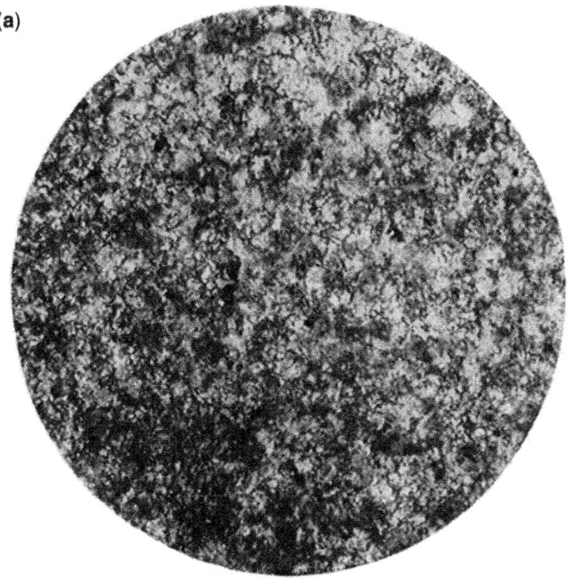

(b)

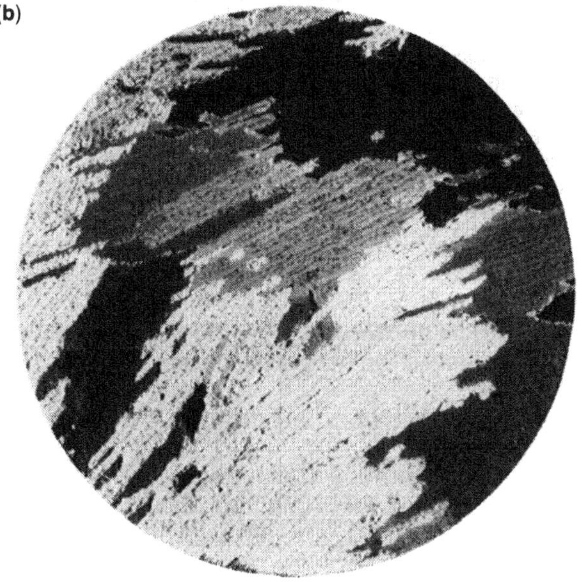

Fig. 2.5 (a) Photograph of typical granular sea ice, taken through crossed polarising filters to identify individual crystals. Crystals are about 1 mm diameter, and randomly oriented (photograph courtesy of BP Petroleum Development Ltd). (b) Horizontal section of columnar sea ice (Beaufort Sea), photographed through crossed polarising filters. The large units of different shades are columnar crystals a few centimetres across. The substructure of the platelets can be clearly seen, since platelets are separated by darker stripes due to lineated brine pockets. This sample shows preferential, though not perfect, alignment of the *c*-axes (photograph courtesy of BP Petroleum Development Ltd).

sea ice has a gross salinity, if melted down, of about 5–10 ppt. A more detailed account of brine and gas volume is given in the next section.

2.3.2 First-year ice

2.3.2.1 General

Once a solid layer of ice has formed at the sea surface, a different freezing process begins. Under mobile environmental conditions dynamic action may lead to fracture and rafting of young ice, but if stable conditions prevail further crystal growth occurs. No longer does ice form as a conglomerate of individual crystals and fragments thrown randomly together, but instead it grows in a stable fashion from the solid surface layer downwards. The growth process is slow, typically about 1 cm per day, enabling the formation of large regular crystals of size about 1–10 cm and sometimes larger. The crystals may be observed in their early stages of growth as thin plates extending down beneath the lower surface of the solid ice cover, when they are known as *underwater ice*. The crystals are elongated in the vertical direction and have a substructure consisting of platelets approximately 1 mm thick with horizontal *c*-axes. The predominance of horizontal *c*-axes is due to preferential growth rates for crystals oriented in this direction: the direction of heat flux is vertical, and crystals grow with their maximum thermal conductivity aligned with heat flux. The maximum conductivity of a crystal is perpendicular to the *c*-axis (Weeks and Ackley, 1982). As the fresh ice crystals form, salt is expelled at the ice–water interface of growing platelets, and cold briney streamers fall from the surface, to be replaced by warmer and fresher water from beneath. The ultimate salinity of sea ice is due to salt trapped between platelet boundaries as they coalesce to form a coherent solid. Ice formed in this way, with predominantly horizontal *c*-axes, is known as *columnar ice*. A horizontal section through columnar sea ice is shown in Fig. 2.5(b), photographed through crossed polars. The brine pockets between ice platelets can be clearly seen.

If such crystals grow as the young ice drifts freely around, or in an environment where sea currents are highly variable, the columnar crystals grow without any preferred direction in the horizontal plane. The ice is transversely isotropic – that is, its properties are the same in all horizontal directions.

However, in landfast ice conditions, where ice is stable relative to the sea bed and where uniform

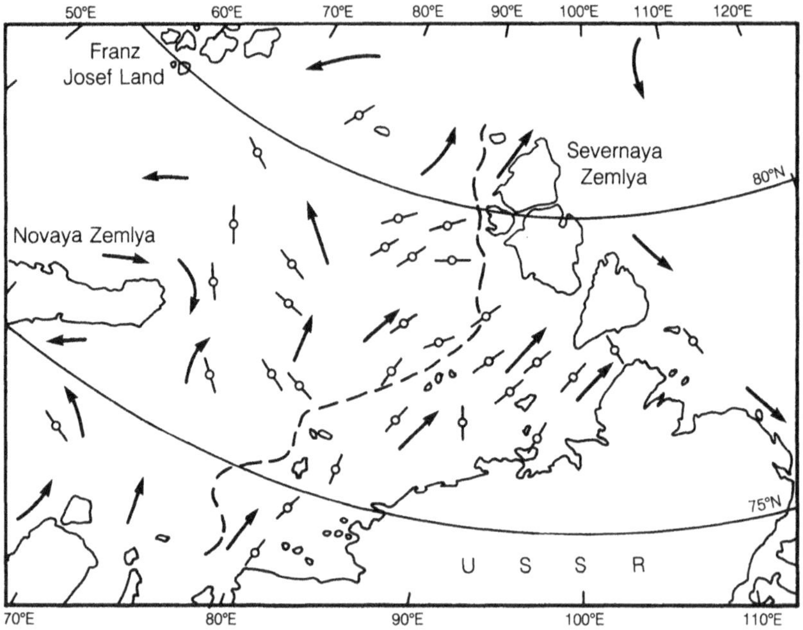

Fig. 2.6 Correlation of *c*-axis orientation for sea ice in the Kara Sea. Measurement sites are indicated by circles, and *c*-axis orientation by lines through the circles. The *c*-axis orientation is correlated with prevailing sea currents, which are indicated by arrows. The dashed line marks the fast-ice boundary (after Cherepanov, 1971; Weeks and Gow, 1978).

directional sea currents occur, preferred orientations of columnar crystals develop. (A preferred, though far from perfectly aligned, orientation of columnar crystals can be discerned in the section of Fig. 2.5(b).) The occurrence of this has been documented by Cherepanov (1971) and Weeks and Gow (1978, 1979), who mapped striking long-range correlations of *c*-axis direction in first-year ice in the Kara Sea (Fig. 2.6) and offshore Alaska. Oceanographic measurements and subsequent laboratory experiments (Weeks and Gow, 1978; Langhorne, 1983) have demonstrated that this must be due to prevailing nearshore sea currents, which lead to preferential growth of crystals with their *c*-axes aligned with the current direction. The mechanism for this to occur is not yet thoroughly understood. The degree of alignment of *c*-axes generally improves with depth through the ice cover, and a typical landfast first-year ice cover consists of about 30 cm of granular ice, followed by about 50 cm of more or less randomly oriented columnar ice, followed finally by preferentially oriented columnar ice. The lower columnar crystals frequently have *c*-axes aligned to better than $\pm 5°$.

Further offshore, where growing ice is less stable with respect to sea currents, there is generally less preferred orientation of columnar crystals. Indeed, in more dynamic environments, mechanical action and ice rafting processes lead to an altogether more confused picture of first-year ice structure (Fig. 2.7). Inclusions of frazil ice and fragments of columnar ice which have not grown *in situ* may occur at any depth, and become incorporated during the freezing process, creating a complex composite material. In addition, flooding of surface snow cover by sea water may occur, leading to large amounts of *infiltration ice* being formed. This closely resembles frazil ice, being composed of small ice crystals seeded by snowflakes, and is especially prevalent in the Antarctic Ocean, where snow cover on sea ice is often sufficiently deep (about 25% of the ice thickness) to depress the surface below sea level (Fig. 2.7(c)). Recent studies

(Gow *et al.*, 1982) have indicated that as much as 50% of the ice of the Weddell Sea may be of this type.

Much of the early work on first-year ice structure was performed on nearshore landfast ice in the Beaufort Sea, which is not necessarily representative of conditions further out to sea, or in the rather different growth conditions of the Antarctic. In fact, as we have seen, first-year ice may take a variety of forms, depending mainly on the degree of mechanical action it has been subjected to.

Near to the coast the ice is typically quite flat and uniform for miles on end, the uniformity broken only by occasional long-range tensile cracks (Fig. 2.8(a)). This leads to the uniform crystal structure shown schematically in Fig. 2.7(a). Further offshore, where wind and currents keep the ice constantly in motion, ridging and rafting occur at all stages of development of the ice cover, leaving a highly irregular field of flat pans of ice of variable thickness and crystal composition, separated by blocky ridges of fragmented ice (Fig. 2.8(b)). The internal structure of such ice is shown schematically in Fig. 2.7(b).

Sea ice is sometimes found not to be simply pure frozen sea water, but to contain significant concentrations of entrained sediments. This generally shows up as discrete visible layers within the ice, often tens of centimetres thick, which are discoloured brown or yellow and consist of fine suspensions of clay, silt or sand. Occasionally, coarser material such as gravel, dense mud or kelp may be found. The origin of this sediment-laden ice is not well understood, but is probably associated with one or all of the following:

- fluvial action and sediment transport near river deltas
- seabed gouging or turbulent seawater action in vicinity of major ice features
- flotation of seabed sediments by *anchor-ice* formation – a rare and poorly understood phenomenon by which ice accretes at the sea bed around rock or vegetation and, at length, acquires sufficient buoyancy to float to the surface

Sediment-laden ice is not sufficiently prevalent to affect the general properties of ice covers, and will

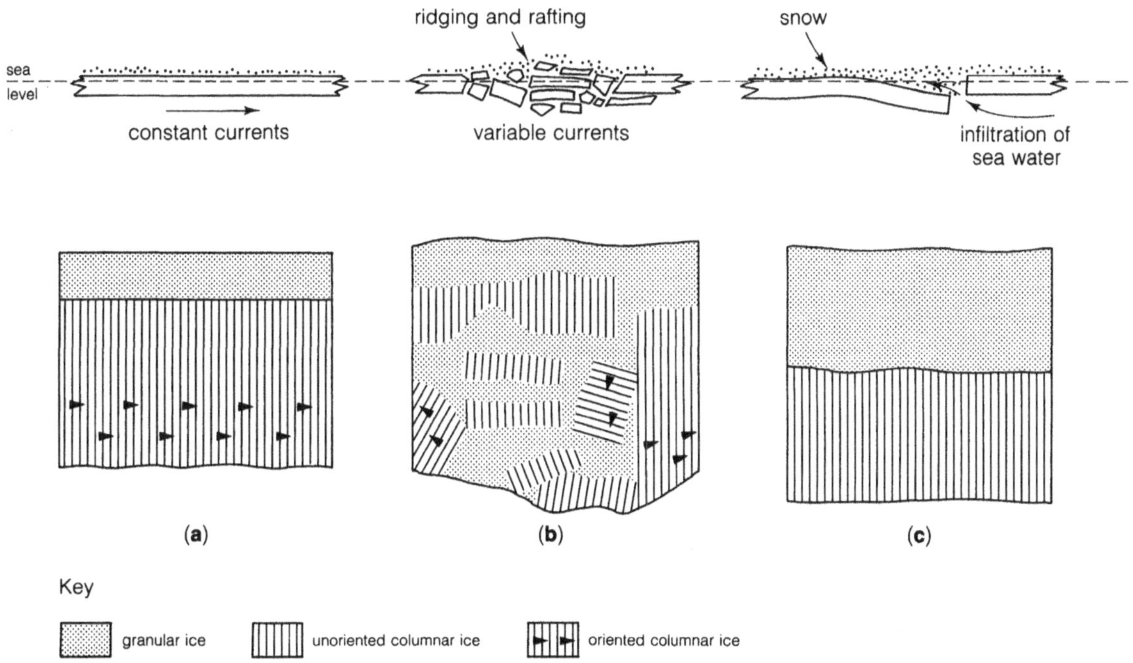

Fig. 2.7 Schematic diagrams of first-year ice internal structure and its genesis. (a) Stable landfast ice growing in the presence of sea currents, developing lower layers of oriented columnar ice crystals. (b) Ice formed further offshore in a more dynamic environment, showing evidence of ridging and rafting. (c) Infiltration ice, formed when the amount of snow cover is sufficient to depress the ice surface below sea level. This is particularly common in Antarctica.

(a)

(b)

Fig. 2.8 (a) Flat landfast ice which has recently been split by a long-range tensile crack. The freeboard of the floating ice can be seen here, and is about 0.25 m (photograph courtesy of BP Petroleum Development Ltd). (b) First-year ice in mobile conditions, showing ridging and heterogeneity (photograph courtesy of BP Petroleum Development Ltd).

not be further considered here. The subject has been reviewed by Osterkamp and Gosink (1984).

2.3.2.2 Temperature profiles through first-year ice

Temperature profiles through sea ice are generally approximately linear, at least during winter: the upper surface of the ice, within some 10–20 cm of the ice–air contact, generally remains within about ±10°C of the mean daily air temperature, while the bottom, in contact with sea water, remains steadily at −1.9°C. This leads to typical profiles as shown in Fig. 2.9 (dashed line), with a gradient of about 12 deg m^{-1}, from a surface at −20°C through ice of thickness 1.5 m.

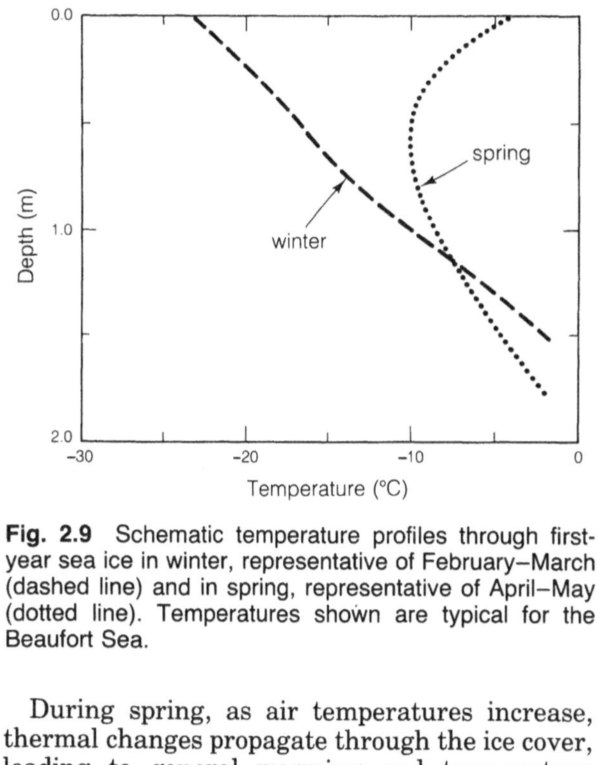

Fig. 2.9 Schematic temperature profiles through first-year sea ice in winter, representative of February–March (dashed line) and in spring, representative of April–May (dotted line). Temperatures shown are typical for the Beaufort Sea.

During spring, as air temperatures increase, thermal changes propagate through the ice cover, leading to general warming and temperature profiles of the type shown in Fig. 2.9 (dotted line).

2.3.2.3 Brine and gas pockets

The salinity of sea ice is almost entirely due to trapped pockets of briney water, with negligible amounts of the salt going into solid solution with the ice crystal. Gas, mainly air, may also exist in the form of bubbles trapped during the freezing process. This may occur in several ways: during dynamic action at the surface; as a result of dissolved gases in sea water; or as a result of gases rising from plant and animal life, living or decaying, and becoming trapped underneath the growing ice sheet. Typical gas concentrations are in the range 0.5–5% by volume (Zubov, 1945; Nakawo, 1983) and the composition of the gas involved is generally close to that of pure air: analyses of ice from the Barents Sea (Zubov, 1945) showed composition of about 82% N_2, 17% O_2, 0.4% CO_2 etc. This is lower in oxygen than pure air, and higher in carbon dioxide. In the lower consolidated layers of ice cover, gas porosity of only about 0.5% is typical, while nearer the surface (above sea level, in the upper 0.2 m) gas porosity is normally higher, in the range 1–5%.

Because of the trapped brine and gas, sea ice may have a different density from that of pure bubble-free freshwater ice (0.917 Mg m^{-3}). In practice, however, the density effects of brine and gas tend to cancel each other out, and a typical density for sea ice is 0.915–0.920 Mg m^{-3} (Nakawo, 1983), very close to that of pure ice. In the upper 0.2 m of ice cover the density may be lower, in the range 0.89–0.92 Mg m^{-3}.

The amount of salt water trapped in sea ice as it forms is principally dependent on the rate of freezing. At the surface, in the granular layer, salinities may be relatively high (8–12 ppt), since freezing has occurred by rapid congelation of a surface skim of water and ice at a critical freezing temperature. However, lower down in an ice cover freezing rates are typically rather slower and brine is more efficiently expelled from the ice as it freezes. This results in salinities of the order of 5–8 ppt. The relationship between freezing rate and salinity is shown strikingly in Fig. 2.10 (Nakawo and Sinha, 1984).

The entrapped brine occurs in the form of fine pockets of fluid between platelets of pure ice, as shown in Fig. 2.5(b). The fluid brine exists in a complex stable equilibrium with the surrounding ice: at a given temperature the fluid within the pockets must always be of just such a salt concentration that it coexists in equilibrium with the surrounding pure ice. At −1.9°C this concentration is 35 ppt, but as the temperature is lowered, pure ice crystallises out of solution onto the walls of the brine pocket leaving behind a more concentrated solution of brine. It continues until the brine is sufficiently concentrated to be in equilibrium at the new, lower temperature. The process is thermodynamically reversible. It continues, as temperature drops, until seawater salts begin to

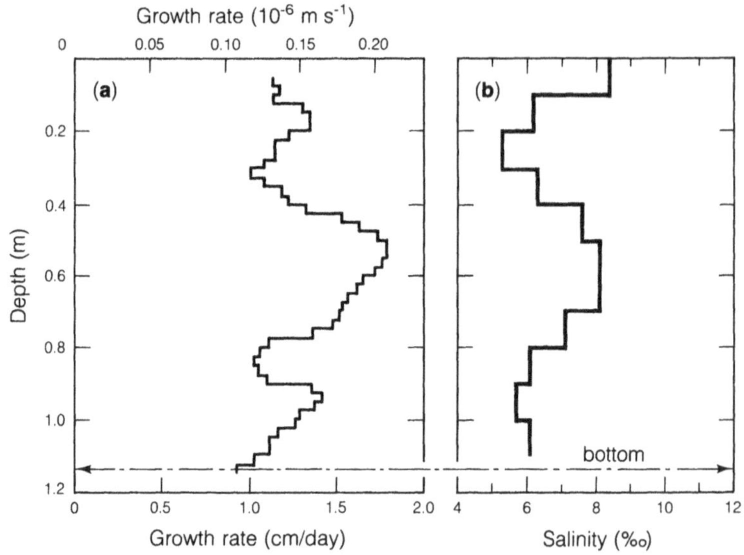

Fig. 2.10 Growth rate (a) of first-year ice (Eclipse Sound, NWT) showing its relationship to (b) salinity profile (after Nakawo and Sinha, 1984).

precipitate out: the first precipitation occurs at −8.2°C, when $Na_2SO_4.10H_2O$ salts come out of solution; next, and most importantly, $NaCl.H_2O$ precipitates out at −22.9°C; and the remaining trace salts precipitate out below −36.8°C. Since NaCl is the predominant salt (85% by weight) sea ice below −22.9°C is effectively solid, and the salt in brine pockets is largely crystalline.

Empirical equations exist which allow the gross volume of the brine pockets to be calculated as a function of temperature and salinity (Frankenstein and Garner, 1967, based on tables by Assur, 1960). The equations make the assumption that the ice is gas-free, and in their fullest form involve three separate equations referring to different temperature ranges. As a general rule, however, it is adequate to use a single equation for the entire range of temperatures normally encountered (−0.5 to −22.9°C). If the temperature of an ice sample is θ (°C) and its gross salinity is S (ppt), the brine volume ν_b (expressed as a fraction) is given approximately by:

$$\nu_b = 0.001S(0.53-49.2/\theta) \qquad (2.1)$$

This means, for instance, that ice of salinity 6 ppt has a brine volume of 0.062 (that is, 6.2%) at −5°C, of 0.033 at −10°C and of 0.025 at −15°C.

Considerably more rigorous analyses exist for the calculation of brine and also gas volumes as a function of temperature, salinity and density (Cox and Weeks, 1982), and the reader is referred to their report for a definitive account. However, the method does not lend itself to statement as a simple equation, and for most purposes equation (2.1) should prove adequate.

The behaviour of brine pockets with time and as temperature rises is of interest, especially to the ensuing discussion of ice types which have survived several seasons. With time, brine pockets tend slowly to migrate as a result of the temperature gradient within the ice cover. Normally the sense of the temperature gradient is from cold surface to warmer ice lower down, and brine pockets migrate downwards by an interesting thermodynamic process which is shown schematically in Fig. 2.11. If a brine pocket exists in a temperature

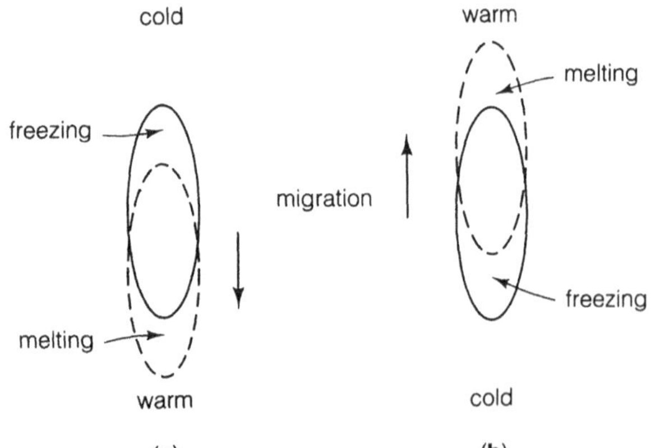

Fig. 2.11 Schematic diagram of the migration of brine pockets along vertical temperature gradient: (a) in winter, when brine migrates downwards; and (b) in spring/summer, when brine near the surface may migrate upwards (adapted from Zubov, 1945).

gradient, it is energetically favourable for ice to melt at the warmer end of the brine pocket and refreeze at the colder end. This results in a net migration of the brine pocket towards higher temperatures. The pockets become larger and longer as they migrate, since they are passing into progressively warmer surrounding ice, and they may coalesce to form major channels of diameter 0.1–1.0 cm diameter. The process is aided by brine convection currents when the temperature gradient is as shown in the figure (cold surface, becoming warmer with depth), but may occur also when the temperature gradient is reversed. This generally occurs during spring and early summer, at which time the upper and lower surface of the ice are both warmer than the interior (Fig. 2.11(b)). The result is that brine slowly becomes expelled from the sea ice, to both top and bottom surfaces, and the gross salinity of the ice decreases with time. A typical rate of decrease in salinity for first-year ice is approximately 0.5 ppt per month (Nakawo and Sinha, 1981). A schematic diagram of brine channels in first-year ice is shown in Fig. 2.12 (after Lake and Lewis, 1970).

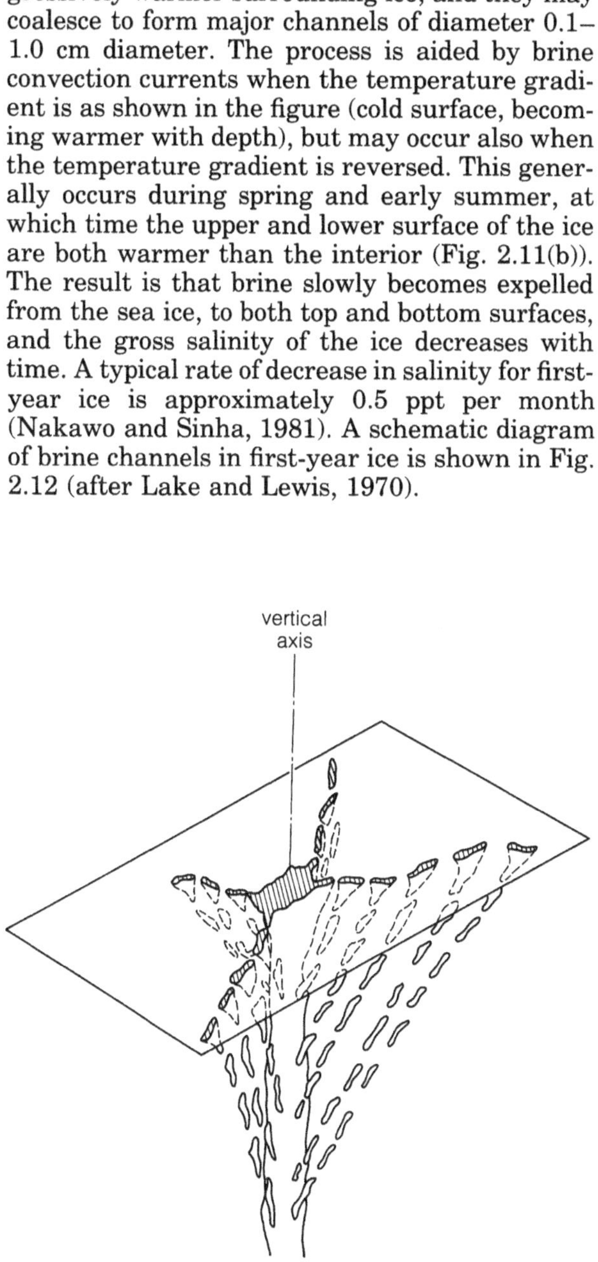

Fig. 2.12 Schematic diagram of brine channels in first-year ice, after migration processes have set in (after Lake and Lewis, 1970).

2.3.2.4 First-year ridges

The dynamic action to which first-year ice may be subjected leads to formation of several distinct forms of ice feature, shown schematically in Fig. 2.13.

A *compression ridge* is formed as two sheets of ice are driven towards each other. This may occur either at the boundary of two plates already separated by the growth of tensile cracks, or it may occur spontaneously in undeformed ice as the result of excessive compressive stress (Fig. 2.13(a)). A ridge formed in this way, by a process of crushing and flexural failure, is generally highly irregular in direction and in vertical height. Examples are shown in Fig. 2.8(b). When first formed, a ridge is simply a pile of unconsolidated rubble, but in due course it may become to some extent consolidated by refreezing processes, especially below the water line. Compression ridges generally have surface "sails" about 1–4 m above sea level and underwater "keels" about 5–10 m below sea level. They may be much deeper than this, and have been known to reach the sea bed in water of depth 30 m. The magnitude of a ridge is a function of the thickness of the ice from which the ridge formed (which influences the block size) and of the amount of ice which had to fail during formation of the ridge. A discussion of the forces required to form a ridge is given by Parmerter and Coon (1973): it involves a combination of forces required physically to fracture the ice and forces required to uplift or downthrust the blocks so formed against the forces of gravity and buoyancy.

A *shear ridge*, on the other hand, is formed by left or right lateral movement. It is normally strikingly straight, since the process of shear necessarily removes obstacles. Shear ridges may be several kilometres long and generally have near-vertical walls consisting of finely pulverised ice. They are rarer than compression ridges. A shear ridge may form either from relative movement of two separate sheets of ice or from a compression ridge if the direction of relative motion changes after initial formation (Fig. 2.13(b)). It is unlikely that a shear ridge can ever develop spontaneously from undeformed ice.

For completeness, Figs 2.13(c) and (d) show the process of *rafting*, in which thin plates of ice override and underride each other without significant rubble formation. In young ice of about 30 cm thickness alternating upthrust and downthrust rafts are commonly observed (Fig. 2.13(d)), a feature which is known as "finger-rafting". It is

(a)

(b)

(c)

(d)

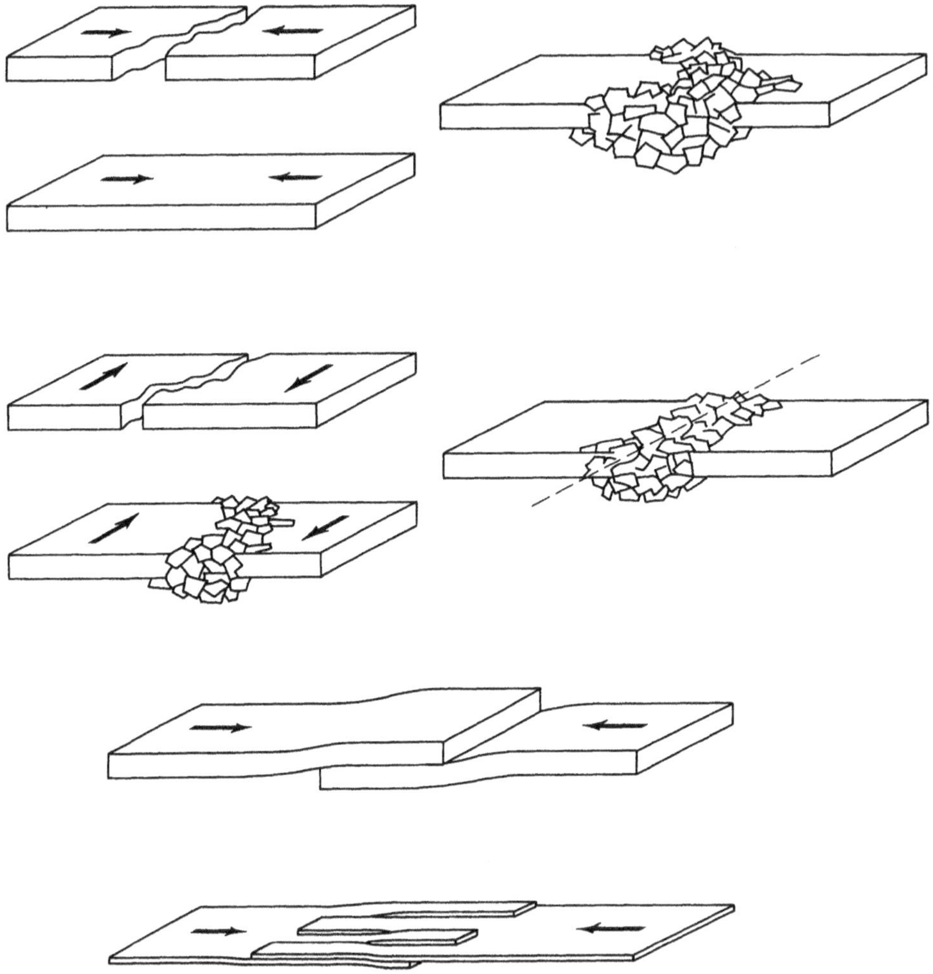

Fig. 2.13 Ridge formation in first-year ice. (a) Compression ridge, formed either from two separate sheets of ice or from a single undeformed sheet. (b) Shear ridge, formed by left or right lateral movement of two ice sheets. The line of strike is generally very straight. A shear ridge may develop from an initial compression ridge if the relative motion alters. (c) Rafting, formed as thin ice plates over- and underride each other. (d) Multiple alternating rafts, known as "finger-rafting", are common in young ice of the order of 30 cm thick.

likely that the process of rafting is associated with an initial elastic buckle formation. As we shall see later in Section 6.2.1, elastic buckling is strongly dependent on ice thickness, and is not normally observed in ice of thickness greater than about 1 m: this may be the reason why rafting seems to be restricted to thin ice.

The formation processes of first-year ice and its subsequent development into multi-year ice are summarised in Fig. 2.14.

2.3.3 Second-year ice

First-year ice generally grows to a thickness of 1.0–2.5 m, reaching its maximum thickness in April–May (Northern Hemisphere) or October–November (Southern Hemisphere).

As summer temperatures rise, further freezing is arrested and deterioration begins: rising sea temperatures begin to dissolve the lower surface of the ice cover and melt pools form on the surface,

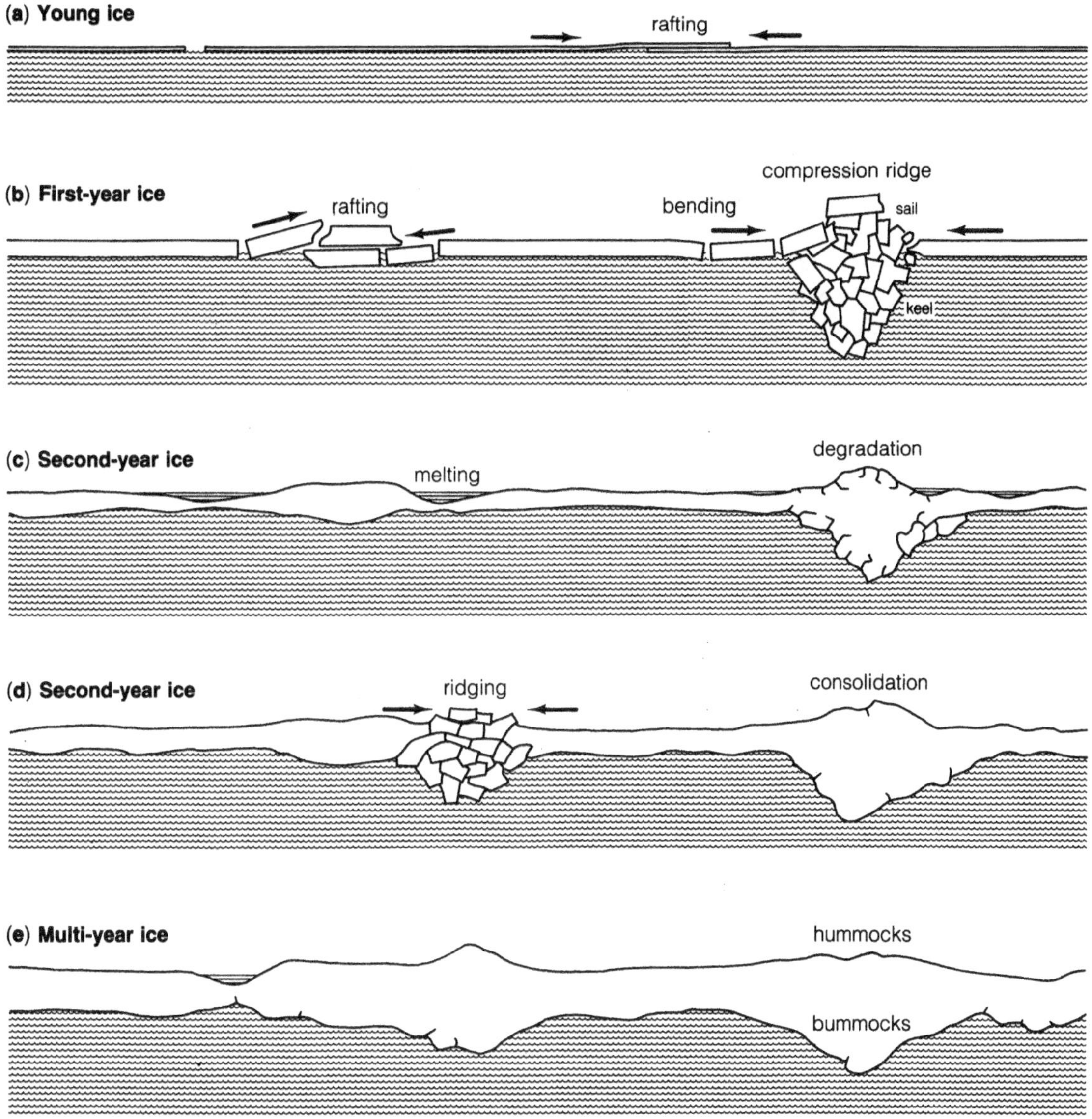

Fig. 2.14 Development of an ice cover. (a) Young ice, 30 cm thick, in early winter. (b) First-year ice, 1.5 m thick, in mid-winter/early spring; ice action causes ridging and rafting. (c) Second-year ice, of variable thickness, in late summer; surface and bottom melting has occurred. (d) Second-year ice during its second winter, undergoing refreezing, accretion and further ice action. (e) Mature multi-year ice, 4–6 m thick, 5–10 years old, containing old smoothed ridges, hummocks and bummocks. This diagram shows normal conditions: extremes may exceed quoted thicknesses.

often in extraordinary lineated patterns (Fig. 2.15). Surface melting is accelerated as reflective snow is removed and also as dirt becomes exposed, lowering the surface albedo. Floes break free from the stable cover and begin to drift around with wind and sea currents. Most of this deteriorating first-year ice melts entirely during the summer months. However, at high latitudes some survives and still exists, in a warm and rotten state, at the onset of freezing the following winter. Ice which has survived one summer season is known as second-year ice. It is fresher than first-year ice since brine has generally been expelled during the summer warming process. Typical summer

Fig. 2.15 Highly deteriorated first-year ice, in July in the Canadian Beaufort Sea. If this ice survives the summer season it will become the basis for second-year ice. In the background is Dome Petroleum's Single Steel Drilling Caisson (SSDC). (Photograph by the author, courtesy of BP Petroleum Development Ltd.)

salinities are in the range 1–4 ppt. Salinity near the surface may be particularly low (less than 1 ppt), and the ice is often quite porous and granular as a result of recrystallisation during successive freeze–thaw cycles. Brine expelled to the surface collects into discrete pools, which may run off the surface if appropriate channels exist, or may in some cases melt right the way through the ice thickness. The process of growth of a meltpool is of interest, and is aided by albedo and density effects. A pool of meltwater, being typically blue or turquoise, has a significantly lower albedo than snow or ice, and hence absorbs radiation quite efficiently. Further, the density inversion characteristics of fresh water (as discussed in Section 2.3.1, it needs only to be fresher than 25 ppt) mean that as the meltwater is warmed above zero it becomes more dense. This sets up efficient inverted convection currents which aid melting at the lower ice surface, even if the upper surface of the meltpool remains solidly frozen (Fig. 2.16).

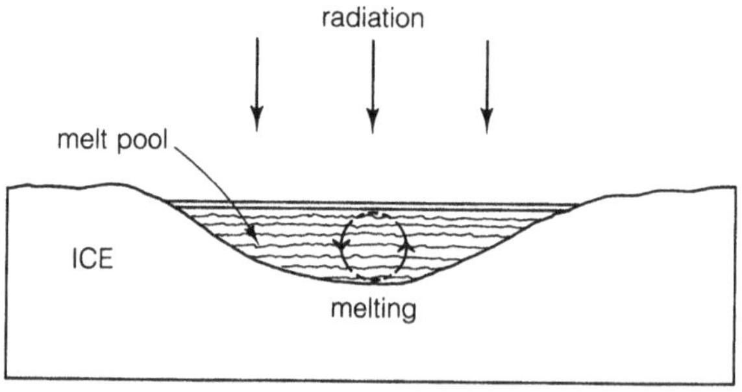

Fig. 2.16 Schematic diagram of the process of melting in a meltwater pool. The density–temperature inversion characteristics of fresh water (less than 25 ppt) allow inverted convection currents in which warmed water sinks. Eventually the pool may melt its way entirely through the ice cover.

During the succeeding winter, second-year ice undergoes freezing and dynamic action (Fig. 2.14(c)), and grows to a thickness generally in the range 2–3 m. However, second-year ice has not been intensively studied, largely because of the difficulty of unambiguously distinguishing it from multi-year ice.

2.3.4 Multi-year ice

2.3.4.1 General

Multi-year ice is normally defined to be ice which has survived at least two summer seasons (World Meteorological Organisation, 1970). Much of the ice drifting in the polar pack is 5–10 years old. The terms "multi-year ice" and "old ice" are, however, frequently used more loosely, to cover second-year ice as well. This is largely because it is often not possible to distinguish second-year ice from ice which is older.

Multi-year ice is formed from second-year ice by continuing dynamic action and by melting and refreezing (Fig. 2.14). New ridges are formed by compressive stresses, while old ridges become smoothed by melt action into surface hummocks and subsurface bummocks. Surface meltpools collect in depressions and may drain away or refreeze the following winter season; surface snowdrifts may flood and freeze; and if the winter season is cool enough and the ice thin enough, continuing additional freezing may occur at the bottom surface of the ice.

The interior of the Arctic polar pack consists almost entirely of multi-year ice, more or less consolidated, but broken by extensive systems of cracks and open water leads. From the point of view of nearshore hazards, however, multi-year ice is normally encountered in the form of discrete floes of order 1 km diameter. An example is shown in Fig. 2.17.

Multi-year ice in the Arctic generally reaches an equilibrium thickness of about 2–6 m, when annual melting just equals annual freezing. The ice often appears, on close examination of cores, to show a recognisable annual layer structure, much like the rings of a tree, especially if it has grown in quite stable conditions. Annual layers

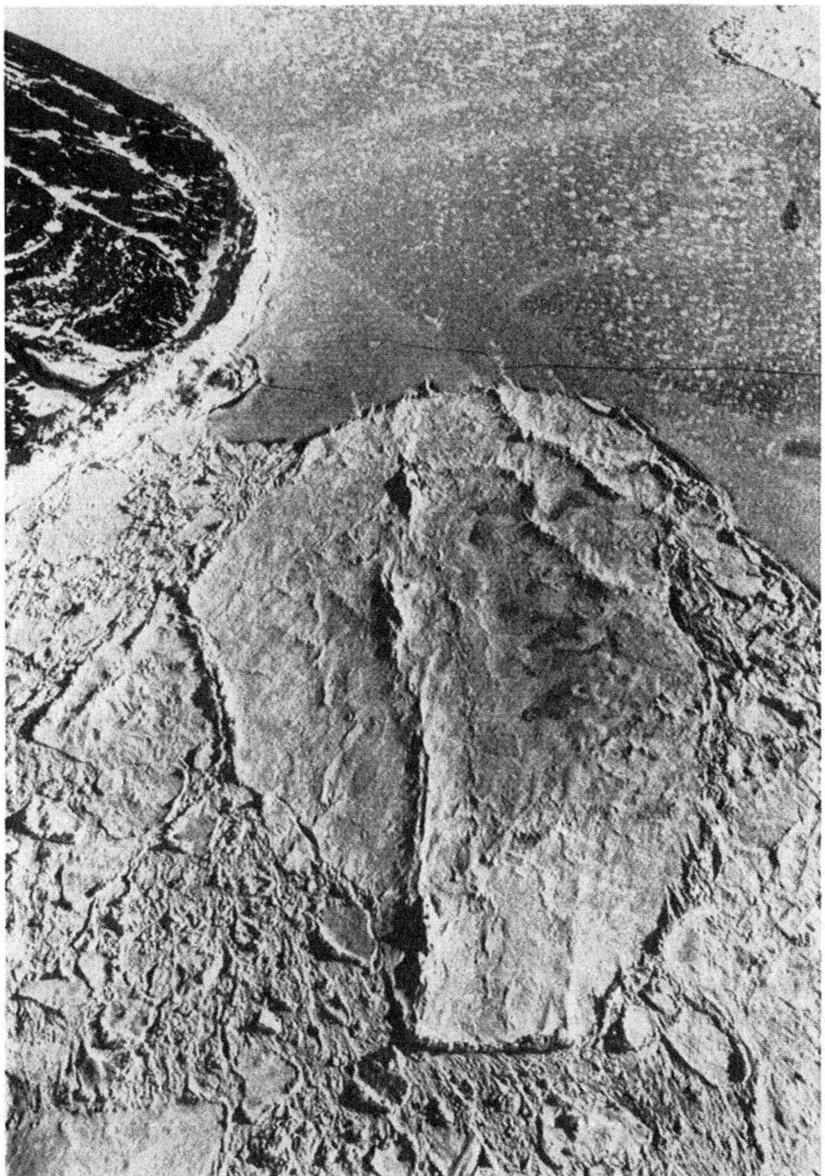

Fig. 2.17 A multi-year floe, in summer just before break-up, bordering with much flatter first-year ice. Note the radial cracks extending from the shoreline of the island. (Photograph by the author, courtesy of BP Petroleum Development Ltd.)

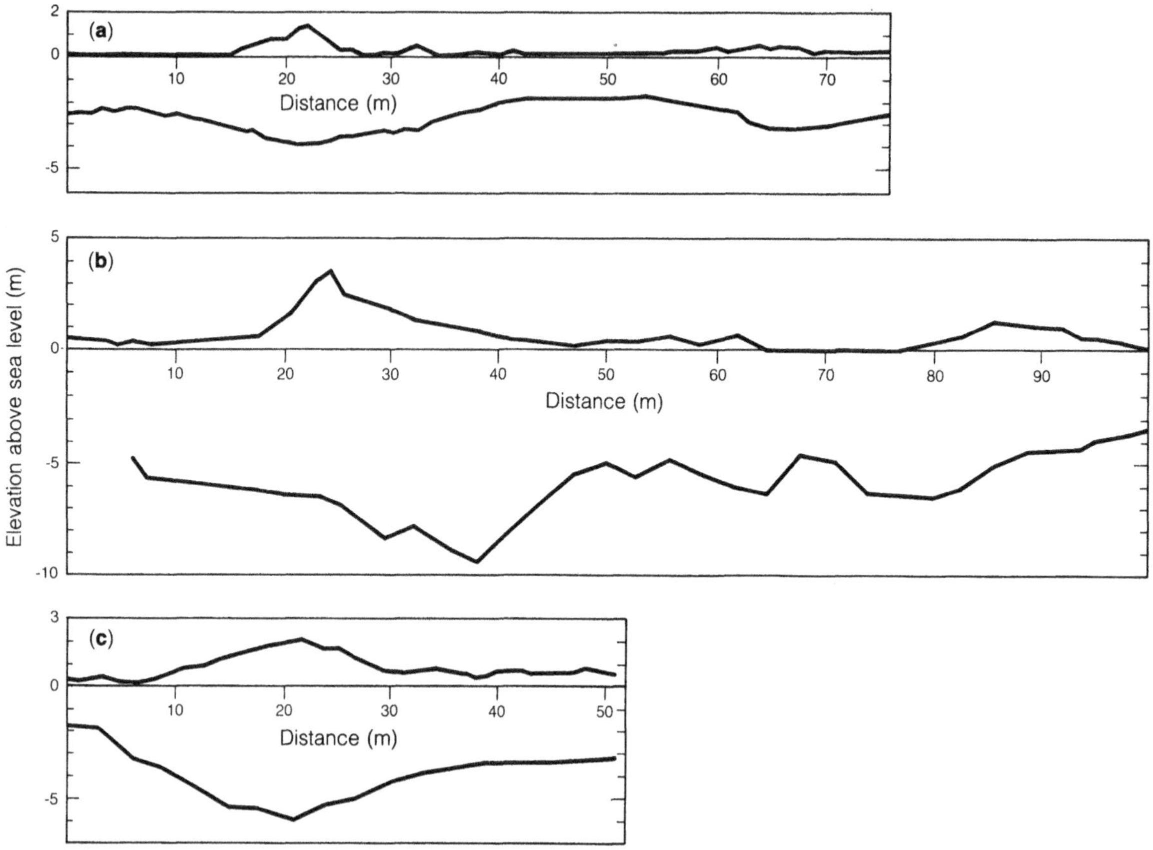

Fig. 2.18 Cross sections through mutli-year ridges in the Beaufort Sea (after Kovacs and Mellor, 1974; Kovacs, 1983) showing height and depth relative to sea level. The vertical scale is exaggerated by a factor of two, and snow cover has been omitted.

formed in this way are generally about 30–50 cm thick, and a typical undeformed floe might contain about 10 annual layers. Multi-year floes are, however, highly irregular and variable in character, and it is hardly meaningful to talk of a typical multi-year floe. If the floe has been through intense dynamic pack conditions then it may show no stable stratification of annual layers and is likely to display dramatic variations in thickness. Ridges on a multi-year floe may have surface sails of height 5 m or more above sea level and keels of 20 m depth below, while at the same time on the same floe thin areas only one or two metres thick (the thickness of first-year ice) may coexist. In contrast, large areas of regular and extremely thick landfast multi-year ice are known to exist in certain regions, particularly in the fjords and bays of the Queen Elizabeth Islands of Canada. This ice, which grows stably *in situ* without appreciable dynamic action, may attain an estimated thickness of 15–30 m (Marcellus and Morrison, 1982) and its surface is irregular only on account of melting and drainage processes.

The salinity of multi-year ice is, as a result of continuing expulsion of salt during the growth process, generally found to lie in the range 0.5–4 ppt.

2.3.4.2 Multi-year ridges

Much time and effort have been devoted to the study of multi-year ridges. Some cross sections through old ice floes in the Beaufort Sea are shown in Fig. 2.18 (Kovacs and Mellor, 1974; Kovacs, 1983). Some of these ice features may be second-year rather than multi-year, owing to difficulties of determining the age of the ice, but the distinction is not important here. The profiles include a ridge of thickness more than 10–12 m in surrounding multi-year ice of average thickness about 6 m.

The horizontal extent of the keel is normally greater than that of the sail. Notice that thin areas of only 2–3 m thickness are regularly encountered on the same floe. An example of a major multi-year ridge offshore north-west Greenland is shown in Fig. 2.19, both from the air and from the ground. This ridge was a distinctive feature in the middle of a multi-year ice floe of estimated average thickness 6 m. The ridge itself had a freeboard of approximately 4 m and a depth below sea level estimated to be 12–15 m. As a general rule, studies (in the Beaufort Sea and Chukchi Sea) have shown that the depth of the keel below sea level is approximately 3.3 times the height of the sail above sea level (Fig. 2.20). Studies in other regions of the Arctic, for instance the Queen Elizabeth Islands (Dickins and Wetzel, 1981), have yielded values as high as 1:5.6. This may well be due to formation under a different (colder) thermal regime, but the process is not understood in detail.

Considerably more detailed accounts of statistics on multi-year floes are given in the next chapter under regional headings, especially Sections 3.2.2 and 3.2.3 on the Arctic Ocean and the Beaufort Sea.

2.4 POLYNYAS

A polynya is not a type of ice, but is the absence of it: it is an area of open water surrounded by ice. There are a number of locations in the Arctic where recurring polynyas appear consistently from year to year. These vary considerably in shape and size and rely for their formation on a combination of currents, tides, upwellings and winds. A map of polynyas in Alaskan and Canadian waters is shown in Fig. 2.21.

The mechanism of formation of polynyas is not thoroughly understood, but they are in practically all cases close to land (within 100 km) and are associated with divergent flow of ice. The divergent flow field generally arises as large-scale pack ice motion, such as the drift of the Beaufort Gyre (see Section 3.2.2.), tries to force its way past an irregular coastline. This appears to be the reason behind the formation of the polynyas of Cape Bathurst and Point Hope, and the divergent flow fields can be seen clearly on a map of sea currents (Fig. 3.2(b), Chapter 3). In the Bering Sea, a polynya regularly occurs to the south of St Lawrence Island in the lee of the prevailing ice drift southwards. The celebrated North Water polynya of northern Baffin Bay appears to be caused by "bottlenecking" of ice flow at the southern outlet of the Kane Basin; an ice bridge appears to form across the straits and ice to the south of it is swept away southwards by sea current. It has been also suggested that current upwelling plays a part in the formation of North Water polynya (Smith and Rigby, 1981).

Polynyas are not of great importance to the subject of ice hazards, but are of significance to ice-cover conditions, since they may play an important role in the process of ice generation, particularly in the Bering Sea (Section 3.2.5). They are important also to wildlife: the polynyas and leads along the Chukchi Coast between Point Hope and Point Barrow, for instance, allow migrating whales to reach the Beaufort Sea.

2.5 ICEBERGS

2.5.1 Sources

Icebergs are not frozen sea water, but are composed of freshwater ice from land-based glaciers flowing off the land into the sea (Fig. 2.22). Glacier ice develops from successive snowfalls of pure freshwater snow which compress under their own weight until they become solid ice. The initial density of freshly precipitated snow is very low, about 0.4 Mg m^{-3}, but density increases systematically with depth and overburden pressure, reaching about 0.6 Mg m^{-3} at depth 10 m below surface. At depth 50 m the density reaches about 0.82 Mg m^{-3}, and below 200 m the ice density exceeds 0.91 Mg m^{-3}, approaching that of pure ice.

The principal sources of icebergs are the Antarctic ice cap in the Southern Hemisphere and the Greenland ice cap (particularly the West Coast) in the Northern Hemisphere. These ice caps are fed by snow accumulation in their interior and build up to such a size and shape that ice flows outward under its own weight, much like any viscous fluid. Ice then calves off at the perimeter into the sea, and the ice cap eventually reaches a state of dynamic equilibrium between accumulation of snow in the interior and outflow of ice at its perimeter.

In addition to the two major ice caps of Greenland and Antarctica, a number of smaller valley glaciers and ice caps give rise to rather smaller numbers of icebergs. These include the glaciers of Franz Josef Land, Svalbard and Severnaya Zemlya and of the Queen Elizabeth Islands, particularly the ice caps of Ellesmere, Devon and Baffin Islands. A few very small icebergs are calved from the glaciers of Southern Alaska.

(a)

(b)

(c)

Fig. 2.19 Multi-year ridges. (a) Aerial view of a multi-year compression ridge offshore Greenland (Kennedy Channel), showing how the ridge has been split in two after formation, by a shear crack. (b) View from the ice surface of the same compression ridge as shown in (a), showing a section through the ridge. The ridge has a freeboard of about 4 m. (c) Multi-year floe bounded by a multi-year shear ridge at its left-hand edge: shear ridges are always straight. The ridge has a freeboard of about 1.5m. (Photographs by the author, courtesy of BP Petroleum Development Ltd.)

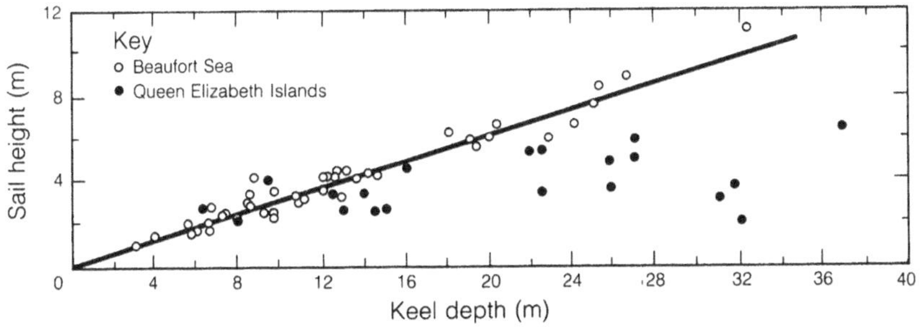

Fig. 2.20 Relationship between pressure ridge sail height and keel depth for the Beaufort Sea (open circles, after Kovacs, 1983, from several data sources). The average sail to keel ratio is 1:3.3. Values obtained by Dickins and Wetzel (1981) for the Queen Elizabeth Islands are superimposed (solid circles).

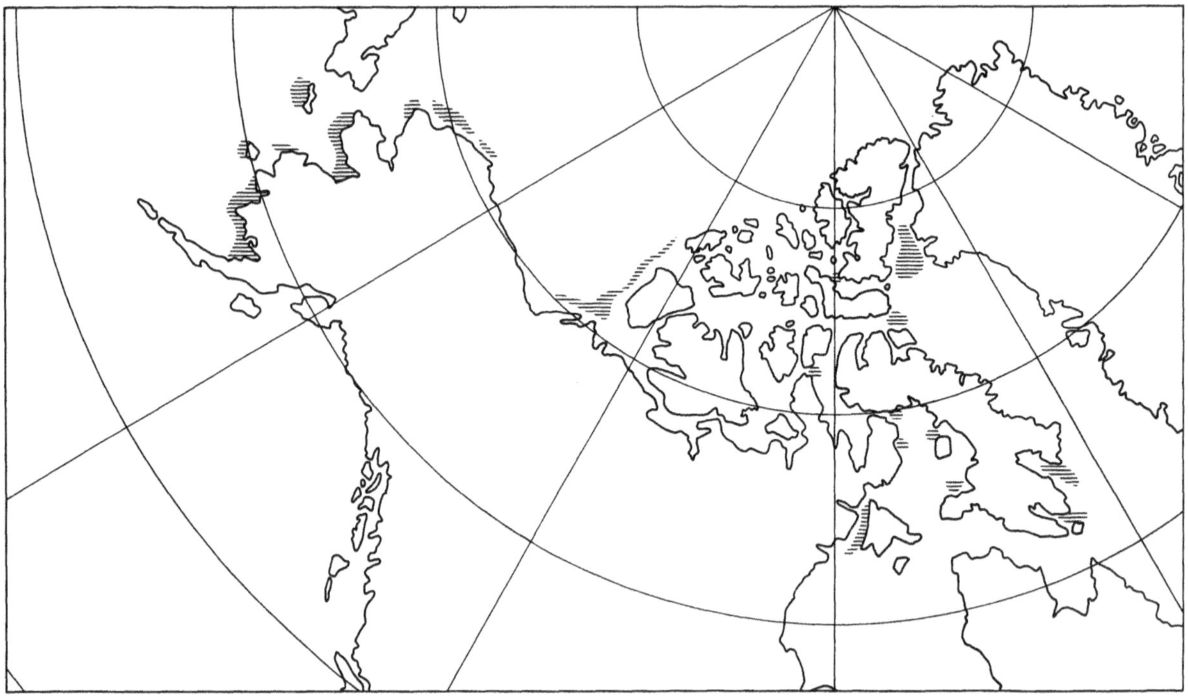

Fig. 2.21 Principal recurring polynyas (hatched areas) in the Alaskan and Canadian Arctic (compiled from Stirling and Cleator (1981) and LaBelle *et al.* (1983)).

2.5.2 Types of iceberg

There are two principal types of iceberg.

(1) *The blocky iceberg*. This is generally formed by ice breaking directly off from a land-based ice cap or glacier into the sea (Fig. 2.22). Such icebergs are generally approximately cubical (diameter to thickness ratio of between 1:1 and 1:2) and have diameters in the range 50–500 m. A typical blocky iceberg is shown in Fig. 2.23(a).

(2) *The tabular iceberg*. Where glaciers flow from the land into an embayment they frequently form an extensive area of landfast floating glacier ice, known as an *ice shelf* (Fig. 2.22). The largest of these in the world is Ross Ice Shelf of Antarctica, which is of areal dimensions 1000×600 km (approximately the size of France) and is 400–500 m thick. These ice shelves give rise to huge floating flat slabs of ice, referred to as tabular icebergs, which may be several hundred metres thick and many kilometres across. In general, they have a length to thickness ratio of up to 1:10, although the largest ever observed was of length 160 km and thickness 500 m (a ratio of 1:300).

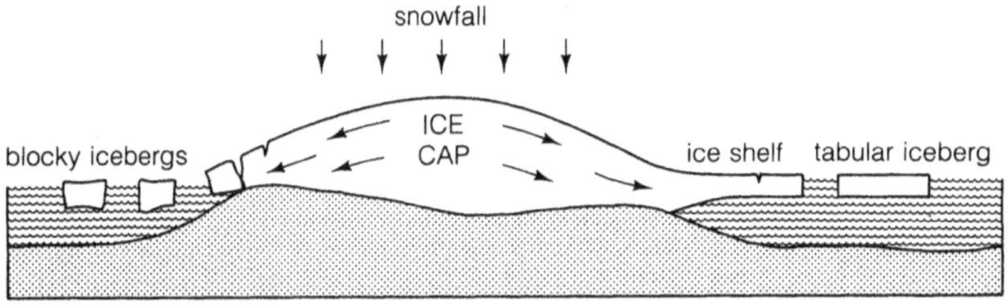

Fig. 2.22 Schematic diagram of an ice cap generating icebergs. Where the glacier flows directly into the sea, blocky icebergs are formed. Tabular icebergs result from ice shelves.

A modest-sized example from north-west Greenland is shown in Fig. 2.23(b).

Icebergs may last for several years, and they undergo major melting at their lower and upper surfaces as they drift into warmer waters and higher air temperatures. This leads to two further classifications of icebergs, as follows.

(3) *A dome iceberg* is generally the result of immersed surfaces melting so severely that the iceberg becomes unstable and inverts, exposing rounded, smoothly melted surfaces. An example, which has recently inverted, is shown in Fig. 2.23(c).

(4) *A drydock iceberg* is heavily eroded at the upper surface by sun and wind, and generally exhibits pinnacles and a highly irregular spikey surface. An example is shown in Fig. 2.23(d).

In their later stages of melting and deterioration, icebergs and fragments of icebergs are known as:

- *bergy bits* – about 10 m diameter, roughly the size of a small house, generally showing about 1–5 m above sea surface;
- *growlers* – a few metres across, roughly the size of a grand piano, generally showing less than 1 m above sea level.

Iceberg size classifications, with approximate masses, are summarised in Table 2.1.

What we see above sea level is of course only the tip of the proverbial iceberg. Most of the iceberg lies beneath the surface. However, the ratio of height to draft depends greatly on the shape and density of the iceberg. The average density of an iceberg is generally in the range 0.85–0.91 Mg m^{-3}, and varies systematically from region to region. Antarctic icebergs are generally composed of lighter snow and firn in their upper levels and may have average densities as low as 0.83 Mg m^{-3}. Greenland icebergs typically have little or no snow cover or permeable ice, since the surface is normally extensively melted and refrozen before calving. The shape also influences the average sail height to keel draft ratio. Typically, this ratio is in the range 1:4 to 1:6 but values between 1:1 (highly unstable drydock iceberg) to 1:11 (flat, dense tabular iceberg) have been reported (Kovacs, 1979).

The drift of an iceberg is difficult to predict. The forces acting on them are not as simple to analyse as, for instance, those acting on a floe of sea ice. A large floe of sea ice experiences wind drag at its upper surface and current drag at its lower surface, and the drag forces involved are essentially proportional to the surface area of the floe. Because the floe is very large, edge effects can be ignored and the area available for drag is the same at the upper surface as it is at the lower surface. In contrast, the drift of an iceberg (especially a blocky one) is dominated by edge effects, since its submerged sides present a very considerable area to the effects of current. The motion of an iceberg is generally determined by local sea currents in the top 100–200 m of the sea, but is also influenced by wind conditions. Winds may contribute substantially to the velocity of icebergs and appear to be capable of driving an iceberg at a speed some 10–30 cm s^{-1} in excess of local current speeds (El Tahan *et al.*, 1983). Figure 2.24 shows exceedance probability for iceberg velocity off the east coast of Canada, compared with measured currents at 15 m and 50 m below the surface. The iceberg velocity is apparently about 1.5 times the current velocity at 15 m, and twice the current velocity at a depth of 50 m. The existence of the Coriolis effect also imposes a transverse deflection on their motion (to the right in the Northern Hemisphere and to the left in the Southern Hemisphere), meaning that their motion frequently corresponds neither to wind nor to current. Observed trajec-

TABLE 2.1 Classification of icebergs by size (as defined by International Ice Patrol)

Description	*Height above sea level (m)*	*Length (m)*	*Approx. mass (tonnes)*
Growler	< 1.5	< 5	100
Bergy bit	1.5–5	5–15	1 000
Small berg	5–15	15–60	100 000
Medium berg	15–50	60–120	2 000 000
Large berg	50–100	120–220	10 000 000
Very large berg	> 100	> 220	> 10 000 000

(a)

(b)

(c)

(d)

Fig. 2.23 Types of iceberg. (a) Blocky iceberg offshore Newfoundland. The iceberg is being towed by the *Hudson Service* supply vessel (photograph courtesy of BP Petroleum Development Ltd). (b) Tabular iceberg in Bessels Fjord, northwest Greenland. Note the figure (Dr R. Frederking) standing at the right-hand edge of the photograph (photograph by the author, courtesy of BP Petroleum Development Ltd). (c) Dome iceberg near Adelaide Island, Antarctica. The smooth rounded surface results from melt action while submerged: this leads to instability and the berg inverts. This particular berg had only just inverted during the past 5 minutes before the photograph was taken (photograph by the author). (d) Drydock iceberg in the waters of the Antarctic Peninsula, heavily and fantastically eroded by sun and wind (photograph by the author).

tories of icebergs can be quite bizarre, especially in the presence of fluctuating tidal currents, which often give rise to looping cusp-like trajectories; also, icebergs within a few kilometres of each other may not necessarily show a coherent drift pattern (Fig. 2.25).

Although the amount of ice released per year by the glaciers of the Arctic and the Antarctic remains approximately constant in the long term, large annual and seasonal variations occur in the number of icebergs sighted drifting freely in the oceans. Seasonal variations occur mainly because many icebergs during winter are locked in landfast sea ice and suddenly become mobile during the summer months. Statistics on iceberg frequency are treated regionally in Chapter 3.

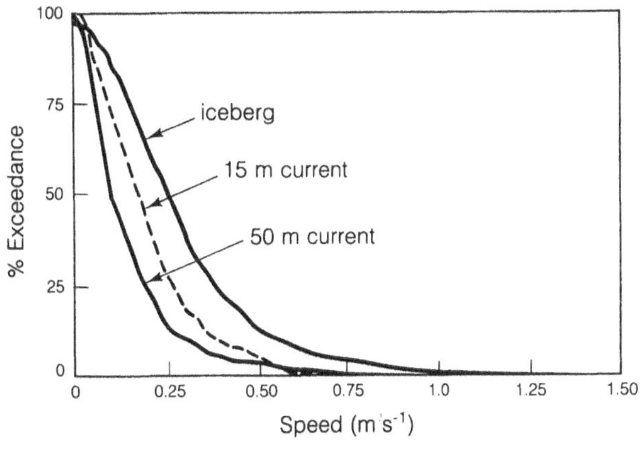

Fig. 2.24 Iceberg drift velocity compared with current velocity at 15 m and 50 m below sea level. The iceberg velocity is roughly twice the current velocity at 50 m depth (after El Tahan *et al.*, 1983).

2.5.3 Ice islands

Ice islands are an Arctic curiosity. They are large masses of flat, low-salinity ice some 50 m thick

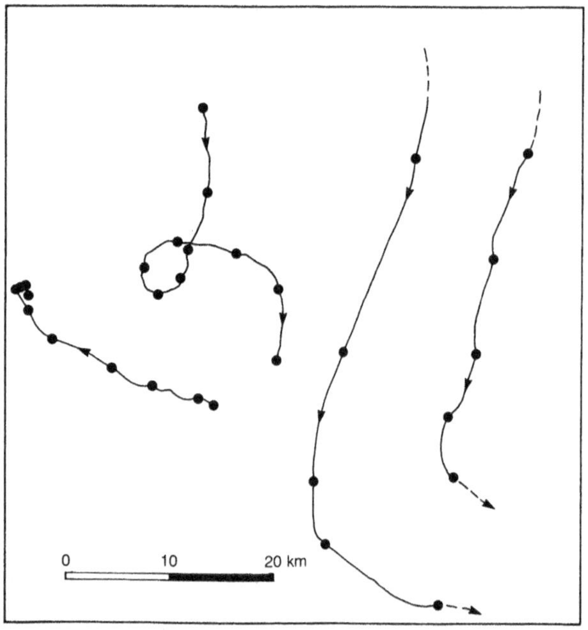

Fig. 2.25 Trajectories of icebergs acted on by currents, tides and winds during a 53-hour period in July 1974 offshore Labrador (after Gaskill and Rochester, 1982, unpubl.) Tracks are marked at 6-hourly intervals by circles, and direction of motion is indicated by arrows (not all bergs were monitored throughout the period).

and up to 40 km in diameter, which have calved from the ice shelves of Northern Ellesmere Island, notably Ward Hunt Ice Shelf (Fig. 2.26). These are not true ice shelves in the strict sense of the term, since they are not fed, at least at present, by land-based glaciers. It is possible that true glacial ice shelves existed in these bays during the last Ice Age but, if so, they exist no longer. The present-day ice shelves can in most respects best be regarded as landfast multi-year ice. They are at present of thickness between 40–90 m and have in general attained an equilibrium with the sea and the atmosphere, undergoing balanced melting and accretion. Isotopic profiles through Ward Hunt Ice Shelf indicate that substantial portions of the interior are as much as 1600–4200 years old (Sackinger *et al.*, 1985).

Much of the surface of these ice shelves, and hence of the ice islands which calve from them, consists of snow ice and lake ice, which are the result of densification of snowfall and refreezing of meltpools. However, beneath the surface much of the ice is either sea ice or brackish ice, often with extraordinarily large crystal size. This appears to be the result of extremely slow bottom freezing. Over the past few thousand years, the ice shelves have developed into a unique hybrid

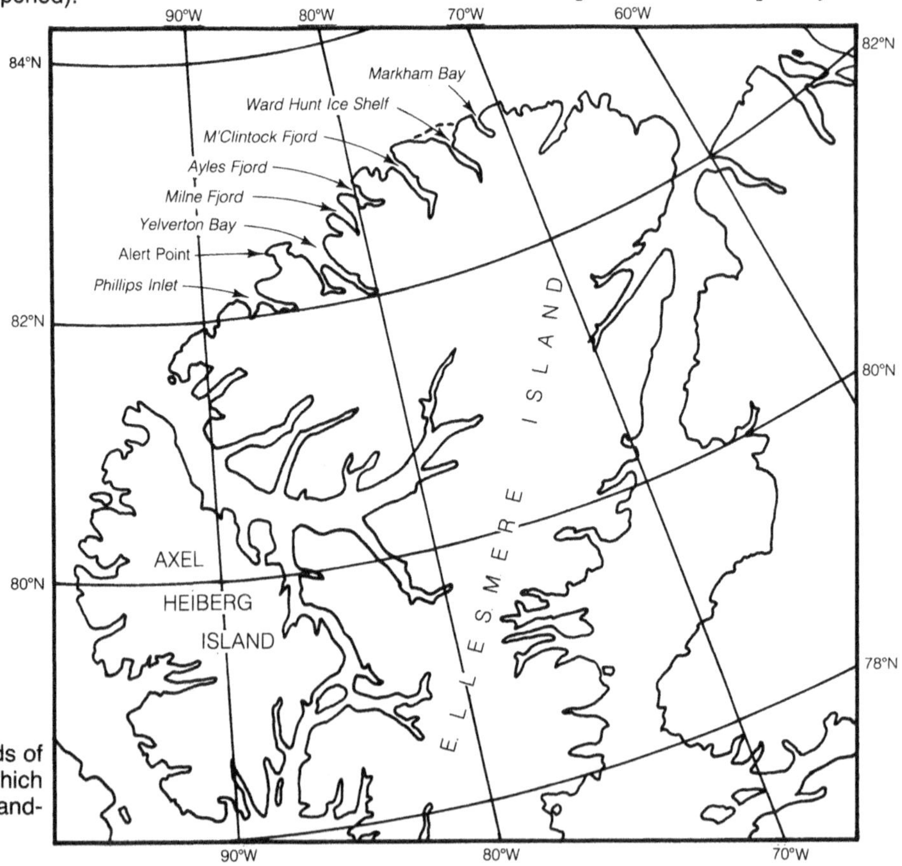

Fig. 2.26 Map of principal fjords of Northern Ellesmere Island, which give rise to ice islands and thick landfast multi-year ice.

Fig. 2.27 Two ice islands grounded in 27 m of water, 50 km north of Prudhoe Bay, Alaska. The islands were captured and then sprayed with water to give them additional stability. They were used for research purposes in 1969 (photograph by permission of The British Petroleum Company p.l.c.).

between glacier ice and multi-year sea ice. The ice shelves are apparently not at all uniform in composition from place to place: core studies on western Ward Hunt Ice Shelf show that a large body of old sea ice exists within the shelf, while recent core studies on a 42 m ice core taken through an ice island calved from eastern Ward Hunt Ice Shelf show the entire ice column to be freshwater ice (Sackinger *et al.*, 1986). The total area of these ice shelves of Northern Ellesmere Island is at present approximately 1500 km^2.

Periodically, calving episodes occur from the ice shelves of Northern Ellesmere Island, releasing large masses of tabular ice islands into the Arctic Ocean. The largest recent calving episode occurred in the winter of 1961/2, when a total of 570 km^2 broke away from Ward Hunt Ice Shelf, and minor episodes have occurred about every 3–4 years since. Ice islands were well known prior to 1961, however, and the most famous of these, which was named T-3 (Fletcher's Ice Island), was discovered in 1950. It was initially of thickness 70 m and was used as a scientific research station for more than 20 years as it drifted clockwise in the Beaufort Gyre. It is estimated (Spedding, 1977) that approximately 3200 km^2 of ice islands were produced or discovered during the period 1946–76. An example of ice islands in the Beaufort Sea, here being used for research, is shown in Fig. 2.27.

There are no definitive statistics on the number and size of ice islands in the Arctic Ocean. However, in 1972 a total of 433 ice islands were sighted along the Beaufort Coast. Of these, 117 had lateral dimensions exceeding 80 m and one was more than 1600 m across (Sackinger *et al.*, 1985). Three years later, in 1975, no sightings at all were reported. The unusually high number present in 1972 was, as we shall discuss later (Section 3.2.3.4), probably the result of a single major calving episode some 10 years previously. It is estimated (Spedding, 1977) that during the late 1970s, and possibly still today, there are approximately 40 ice islands with diameters in excess of 400 m drifting around with the Beaufort Gyre. Of these, perhaps 10 have diameters of several kilometres, and they may periodically be released from the pack ice into the surrounding waters of the Beaufort Sea, the Canadian archipelago or Northern Greenland.

FURTHER READING

Hobbs, P. V. 1974. *The Physics of Ice*. Clarendon Press, Oxford. This is the principal reference work for atomic/molecular structure of ice and physical properties of pure ice. It also contains a detailed account of the properties of other polymorphs of ice besides ice Ih.

Weeks, W. F. and Ackley, S. F. 1982. *The Growth, Structure, and Properties of Sea Ice*. CRREL Monograph 82-1. This gives an excellent account of sea-ice genesis, morphology and mechanical properties.

Chapter 3

Worldwide Occurrence of Ice

3.1 INTRODUCTION

In this chapter we look again at the ice types discussed in Chapter 2, and investigate their regional occurrence worldwide. Wherever possible I shall present detailed maps and statistics, but for a more specialised analysis of, for instance, month-by-month areal coverage statistics, the reader should refer to the various comprehensive charts and atlases available. The principal data sources used here are summarised at the end of this chapter. I have made an effort to collate the data in such a way that statistics for all regions are presented in comparable format. This has not always been possible.

Information is given on the following principal parameters:

- Boundaries of average ice concentrations (0/10, 3/10, 8/10 and 10/10 by area) for significant months: for instance, months of maximum and minimum ice coverage, or months at the beginning and end of the winter season.
- Statistics on the probability of encountering certain concentrations of ice coverage during critical months of the year. The statistics and months selected vary from region to region according to their importance. In general the charts show the boundaries of 20% and 80% chance of encountering ice of more than 3/10 coverage (or in some cases, more than 8/10 coverage). A 20% chance indicates that in only 1 year out of 5 will the ice coverage in question be exceeded; an 80% chance indicates that it will be exceeded in 4 years out of 5.
- Average boundary of winter landfast ice. This is sometimes difficult to define precisely, but in general the boundary mapped refers to the month of March (Northern Hemisphere), and should be taken only as a broad indication of stable consolidated ice. The trouble is, definitions of fast ice vary considerably according to the application, and one man's fast ice may be another man's mobile ice: for instance, fast ice which is sufficiently stable for safe travel by surface vehicles may not be sufficiently stable for constructing an ice platform to drill from.

The data set used here is a 20-year set covering the years 1955–74, digitised in 111×111 km grid elements (Walsh, 1978). The statistics on ice coverage refer to the areal proportion of a grid square covered by ice, expressed to the nearest 1/10. Thus, 10/10 coverage implies only that the coverage exceeds 95%. The charts are derived principally from an analysis of these data by Kelly *et al.* (unpubl., 1982), with additional information, particularly on extent of landfast ice, incorporated as appropriate from other sources.

For the present purpose, which is to provide a rapid guide to general ice conditions worldwide,

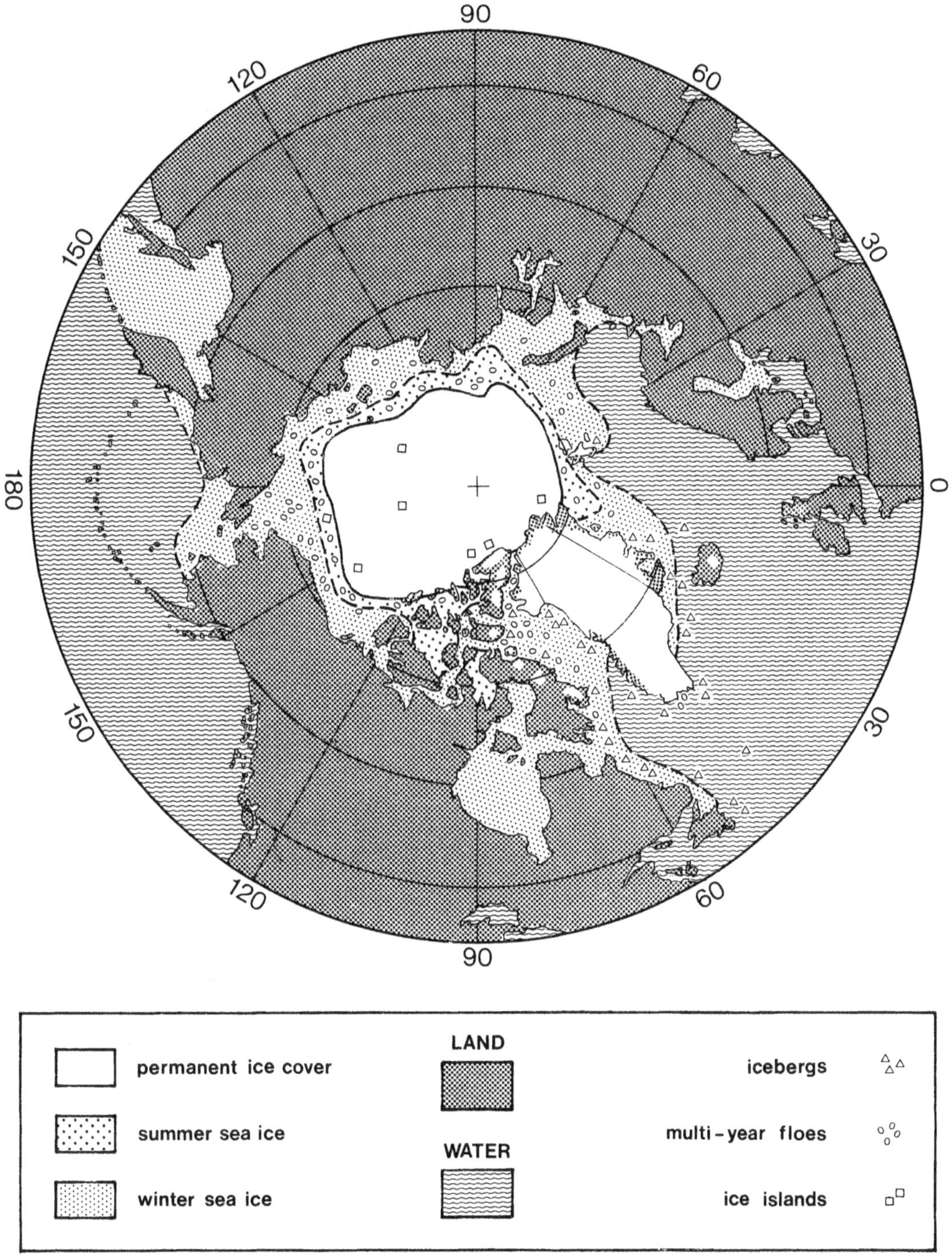

Fig. 3.1 Schematic map of Arctic ice conditions. Dashed lines represent average maximum and average minimum extent of 8/10 sea ice cover. Permanent ice cover is shown as pure white. Ovals represent multi-year floes, triangles represent icebergs and squares represent ice islands. (Representation of these discrete ice features is of course entirely schematic.)

we shall look in turn at the Arctic and the Antarctic. The Antarctic will be treated as one single region; not because it is homogeneous, but because it is unlikely to be the principal area of interest for those reading this book. The Arctic, on the other hand, will be subdivided into eight distinct geographical regions.

We begin with a general discussion of Arctic ice conditions.

3.2 ARCTIC ICE CONDITIONS

3.2.1 The Arctic regions: general

A general schematic map of ice conditions in the Arctic is shown in Fig. 3.1. Principal geographical locations and place names are shown in Fig. 3.2(a).

The extent of ice cover is controlled by a complex combination of air conditions and sea conditions, but the influence of sea currents is particularly important. A general map of surface sea currents is shown in Fig. 3.2(b). Each winter, most of the seas north of latitude about 60°N are frozen. The notable exception to this is the north-east Atlantic Ocean, especially the Norwegian Sea, which is kept largely ice-free by the Gulf Stream. The east coast of Canada, on the other hand, experiences more southerly encroachment of first-year sea ice as a result of prevailing northerly winds and the cold Labrador Current from Greenland and the Arctic Islands. Sea ice and icebergs are common offshore Newfoundland, and yet it is the same latitude as Paris.

First-year ice conditions are highly seasonal and vary widely throughout the Arctic: offshore Newfoundland, ice is present only for several weeks, during the February–April period, while in the

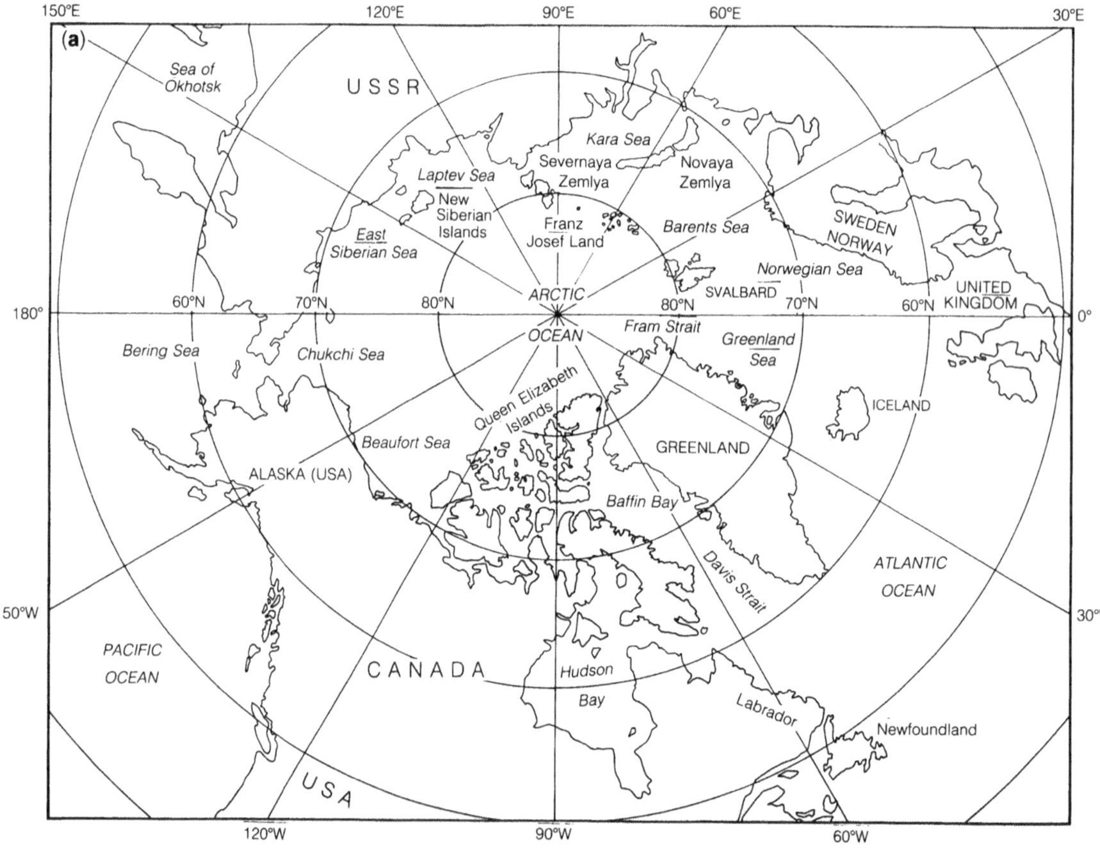

Fig. 3.2 (a) General map of the Arctic, showing principal locations and place names. (b) Map of sea current patterns in Northern Seas.

Beaufort Sea it is present for most of the year, disappearing only for about two months between August and October.

Multi-year ice is present to some extent throughout most of the Arctic regions, but is exceedingly rare south of the Bering Straits and is seldom seen in the Labrador Sea. In the central Arctic Ocean, multi-year ice is present throughout the year (Fig. 3.2(b)), but is in a continual state of motion, even during winter.

Large icebergs are, in general, concentrated around the Greenland Coast, and are carried southwards by the Labrador Current towards Newfoundland and the Gulf Stream (Fig. 3.2(b)). On reaching the Grand Banks of Newfoundland, an iceberg has typically already wasted away to about 10–15% of its original size, and once it encounters the warmer waters of the Gulf Stream it rapidly disintegrates. This means that the average southern limit of iceberg drift can be quite well defined. However, rogue icebergs have been sighted in the most extraordinary places: for instance, offshore southwestern Ireland, and even the Azores. Some of the more unusual sightings in the North Atlantic are shown in Fig. 3.3.

We turn now to a regional summary of Arctic ice conditions. For convenience we divide the Arctic into the following regions:

- Central Arctic Ocean
- Beaufort Sea (Alaska and Canada)
- Chukchi Sea
- Bering Sea
- Canadian Arctic Islands
- Labrador, Newfoundland and West Greenland
- East Greenland and Svalbard
- Soviet Union

The overview is generalised and not definitive. It is intended simply to give a representative description of the known ice properties in each region, and the reader will notice liberal use of imprecise terms such as "typically" and "in general". Wherever possible I have supplied precise statistical data, but serious calculation for design purposes in a particular region should be based on a more

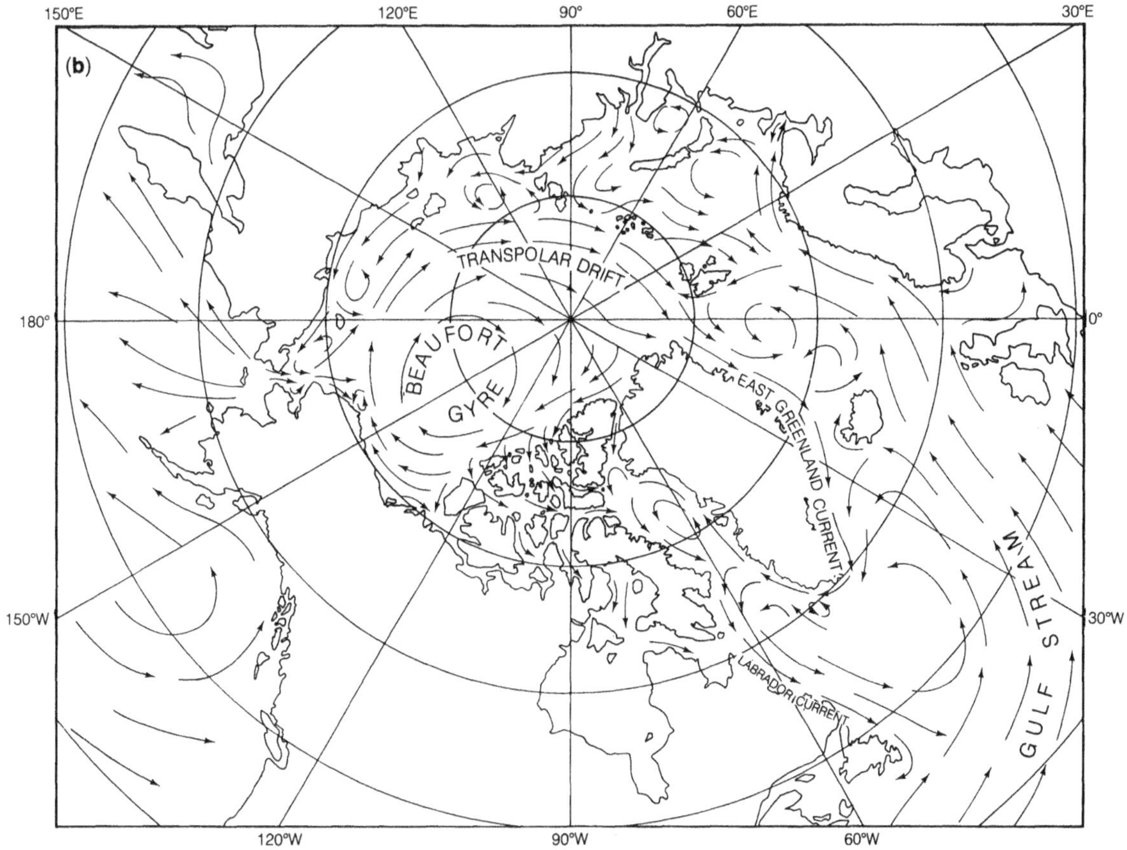

Fig. 3.2 (b)

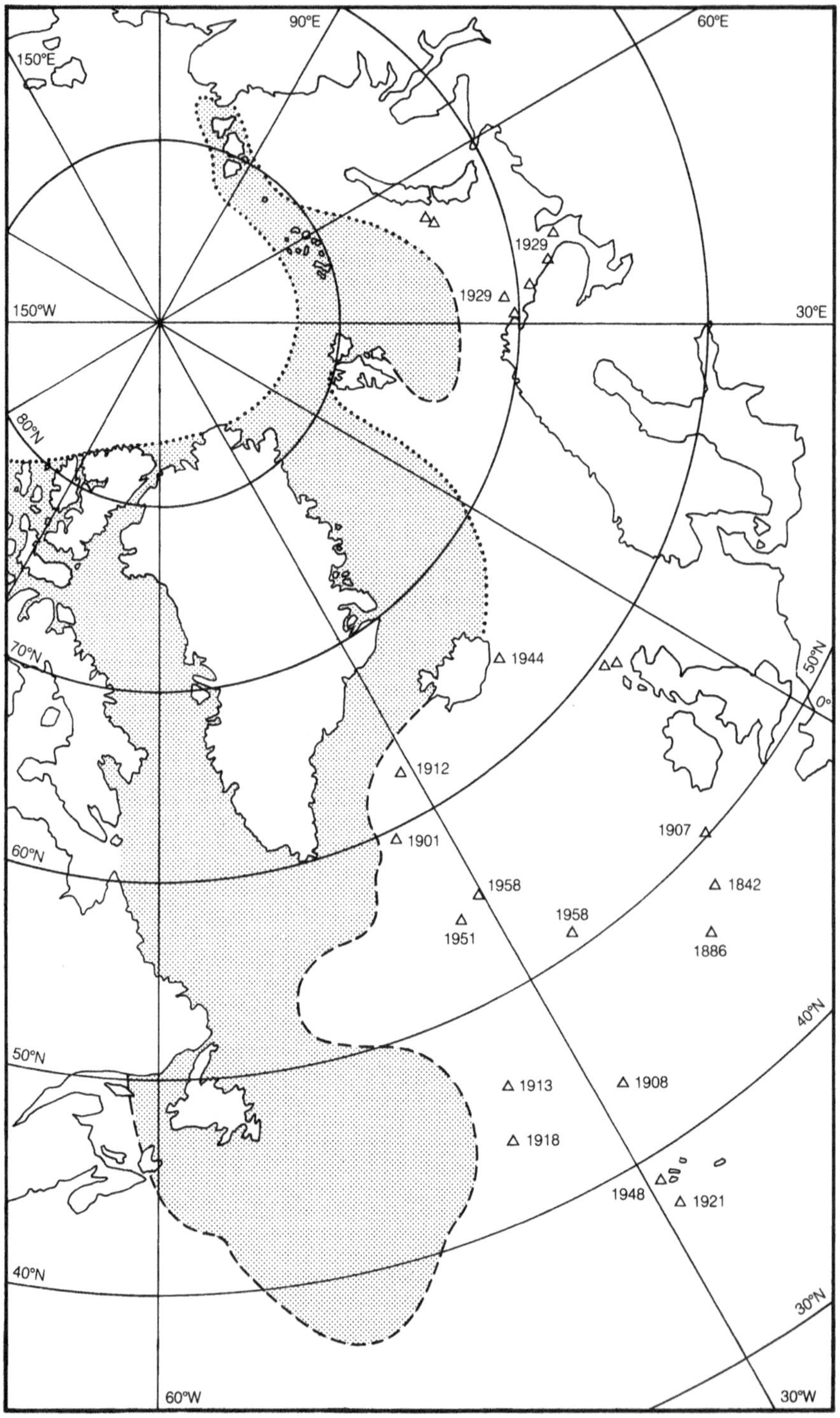

Fig. 3.3 Occurrence of icebergs in northern seas. The dashed line represents the average limit of seasonal drift of icebergs, and triangles indicate (with dates) rare and unusual sightings (data from Kovacs, 1979; Zubov, 1945). Where data on seasonal drift are not available, I have inferred the probable limit from surface current patterns. This is shown as a dotted line.

detailed analysis, and may well involve the collection of additional data.

3.2.2 Central Arctic Ocean

The Arctic Ocean north of about 75°N is permanently covered by sea ice. This is not to say that the ice itself there is the same from year to year, nor that the ice cover is quite total. In fact, the polar pack ice is far from solid and continuous, and at any time as much as 1–5% of it consists of open water or very thin ice less than 30 cm thick (Hall, 1980). It is in a state of dynamic equilibrium. Floes of multi-year ice are constantly being shed at its perimeter, moving southwards to the surrounding open waters, and as new cracks form in its interior the freshly exposed open water leads quickly freeze. This forms new thinner ice cover which will grow into multi-year ice by a process of continuing freezing, dynamic ridging and reworking. Most Arctic ice is thought to be up to 10–12 years old.

Floes of multi-year ice within the polar pack are of highly variable size, showing variations with season and location. Within the central polar pack, especially in winter, the ice is not so much an ensemble of independent floes, but rather a vast broadly coherent field of floating ice broken by intermittent open leads. In winter these open leads are separated by distances in the range 1–100 km and are typically between 10 m and a few hundred metres wide. However, in summer the leads are much more closely spaced, especially in the outermost 100 km or so of the pack, and the ice field may better be described as a group of independent floes drifting through water. Floes in the interior of the polar pack are typically of 10–100 km diameter but undergo wave and swell action as they move towards the edge of the pack, and are generally of no more than 40–80 m diameter if exposed to appreciable wave penetration at the boundary between pack ice and open water.

Statistical information on lead frequency is given by Wadhams (1983) for the M'Clure Strait and the nearby Beaufort Sea and indicates that, during winter, leads of all dimensions are spaced about 10 times further apart than in summer. Further details on multi-year floe size distribution are given for the Beaufort Sea in Section 3.2.3. In other regions data are sparse. The polar pack as a whole has two major movement systems: the Beaufort Gyre and the Transpolar Drift (Fig.

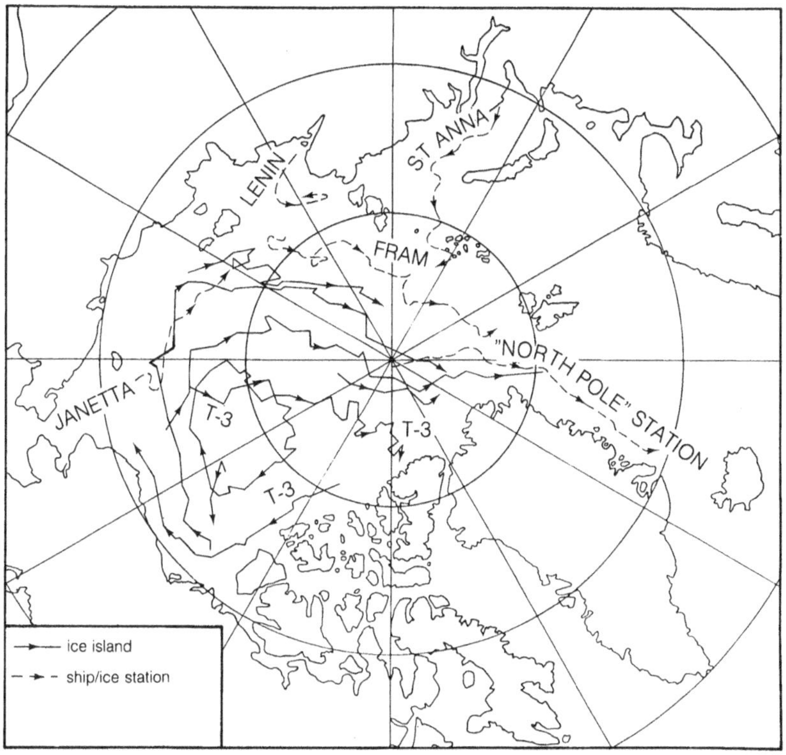

Fig. 3.4 Arctic Ocean drift patterns as determined from drift of ice islands, buoys and ships (compiled from Zubov, 1945; Central Intelligence Agency, 1978).

3.2(b)). In the Amerasian Basin, the ice rotates clockwise in the Beaufort Gyre about a point at approximately 80°N 150°W, under the influence of prevailing wind and current systems. The rotation is not coherent: at the outer edge of the gyre it completes one revolution in about 10 years, while nearer the centre it takes about 3 years. At the southern rim its drift speed is normally 1–2.5 km per day (1–3 cm s^{-1}; Colony and Thorndike, 1985), but this can reach 25 km per day during spring. The Transpolar Drift runs across the Arctic Ocean from Siberia towards Fram Strait between Greenland and Svalbard. This was discovered by Nansen in the historic voyage of his vessel, the *Fram*, in 1893–6 (Fig. 3.4). During this voyage the vessel was locked in ice and drifted freely with the pack while navigational observations were made. The passage from Siberia to Fram Strait took some 3 years, during which 2000 km were covered. This is an average drift rate of about 1.8 km per day.

Numerous other minor and contrary drift patterns exist, and these too have mostly been determined by historic voyages of vessels locked intentionally or unintentionally into the pack ice. Over the past few decades, data have also been collected from ice-island drift and buoy transponders. A compilation of the most important results is given in Fig. 3.4.

Ice of the Arctic pack is generally of 2–6 m average thickness but shows strong regional variations. Average thickness towards the Chukchi Sea is only about 2 m, but this increases towards Northern Ellesmere Island and Northern Greenland, where average thickness exceeds 6 m. The regional distribution of average ice thickness in summer and in winter is shown in Fig. 3.5, on the basis of an analysis by Bourke and Garrett (1987) of 17 submarine cruises between 1960 and 1982.

Within a particular region there is naturally wide local variability as a result of ridging and areas of new ice or open water. The best data source for statistics on ice thickness over the Arctic Ocean comes from submarine upward-looking sonar measurements. Strictly, these measure only draft of keels below sea level, but the measurements can readily be converted to approximate ice thickness by assuming a thickness to draft ratio of 1.12:1. A more sophisticated analysis based on coordinated collection of submarine sonar profiles and aerial laser surface profiles (Wadhams, 1983) yields the alternative empirical formula that total thickness h (in m) is related to submerged thickness (draft) h_s by the formula:

$$h = 0.88h_s+0.36 \qquad (3.1)$$

Typical distributions of ice thickness determined in this way are shown in Fig. 3.6 (Wadhams and Horne, 1980), summarising data collected by the submarine *USS Gurnard* in April 1976 in the Beaufort Sea. A map of the track taken is shown in Fig. 3.6(c). The complete data set, including all keels, has been analysed in probabilistic terms, and Wadhams and Horne show that the probability density function shows an approximately negative exponential form. This means that if we plot probability density on a logarithmic scale we get an approximately straight line, at least for ice of thickness exceeding about 6 m (Fig. 3.7). Note that this is *not* a log-normal distribution. Two data sets are shown in Fig. 3.7: one for Section 1 of the submarine track (Fig. 3.6(c)) in relatively nearshore transition zone conditions of the Beaufort Sea, and the other for the entire track shown.

If we represent the probability density function for draft h_s by $f(h_s)$ then we can write, for keel drafts exceeding about 6 m,

$$f(h_s) = B\exp(-bh_s) \qquad (3.2)$$

Fitting this equation to the straight portion of the plots in Fig. 3.7 yields the following approximate constants (m^{-1}) for equation (3.2):

section 1 (nearshore Beaufort Sea):	$b = 0.32$, $B = 0.66$
whole track:	$b = 0.37$, $B = 0.41$

The data have also been analysed in such a way that major keels are excluded from the distribution (Fig. 3.6(b)), and this distribution thus represents the thickness of broadly undeformed "level" ice (for details of the definition of this concept, see Wadhams and Horne, 1980). The probability function for ice defined in this way shows an interesting bimodal distribution, with major peaks of thickness at around 1 m (draft 0.8 m), probably representing thin young first-year ice, and at thickness about 3.1 m (draft 2.8 m), probably representing average undeformed multi-year ice. It is also possible to speculate that other minor peaks within the distribution can be identified with mature first-year, second year and multi-year ice, but this is not well established.

The deepest keels so far observed in the Arctic have had drafts in the region of 50 m but, in general, a keel exceeding depth 30 m can be regarded as exceptional. Deep keels are normally concentrated in areas of high ridging frequency, and occur at a rate (depending on the definition adopted of a "significant keel") of one every few hundred metres – that is, several keels per kilometre.

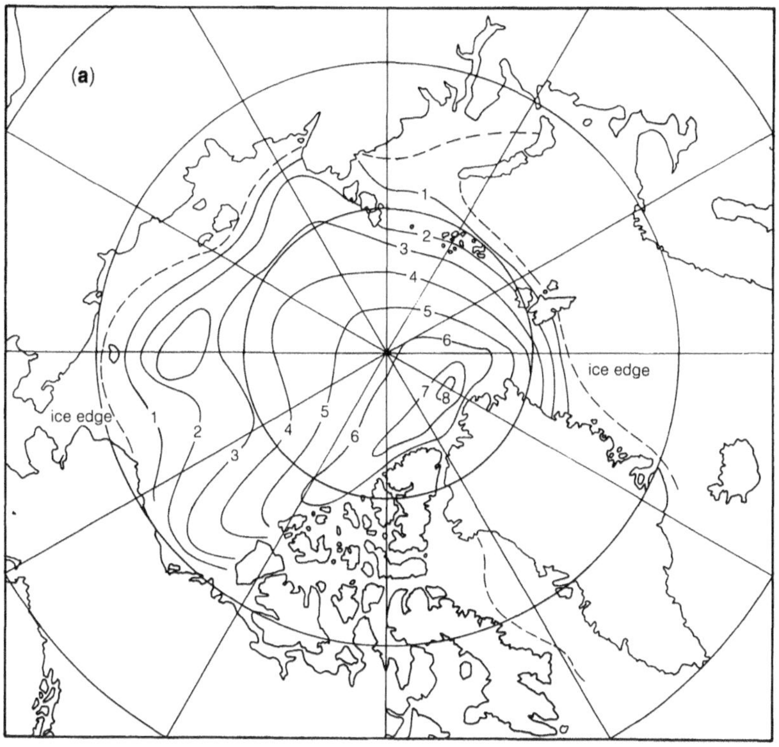

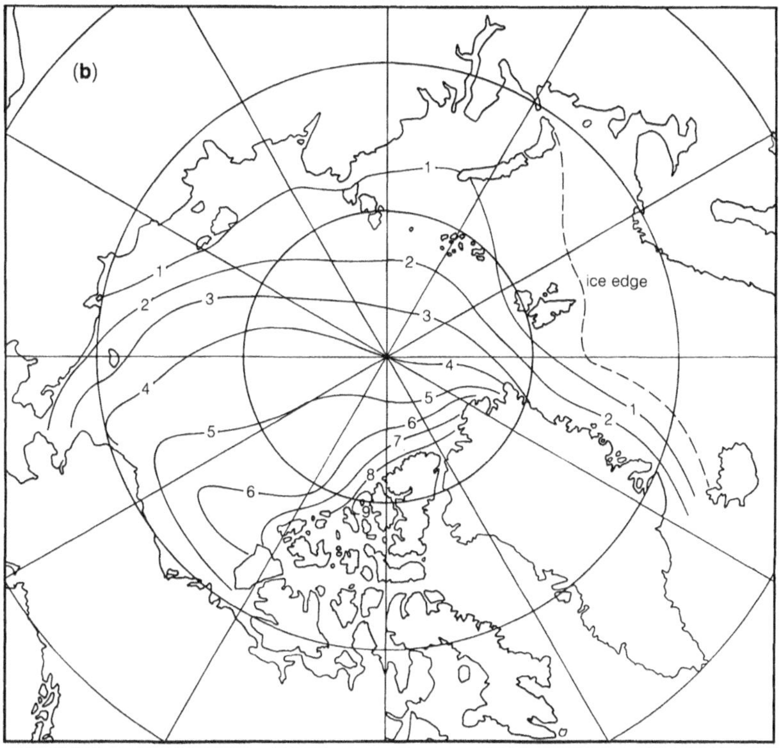

Fig. 3.5 Regional distribution of ice thickness in the Arctic Ocean, based on maps of ice draft compiled by Bourke and Garrett (1987) using sonar data from 17 submarine cruises between 1960 and 1982. Data have been converted from draft to thickness by applying a factor of 1.12 and interpolating the contours. (a) Summer (minimum thickness); (b) Winter (maximum thickness). (Bourke and Garrett also supply maps of ice draft for spring and autumn, which are not shown here.)

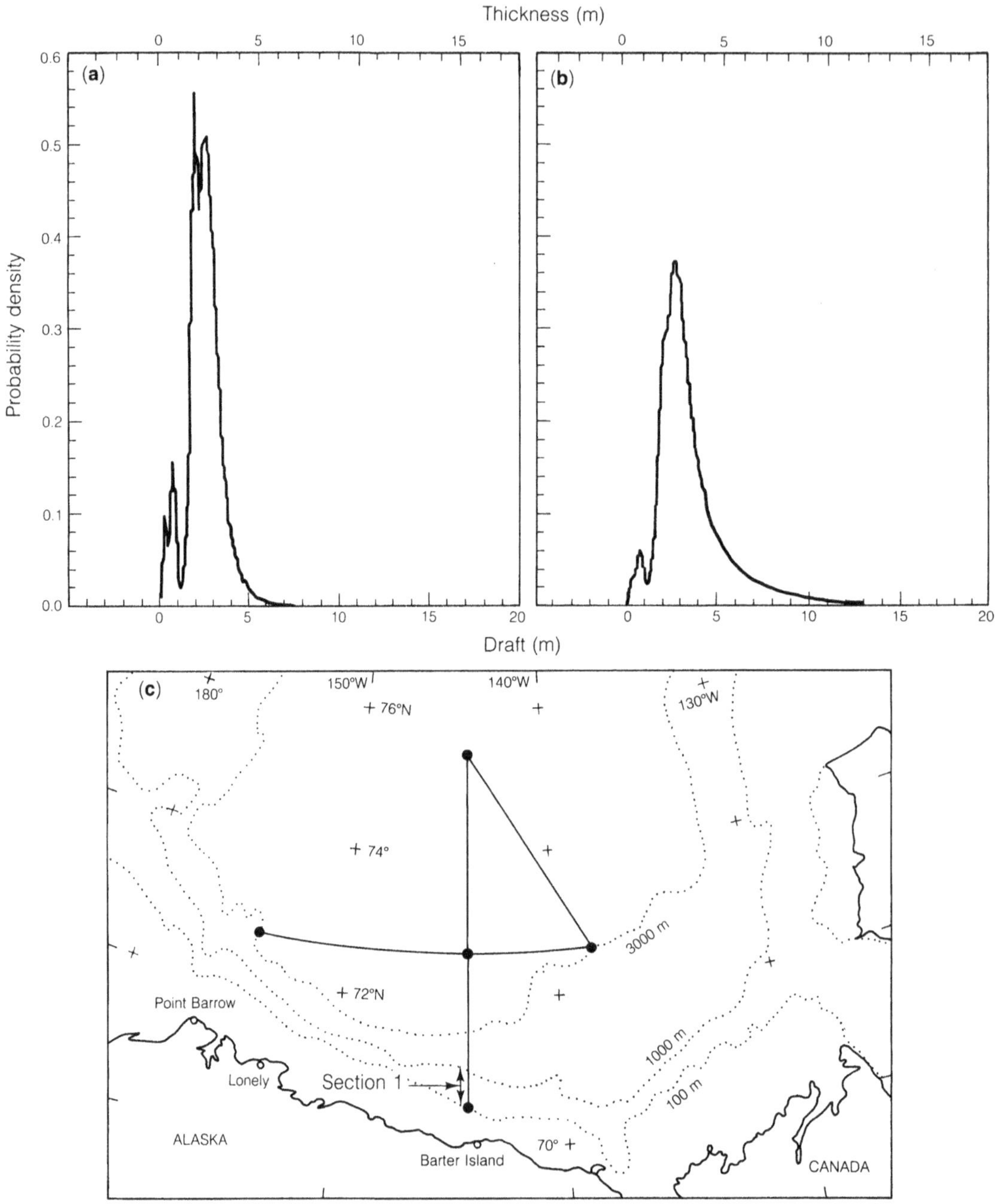

Fig. 3.6 Probability distribution for draft and (inferred) thickness of polar pack ice. (a) Overall distribution, including all keels. (b) Probability distribution for "level" ice only. Note the bimodal distribution, showing first-year and multi-year ice populations. (c) Track of the submarine *USS Gurnard* while collecting these data. (After Wadhams and Horne, 1980.)

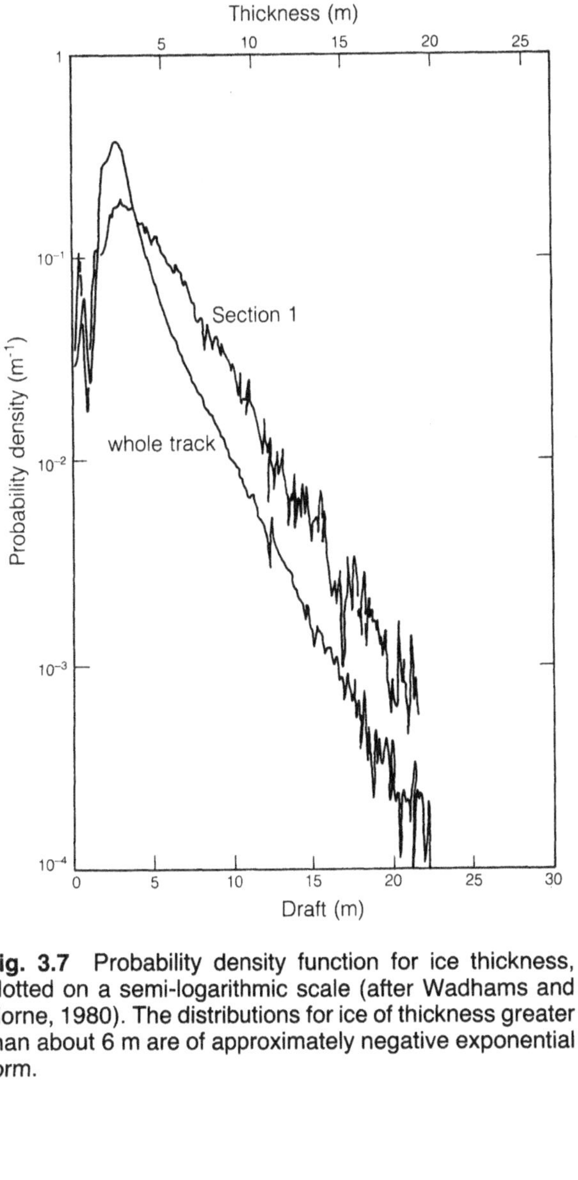

Fig. 3.7 Probability density function for ice thickness, plotted on a semi-logarithmic scale (after Wadhams and Horne, 1980). The distributions for ice of thickness greater than about 6 m are of approximately negative exponential form.

3.2.3 Beaufort Sea

3.2.3.1 General

The Beaufort Sea, as a result of intensive petroleum exploration activities, both in the Alaskan and the Canadian sectors, is the most highly studied region of the Arctic. A plan of the region is shown in Fig. 3.8(a), and a pictorial section through the continental shelf is shown in Fig. 3.9.

Moving out from the coast, a zone of bottomfast ice 2–3 m thick exists during winter, passing without noticeable change into a zone of essentially undeformed stable landfast ice. A striking feature of the Alaskan Beaufort Sea is the existence of a series of shoals or "stamukhi" occurring at approximately the 20 m bathymetric contour. Their depth below sea level is generally less than 11 m. It is common for first-year ice ridges to ground out at this point and provide pinning points which keep the shoreward landfast ice stable during winter. Some of these stamukhi are evidently not stable features but are reworked and shifted by constant ice action (Reimnitz and Kempema, 1984). Nonetheless, they effectively define the winter extent of stable landfast ice. This extent is about 25–50 km from shore. Landfast ice begins to form around mid-October and is generally present until June the next year.

Beyond the landfast ice is a so-called transition zone between stable ice and the constantly shifting polar pack ice. The edge of the pack typically moves at a rate of a few kilometres per day, often intermittently, and gives rise to intense ridging activity and shear in the transition zone, which generally lies between 20 and 60 km from the coast, in a water depth range of 20–100 m. The zone of motion is surprisingly localised at any one time, however, taking place in a narrow "slip-plane" just a few hundred metres wide (Wadhams, 1980). Within this region the ice is chaotically fractured and pulverised.

During the middle of winter the Beaufort Sea is generally 10/10 ice-covered, and during the entire period between 1 November and 15 May the ice coverage consistently exceeds 5/10. A summary of Beaufort Sea ice conditions is given in Fig. 3.8(b)–(f).

Break-up generally occurs in late June and freeze-up in late October/early November (LaBelle *et al.*, 1983). However, this is highly variable, and in 1975 break-up was so late and freeze-up so early that the passage around Point Barrow never became ice-free. A 30-year record of the length of the summer operating season for the Alaskan Beaufort Sea is shown in Fig. 3.10. This represents the length of time during which a continuous passage could be navigated around the Alaskan Coast and Point Barrow to Harrison Bay in ice coverage conditions of 5/10 or less. The value 5/10 is fairly arbitrary, but represents approximately the ice-coverage conditions in which passage by most vessels becomes severely restricted. Note the "concertina" form of the graph: if the operating season begins early it is likely to end late, and *vice versa* (Karas *et al.*, unpubl., 1984). The mean length of the season of 5/10 coverage or less is 57 days, with a standard deviation of 24 days.

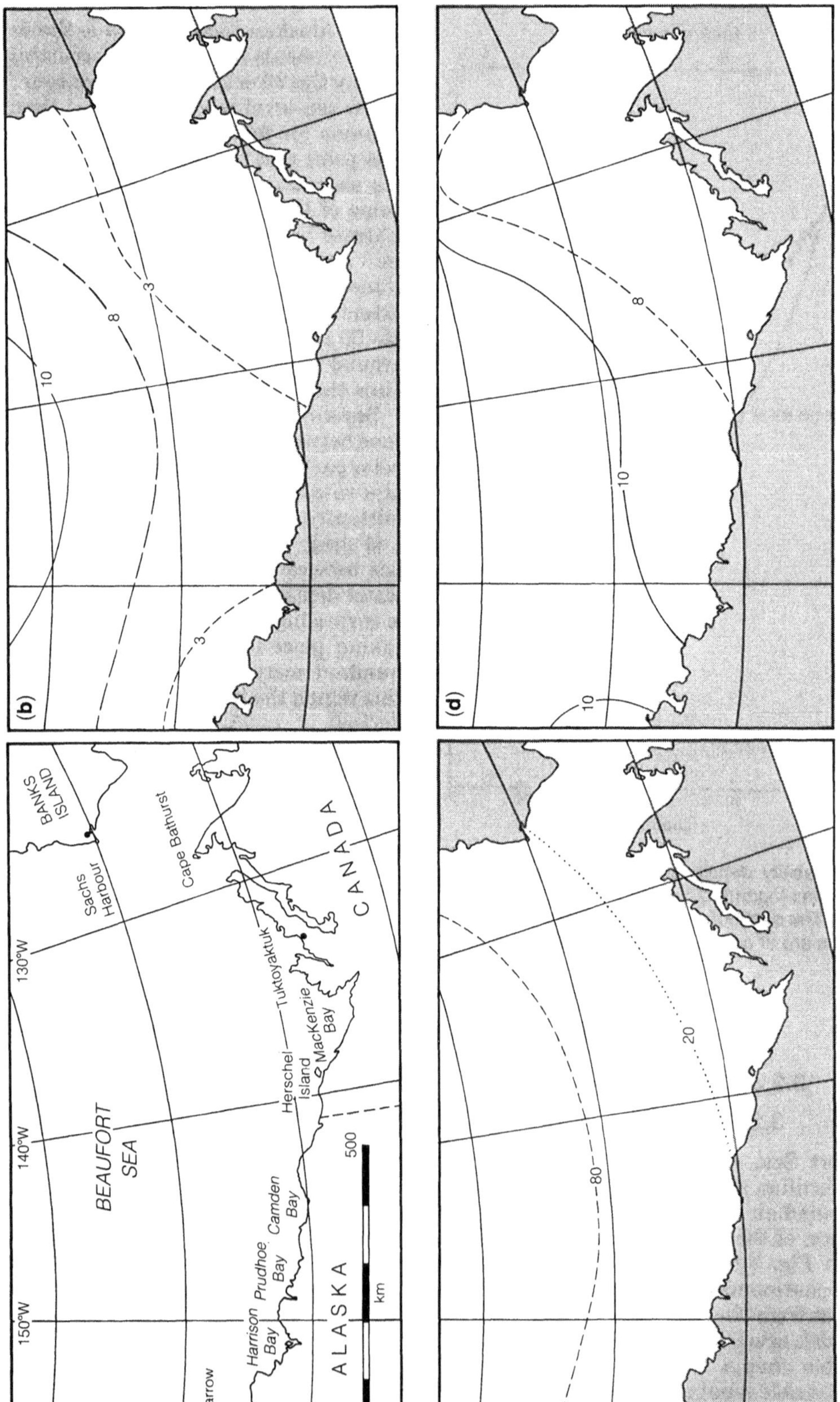
(a)
74°N
72°N
70°N
150°W
140°W
130°W
Point Barrow
Harrison Bay
Prudhoe Bay
Camden Bay
BEAUFORT SEA
Herschel Island
MacKenzie Bay
Tuktoyaktuk
Sachs Harbour
BANKS ISLAND
Cape Bathurst
ALASKA
CANADA
0
500
km
(b)
10
8
3
3
(c)
80
20
(d)
10
8
10
10

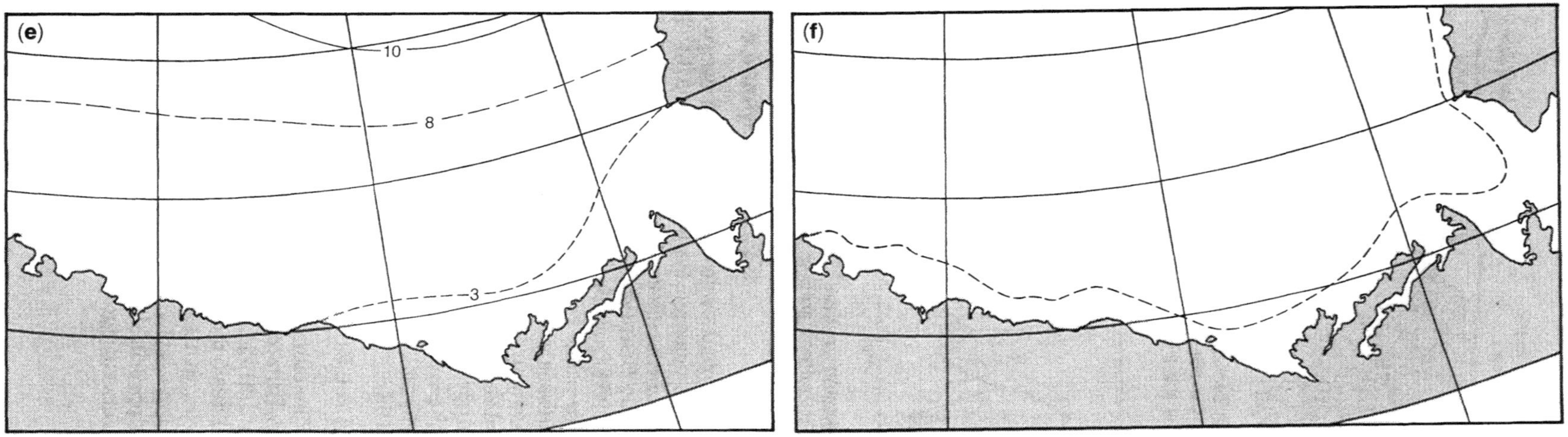

Fig. 3.8 Ice conditions in the Beaufort Sea: (a) general map of region; (b) average ice concentration during August (minimum extent); (c) chance (%) of encountering ice of concentration greater than 3/10 during August; (d) average ice concentration during June (beginning of summer season); (e) average ice concentration during September (end of summer season); (f) edge of landfast ice during winter (February–April). Between October and May the ice cover over this whole region is effectively total (greater than 9/10).

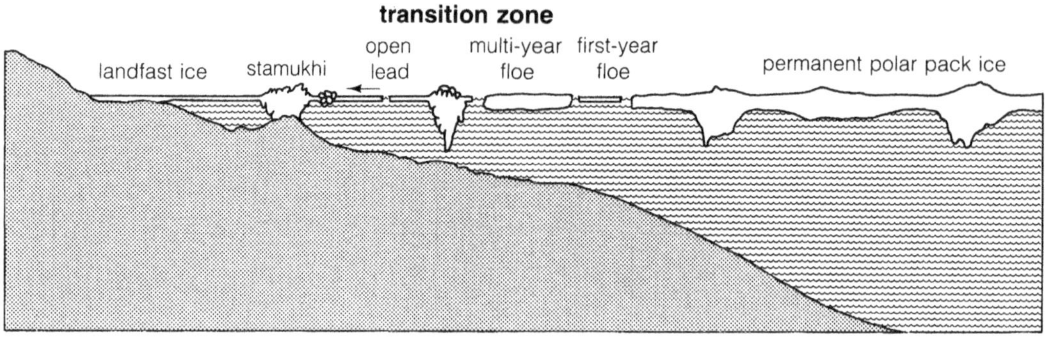

Fig. 3.9 Pictorial section through the continental shelf of the Beaufort Sea, showing typical ice types. Compression ridges grounded on stamukhi shoals provide pinning points for the landfast ice. Between the landfast ice and the moving polar pack lies a zone of intense ice activity known as the transition zone. Not to scale.

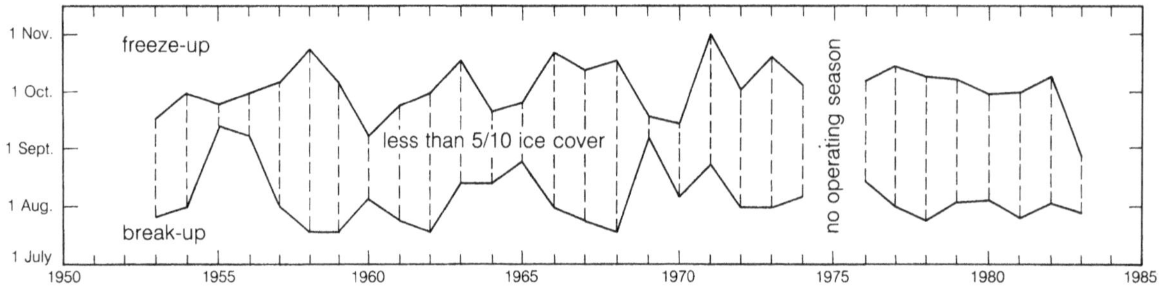

Fig. 3.10 Opening and closing of the 5/10 ice cover "operating season" for the Alaskan Beaufort Sea (Harrison Bay). Note that early break-up is correlated with late freeze-up (after Karas *et al.*, 1984, unpubl.).

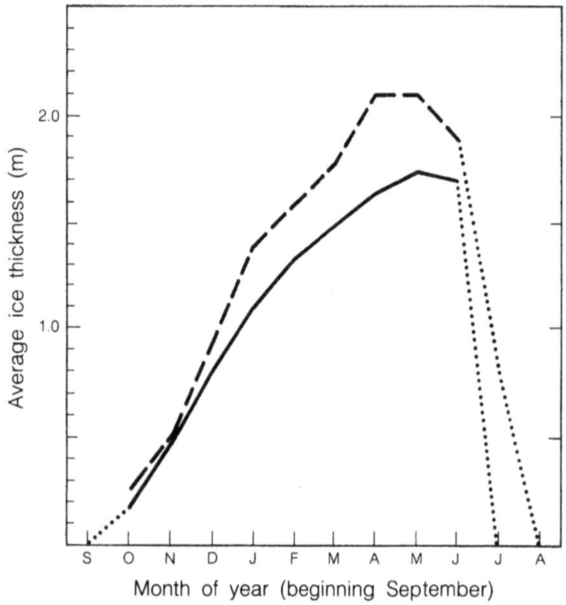

Fig. 3.11 Cycle of average seasonal ice thickness (solid line) for Herschel Island in the Beaufort Sea. Dashed line shows extreme ice thickness (data from Westermann, 1984, unpubl.).

3.2.3.2 First-year ice

A typical cycle of growth and decay of nearshore first-year ice is shown in Fig. 3.11. This is based on data collected at Herschel Island in the Canadian Beaufort Sea and compiled by P. Westermann (1984, unpubl.). In this and later similar figures for other regions, average and extreme ice thicknesses are calculated from a combination of direct measurements and inferences from meteorological observations using the "freezing degree-day" approach, which relates ice thickness h (in m) to the cumulative number of freezing degree-days N_θ (in °C) by the empirical formula (LaBelle *et al.*, 1983):

$$h = 0.01\sqrt{(8N_\theta - 501)} \qquad (3.3)$$

The cumulative number of freezing degree-days is calculated as mean daily temperature below 0°C multiplied by number of days for which it persists. Thus 5 days at a mean daily temperature of −10°C constitutes 50 freezing degree-days. The formula applies well only to fast ice formed *in situ* without appreciable snow cover, but is a serviceable general rule.

The mobility of first-year ice is highly variable. In most years the landfast ice close to shore (< 10 km) is truly landfast and stable during the winter months, moving no more than a few metres (perhaps 1–5 m) during a season. However, at the 20 m bathymetry line relatively large episodic movements are quite common: during periods of high wind, movements of several hundred metres may occur over a period of a few hours. Maximum reported movement rates in nominally landfast ice have been in the range 10–50 m h^{-1} (Agerton, 1981; Spedding, 1983). Further out into the transition zone, movement may be practically continuous at a rate of several kilometres per day and may involve both first-year and multi-year ice.

3.2.3.3 Multi-year ice

During summer, the Beaufort Sea typically contains approximately 5–30% areal coverage of multi-year ice. The coverage in the Canadian Beaufort sector is typically in the range 5–10% (Marcellus and Morrison, 1982), while in the Alaskan sector, particularly towards Point Barrow, the concentrations are frequently in the range 10–30% (Markham, 1981) even during August. After

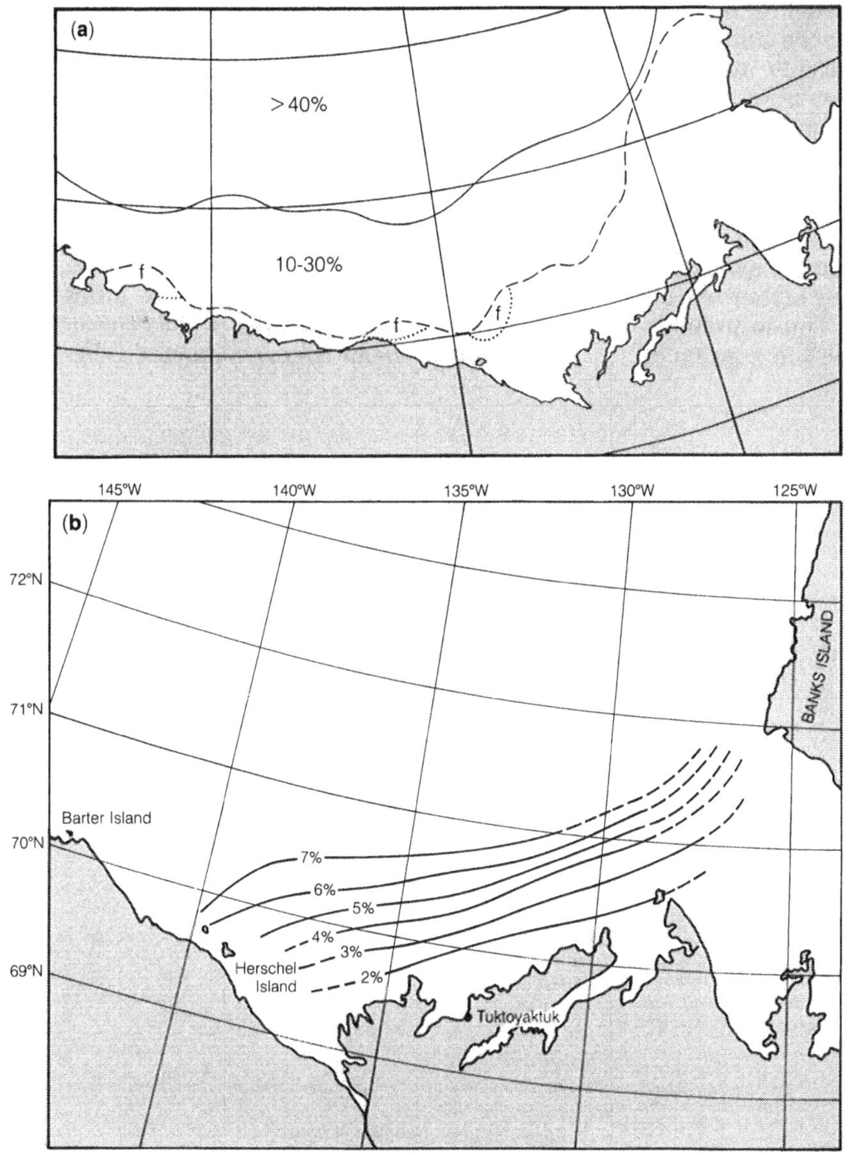

Fig. 3.12 (a) Median amount of multi-year ice in the Beaufort Sea, for end of July (after Markham, 1981). The solid line shows the boundary of 40% coverage or greater; the dashed line shows the boundary of 10–30% coverage; areas denoted "f" and enclosed by dotted lines indicate "a few old floes". (b) More detailed statistics plotted for Canadian Beaufort Sea (after Marcellus and Morrison, 1982). Lines show average concentration of multi-year ice during summer (%).

severe winters, the concentration along parts of the Alaskan coast near Point Barrow may approach 100% throughout the summer. Typical concentrations of multi-year ice are mapped in Fig. 3.12.

The most important other parameters from the point of view of offshore hazard to structures are:

- size distribution of floes
- thickness
- velocity

When combined with ice concentration statistics these enable the calculation of probability of impact occurring and, once impact has occurred, estimation of the likely force exerted on a structure. The parameters – especially the number of floes impacting – are highly variable and site-specific, and the quantities given here are for guidance only. Methods of calculating impact probability are examined later, in Chapter 7.

Size distribution of floes. In the nearshore Beaufort Sea, multi-year floes are typically of diameter up to 1 km. Floe size varies systematically with ice concentration, being higher within areas of high concentration of ice. This is probably due to wave attenuation effects within a group of floes, which decrease the risk of break-up by flexure, and provide safety in numbers: a large isolated floe in open water with an appreciable swell stands little chance of remaining intact.

Multi-year floes of diameters 10–20 km have been reported during the summer season, and are regularly of this size some 200 km offshore (see, for instance, Lindsay, 1981). The best detailed analysis of floe size distribution characteristics is given by Rothrock and Thorndike (1984) using data from LANDSAT imagery, SLAR imagery and aerial photography. The principal results of their analysis are shown in Fig. 3.13(a). This shows cumulative number distribution $n(L)$ of mean caliper diameter L, and is to be interpreted as follows:

(a) "Mean caliper diameter", L, may be thought of simply as floe diameter. Strictly, however, it is the diameter obtained by imagining that we measure the diameter of an irregularly shaped floe using parallel calipers and then take the average caliper reading as the floe is rotated through 360°. For a perfect disc, of course, mean caliper diameter is precisely equal to diameter. The area $\mathcal{A}$ of a typical floe is related to mean caliper diameter L by the

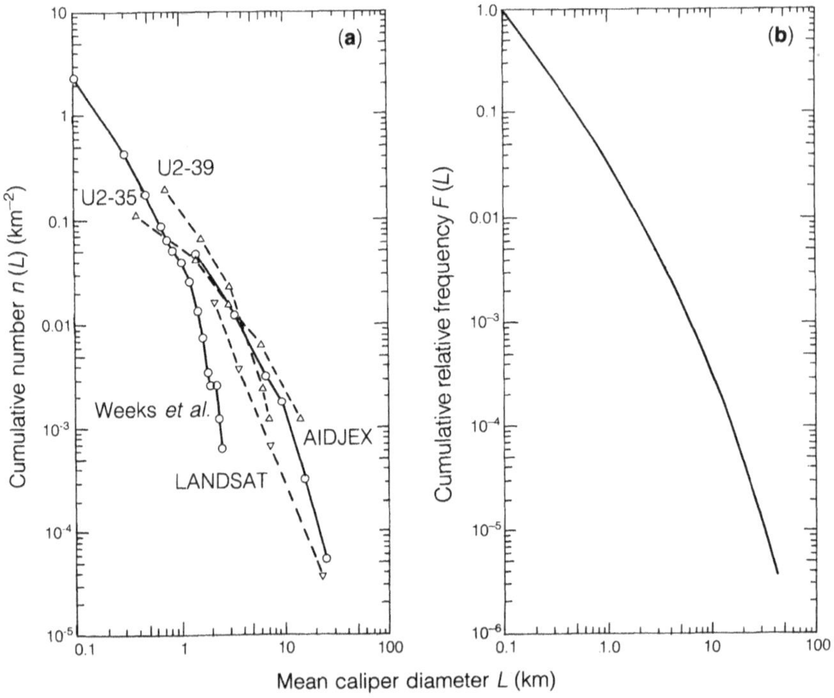

Fig. 3.13 Statistics on multi-year ice floe diameters in the Beaufort Sea and the Arctic Ocean. (a) The cumulative number distribution n (L) of floes per km^2 with mean caliper diameter L or greater (after Rothrock and Thorndike, 1984). The wide divergence between the five plots shown demonstrates the effect of regional and seasonal variability. See text for data sources. (b) The cumulative relative frequency $F(L)$ of floes of mean caliper diameter greater than L, for a representative line drawn through the distributions of Fig. 3.13(a).

approximate formula:

$$\mathscr{A} = 0.66L^2 \quad (3.4)$$

Note that this is slightly less than the relationship

$$\mathscr{A} = \pi L^2/4 = 0.785L^2$$

for a perfect disc, reflecting the irregularity of real floe shape.

(b) "Cumulative number distribution", $n(L)$, represents the number of floes per km^2 with mean caliper diameter greater than or equal to L.

(c) The data sources in Fig. 3.13 are:

- Weeks *et al.* (1980), based on SLAR (Sideways-Looking Airborne Radar) imagery of 1644 km^2 north-west of Barrow and 2375 km^2 north of Cape Simpson, on 23 September 1975.
- AIDJEX (Arctic Ice Dynamics Joint Experiment) aerial photograph mosaic covering 10 450 km^2, some 400 km offshore in the region of the AIDJEX experiment at about 145°W 77°N.
- LANDSAT image covering 28 675 km^2 taken during August 1973.
- Aerial photographs of areas totalling 900 km^2.

All the data broadly come from the Beaufort Sea, but stray into the Chukchi Sea (Weeks *et al.*, 1980) and into the far offshore Arctic Ocean (AIDJEX experiment). For design purposes, it is of course desirable to collect site-specific and seasonally defined data, but very few data of such quality are yet publicly available. The data of Fig. 3.13 show wide variability, particularly in respect of large floes (greater than 1 km diameter), but indicate, for instance, that floes of diameter 1 km or larger occur with a frequency of between 0.04–0.13 km^{-2} (that is, 4–13 per 10×10 km square). Extremely large floes, of diameter 20 km or more, occur in the far offshore Arctic Ocean with a frequency of about 10^{-4} km^{-2} (that is, one such floe in a square 100×100 km). However, in the nearshore environment they are considerably less common: the largest floe diameter observed by Weeks *et al.* (1980) was 3.6 km.

For design load calculations it is therefore not necessarily appropriate to use the number densities provided in Fig. 3.13(a). In Chapter 7 when we come to look at multi-year impact risk, I shall propose a two-fold approach to the problem of floe frequency and diameter distribution:

(1) Define the cumulative relative frequency (rather than cumulative absolute number) distribution for mean caliper diameter. This is done simply by dividing the cumulative number distribution by the total number of floes per km^2.

(2) Use site-specific statistics on mean multi-year ice concentration in order to convert this to a number of floes km^{-2} as a function of size.

In this way, statistics can be generated which are consistent with local observations of ice concentrations. A suitable cumulative relative frequency function (for floes of diameter exceeding 0.1 km) is shown in Fig. 3.13(b) and will be used in Chapter 7: it is simply a middle line drawn through the distributions of Fig. 3.13 and normalised by dividing by the total number of floes per km^2 (in this case 2.2 floes km^{-2}). The ordinate represents the probability $F(L)$ that a randomly selected floe has a diameter exceeding L. The distribution shown has a mean diameter $\bar{L} = 0.34$ km and the mean floe area $\bar{\mathscr{A}}$ is 0.23 km^2.

Now, if we know that the areal coverage of multi-year ice is a certain proportion c, then the mean number of floes per km^2 having a diameter exceeding L is simply:

$$\frac{c}{\bar{\mathscr{A}}}F(L)$$

As we have seen, typical areal coverage of multi-year ice in, for instance, the nearshore Canadian Beaufort Sea is in the range 3–5% during summer. For the floe size distribution given (Fig. 3.13(b)) this represents 0.13–0.22 floes km^{-2} or 13–22 floes 100 km^{-2}. This, as expected, confirms that there are rather fewer multi-year floes in nearshore areas than might be inferred directly from Fig. 3.13(a). Work on seasonal variability of floe density and floe size is in progress (Stringer, 1986).

Thickness. Data on ice thickness for the Arctic polar pack have already been shown in Figs 3.5, 3.6 and 3.7, but generally refer to ice several 100 km offshore. These data were all collected by submarine, generally in water depths between 100 and 3000 m.

Few published data exist on multi-year ice thickness in the nearshore regions, since they lie in water depths of 20–50 m and are not generally accessible by submarine. Data which do exist consist largely of isolated landings on multi-year floes, in which thicknesses were measured by auger holes (see, for instance, Fig. 2.18). In general, these studies are aimed at investigating major multi-year ridges or particularly thick or interesting-looking floes – and the data set is therefore very likely to be distorted. They do not represent a random sample from the floe population and are not necessarily average or typical. However, it might be expected that nearshore open-water floes should, in general, be thicker than the average encountered further offshore: during summer, when the edges of the polar pack break up and are exposed to open-water wave action, it is the thickest floes which stand the highest chance of surviving as they drift towards the shore. This means that the thickness distribution for drifting multi-year floes during summer is not necessarily the same as that derived from submarine profiles of winter pack ice.

There is another problem with directly applying the statistics derived from submarine profiles: we have to decide whether we are interested in statistics "at a point" on the ice cover, or in statistics "for a floe". These two concepts may differ. In collecting statistics from a submarine we measure the variability of thickness along a line travelled by the submarine, and we can derive statistics such as the mean and standard deviation of thickness at a point and the shape of the distribution. Typically, the major variations of thickness occur on a distance scale of several hundred metres: that is, a typical keel normally extends for several hundred metres in the horizontal direction. This means, of course, that if we are interested in statistics for total average thickness of a floe of diameter a few kilometres then we should be taking an average over a significant number of individual points. This kind of statistic is of use in the calculation of average kinetic energies of impacting floes, and the standard deviation of "average thickness of a floe" is naturally rather less than the standard deviation of "thickness at a point".

The difference between these thickness statistics can to a certain extent be quantified, but not without a large degree of quite careful processing of raw submarine sonar data. It involves defining first of all a criterion to delineate "a floe" on the basis of submarine sonar data alone, and then analysing the statistics of average thickness of floes so defined. The problem has been tackled by Marcellus and Morrison (1982) using tracks from the *Sargo* (1960) and the *Seadragon* (1962), but the details are quite sensitive to the manner in which the definition of "a floe" is implemented. Their criterion, for instance, defines a floe as any stretch of ice with draft continuously exceeding 2 m and, additionally, they restrict the analysis to floes with diameter exceeding 500 m. Because of these limitations, and also because the statistics are biased towards the relatively far offshore, I shall not reproduce the results here.

Velocity. Ice movement occurs by a combination of wind and current, but is most strongly correlated with wind. A detailed discussion of driving forces is given in Section 7.5. Movement rates generally lie in the range 0.01–1.0 m s^{-1}, and the average speed of ice floes during summer in the Canadian Beaufort Sea is about 0.18 m s^{-1} (Marcellus and Morrison, 1982). During winter, beyond the transition zone, the average speed is about 0.01–0.05 m s^{-1}. Exceedance plots for the Canadian Beaufort Sea are given in Fig. 3.14.

As a rule of thumb, experience shows that ice floe velocity is typically in the range 2–4% of the wind speed; the assumption of 4% is a conservative upper bound, and for the data of Fig. 3.14 the assumption of 3.5% appears to give the best fit. However, as we shall see later, in Section 7.5.1, that there are good theoretical grounds for maintaining that the terminal velocity of a wind-driven ice floe is about 2.5% of the wind speed. The direction of motion is influenced by the Coriolis effect, and is found generally to lie about 28° to the right of the direction of wind (Zubov, 1945).

3.2.3.4 Icebergs

Icebergs of true glacial origin (as distinct from ice islands) are exceedingly rare in the Beaufort Sea, and have not to my knowledge ever been reported. The reason for this is fairly clear: although icebergs are produced by the minor glaciers of the Queen Elizabeth Islands, they are effectively prevented by the prevailing currents from drifting out into the Beaufort Gyre. All the currents from the Arctic Ocean (Fig. 3.2(b)) pass southwards through the islands and sweep icebergs towards Parry Channel and, eventually, Baffin Bay.

There is, however, no reason totally to exclude the possibility of icebergs being transported into the Beaufort Sea. Icebergs are very common in Nansen Sound (see Fig. 3.18(a)) and it would be surprising if they were not sometimes incorporated into the Beaufort Gyre. As a risk to structures, however, they are invariably neglected.

Ice islands have been observed in large numbers in the Beaufort Sea, though not on a regular basis. The principal incursion of ice islands occurred during the period 1972–5 (Sackinger *et al.*, 1985) and is thought to have been the result of a massive calving event from Ward Hunt Ice Shelf 10 years previously, over the winter of 1961–2. In 1972,

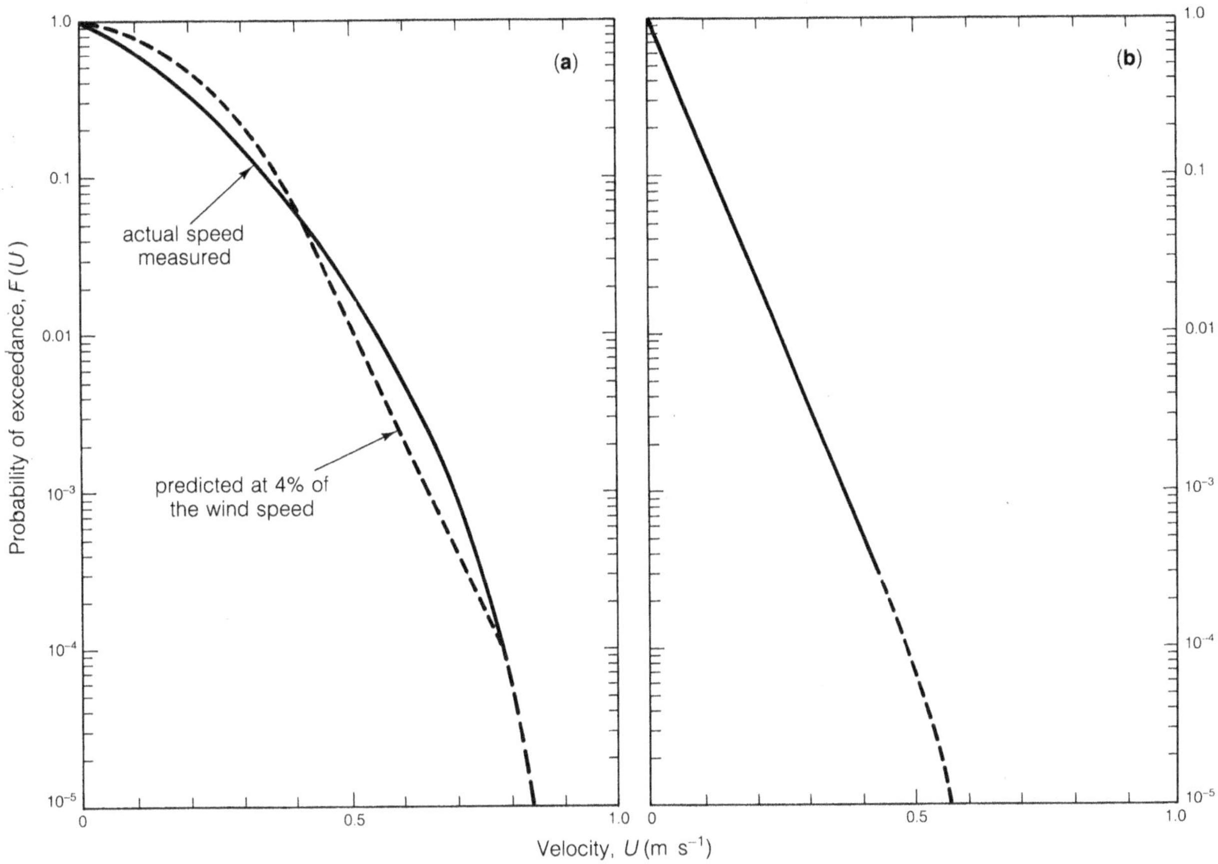

Fig. 3.14 Exceedance plot for velocity of multi-year floes: (a) in summer–the ordinate shows the probability $F(U)$ that the velocity of a floe exceeds U, and the dashed line shows a calculation based on 4% of wind speed; (b) in winter.

433 ice island fragments were observed along the coast of the Beaufort Sea, the largest being over 1.6 km in length, near Flaxman Island. A year later, 299 fragments were observed, of which the majority (65%) were grounded in water depth 25–35 m and the remainder in less than 20 m water depth. Evidently these were fragments only, and not ice islands of the true original thickness 40–60 m. In 1974, only 27 fragments were observed, and in 1975, none. This shows fairly clearly that ice islands appear episodically, and are not a simple stationary stochastic process. With routine surveillance of the ice shelves which generate ice islands, and of the Beaufort Gyre which transports them southwards, their arrival should be predictable and come as no surprise.

Nonetheless, calculations have been performed for the risk of incidence of ice islands. Dunwoody (1983) calculated that a typical structure in the Canadian Beaufort Sea runs an annual risk of about 0.001 of being struck by an ice island – that is, a return period of about 1000 years.

3.2.4 Chukchi Sea

3.2.4.1 General

The Chukchi Sea is bounded to the south by the Bering Straits and to the north roughly by a line joining Point Barrow to Wrangel Island. The Chukchi Sea is largely ice-covered (between 50 and 100%) between mid-November and mid-June, but is generally ice-free, at least in its southern regions, during the months of August and September. A summary of ice-coverage conditions is given in Fig. 3.15.

The area has not been as extensively studied as the Beaufort Sea, and information is lacking, particularly on relative concentration of first-year and multi-year ice. Multi-year ice is, however, evidently quite common in the Chukchi Sea, as it is

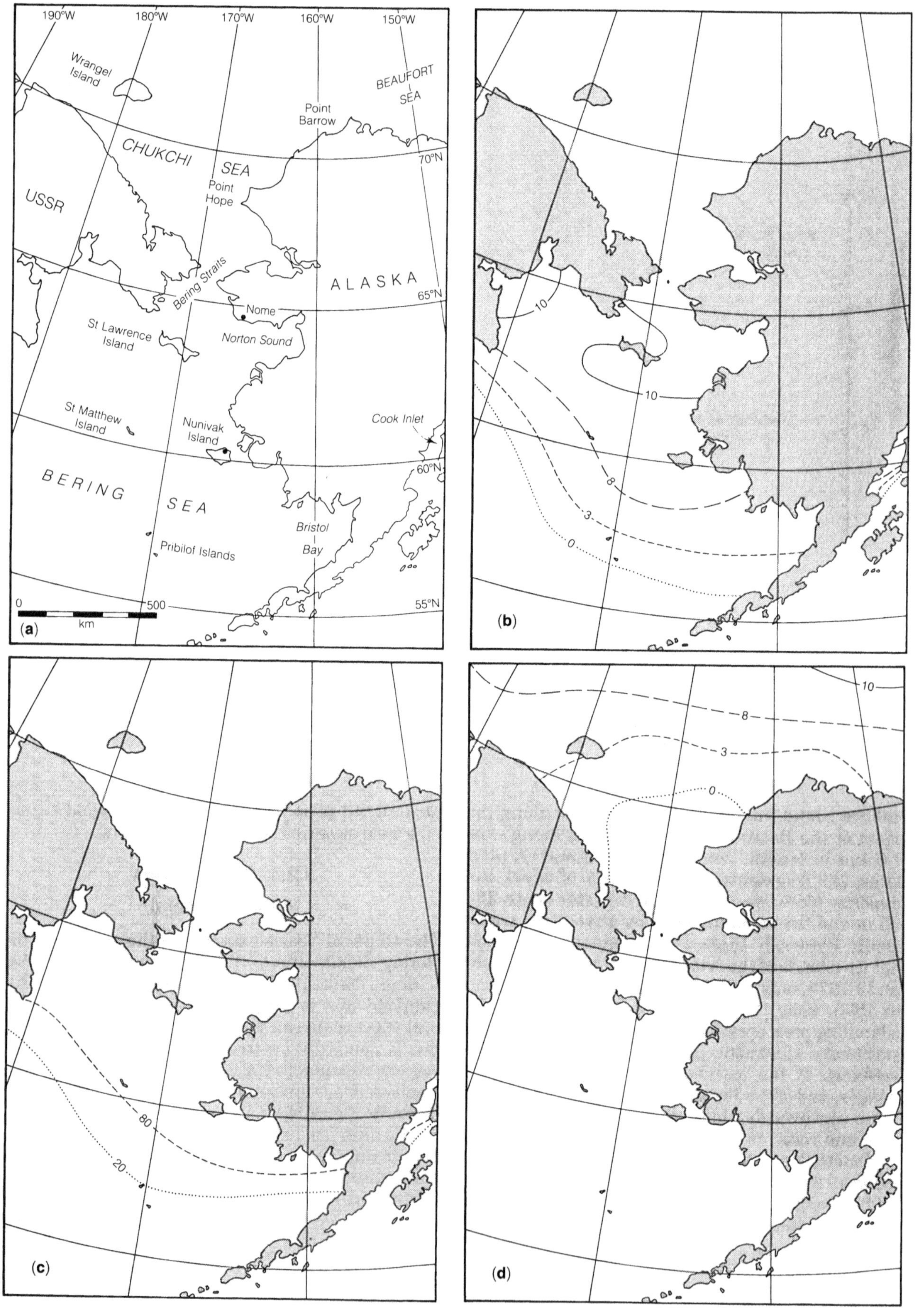
190°W
180°W
170°W
160°W
150°W
Wrangel Island
BEAUFORT SEA
Point Barrow
CHUKCHI SEA
70°N
Point Hope
USSR
Bering Straits
ALASKA
65°N
Nome
St Lawrence Island
Norton Sound
St Matthew Island
Nunivak Island
Cook Inlet
60°N
BERING SEA
Bristol Bay
Pribilof Islands
0
500
km
55°N
(a)
10
10
8
3
0
(b)
80
20
(c)
10
8
3
0
(d)

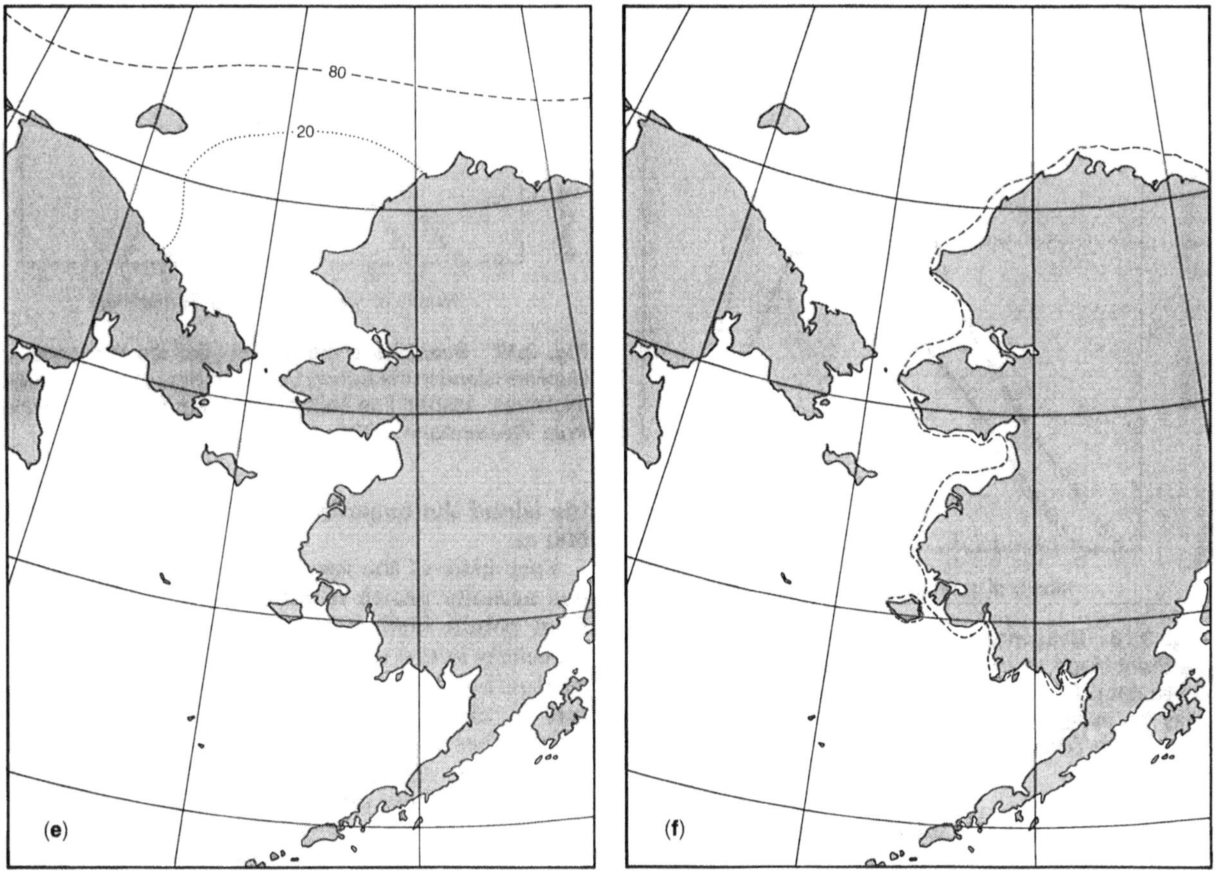

Fig. 3.15 Ice conditions in the Chukchi Sea and Bering Sea: (a) general map of region; (b) average ice concentration during February (maximum extent); (c) chance (%) of encountering ice of concentration greater than 3/10 during February; (d) average ice concentration during August (minimum extent); (e) chance (%) of encountering ice of concentration greater than 3/10 during August; (f) edge of landfast ice during winter (February–March).

readily fed into the area from the polar pack by the motion of the Beaufort Gyre between Barrow and Wrangel Island. However, the currents and winds in the Chukchi Sea are complex, and ice floes are not easily transported south all the way towards the Bering Straits. In general, ice drift patterns show flow towards the south-east along the north coast of the Soviet Union towards Cape Deshneva, but then a sharp turn towards the north on reaching the Bering Straits (Figs 3.2(a) and 3.4). It thus appears likely that there is a rapid diminution in the probability of occurrence of multi-year ice as we travel southwards. Very little reaches the Bering Straits, and even less passes through the straits: sea currents are predominantly south to north through the Bering Straits, and surface currents of up to 1.5 m s^{-1} have been observed during the summer months (Zubov, 1945).

3.2.4.2 First-year ice

First-year ice in the Chukchi Sea generally reaches a maximum thickness of about 1.25–1.4 m during its annual cycle. Seasonal average ice thicknesses are shown in Fig. 3.16 for observation points at Point Hope and Wales (on the Alaskan coast of the Bering Straits). The average extent of the coastal fast ice zone is shown in Fig. 3.14(f). Little appears to be known about the mobility of first-year ice during the winter months. However, extensive ridging of first-year ice is apparent: ridges are observed at a frequency of approximately 3–5 per km of travel, and have average

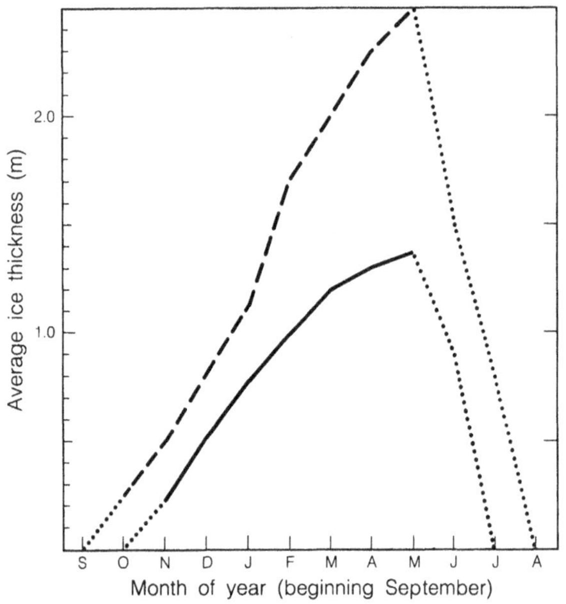

Fig. 3.16 Seasonal average first-year ice thicknesses, for Point Hope on the Alaskan coast of the Chukchi Sea (solid line). Dashed line shows maximum thickness, but may be exaggerated by the presence of multi-year ice. (Data from Westermann, 1984, unpubl.)

sail heights of 1.5 m with a typical range up to 3.7 m (Weeks *et al.*, 1980).

I have been unable to find statistics on concentration of multi-year ice in the Chukchi Sea, and there appear to be no reported sightings of icebergs or ice islands.

3.2.5 Bering Sea

3.2.5.1 General

Ice in the Bering Sea consists almost entirely of first-year ice, with negligible concentrations of multi-year ice. A few multi-year ice floes have been observed in the Bering Straits, and some as far south at St Lawrence Island, but they are extremely rare (Kovacs *et al.*, 1982). I know of no reports of multi-year ice south of this. Ice concentration statistics are summarised in Fig. 3.15, and the average seasonal cycle of thickness is shown in Fig. 3.17. Ice in the Bering Sea is not stable, but is in a constant state of drift, predominantly southwards with the prevailing sea currents. Most of the ice, even in winter, is in the form of modest size floes rather than a continuous sheet. North of St Lawrence Island they are typically of diameter 500 m to several kilometres, but south of

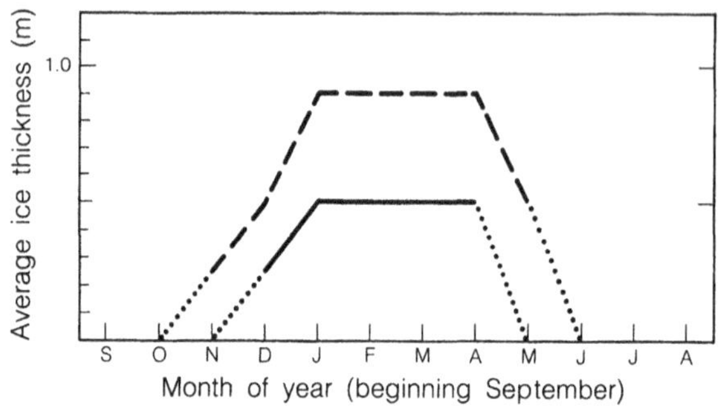

Fig. 3.17 Seasonal cycle of first-year ice thickness at Nunivak Island in the Bering Sea. Solid line shows average thickness; dashed line shows maximum thickness. (Data from Westermann, 1984, unpubl.)

the island the majority are of diameter less than 500 m.

Very little of the ice drifting in the Bering Sea has actually grown *in situ*. Most of it originates from points further north. Recurring polynyas, especially in the region of St Lawrence Island and Norton Sound, are thought to play a significant role in the generation of new ice cover (Pease, 1980). These wind-induced polynyas (Fig. 2.21) are a frequent (though not invariable) feature in winter, especially during periods of strong northerly winds. As ice is driven southwards away from the shoreline, large expanses of cold open water are exposed to cooling winds and freeze quite rapidly and efficiently. The polynya to the south of St Lawrence Island often extends for 160 km to the south. Similar features appear southwards of Nunivak Island and Bristol Bay, but being further south these are probably less efficient in the generation of new ice cover.

First-year ice ridges are a common feature in the Bering Sea, though generally smaller than in the Chukchi Sea owing to the smaller undeformed ice thickness from which they form. Ridges commonly occur with a frequency of about 5 per km and are typically of maximum sail height in the range 1.0–2.4 m.

3.2.6 Canadian Arctic Islands

3.2.6.1 General

The Canadian Arctic Islands consist of the Queen Elizabeth Islands and the series of islands between Banks Island and Baffin Island, which lie to the South of the Parry Channel. In the waters of these islands, all the different forms of floating ice can be found in great abundance.

The sheer complexity of the archipelago and the narrowness of its numerous channels mean that ice in this region is stable and landlocked for most of the year. During the winter months from mid-October to mid-July the entire area is effectively covered with 10/10 cover of first-year and multi-year ice, and many areas do not properly clear from year to year. The area just north of Melville Island, for instance, is frequently covered with 10/10 consolidated multi-year ice for practically the entire year, clearing only, if at all, for a few weeks during September and October. A map of ice coverage conditions is shown in Fig. 3.18. Much of the multi-year ice in the region has in fact grown stably *in situ*, landfast for most of the year, unlike multi-year ice of the Arctic Ocean, which has grown more dynamically while drifting in the polar pack.

Ice islands and icebergs are generated in the Arctic Island, and are a common sight, often locked in landfast ice.

3.2.6.2 First-year ice

Seasonal first-year ice in the Arctic Islands typically grows to a maximum thickness of 2.0–2.2 m during its annual cycle. A typical growth curve is shown in Fig. 3.19. More detailed statistics on winter ice thickness are given by Wetzel (1981). The ice is in most areas effectively immobile for the winter months, and during the months of January to June nearshore ice in the region south of Ellef Ringnes Island typically moves less than 6 m (Baudais *et al.,* 1976). This is not true of all regions, and much of the landfast ice is not even as stable as the landfast ice within the stamukhi zone of the Beaufort Sea: transponder movements south of Cornwall Island, in Belcher Channel, have shown irregular movements of several hundred metres during winter (Wetzel *et al.*, 1974). Furthermore, ice in the principal waterways of the North-West Passage, for instance Lancaster Sound, is frequently highly mobile even in the depths of winter, moving at rates of several kilometres per day. However, most of the ice of the internal Arctic Islands can be regarded as landfast, or at least quasi-landfast, throughout the winter.

3.2.6.3 Multi-year ice

Multi-year ice which has been generated in landfast conditions in the Queen Elizabeth Islands can reach extraordinary thicknesses. Submarine cruises through M'Clure Strait by *USS Sargo* and *Seadragon* in 1960 (McLaren *et al.,* 1984) showed general mean drafts of 4–5 m (inferred thickness 4.5–5.5 m) but extensive areas of abnormally high average thickness. The highest mean draft measured over a 50 km section was 7.8 m, implying an average thickness of about 9 m over a length of 50 km. Extensive areas of thick landfast ice (as distinct from ice shelves, though the distinction may be somewhat artificial) also exist in the embayments of Northern Ellesmere Island (Marcellus and Morrison, 1982), and are estimated to be of 15–30 m thickness.

Floes of multi-year ice may also reach extremely large diameters, owing to the sheltered oceanographic conditions, in which they are protected from waves and sea swell. It is clear from a study of aerial and satellite photographs that floes may exist independently at diameters up to at least 50 km. A drifting floe of diameter 50 km was, for instance, wedged in Byam Martin Channel in late August 1977 (Lindsay, 1981). In general, floe size is limited only by the size of the channels in which the multi-year ice grows.

Once in open water, drift speeds in the range 4–5 km per day are commonly observed. In open water, break-up occurs rapidly.

3.2.6.4 Icebergs

Glacial icebergs in the Queen Elizabeth Islands are mostly produced by the glaciers of Northern Ellesmere Island, Devon Island and Axel Heiberg Island. In addition, icebergs originating from West Greenland frequently drift from Baffin Bay into the waters of Jones Sound and Lancaster Sound.

Icebergs are particularly common in Nansen Sound (between Ellesmere Island and Axel Heiberg Island) and in Jones Sound (between Ellesmere Island and Devon Island). The number of icebergs in these regions at any one time is probably in the region of a few hundred, and the larger bergs amongst them may be 100–200 m in diameter. Sightings over the period 1961–78 are provided in the atlases by Lindsay (1976, 1977, 1981).

Ice islands and ice-island fragments have been reported throughout the Queen Elizabeth Islands; some, surprisingly, as far east as Jones Sound (Lindsay, 1981). It is likely, however, that many of these sightings are not of ice islands from Northern Ellesmere Island but are in fact glacial tabular icebergs from West Greenland. True ice islands have, nonetheless, been tracked through the Parry Channel: in 1959, ice island T-1 was tracked

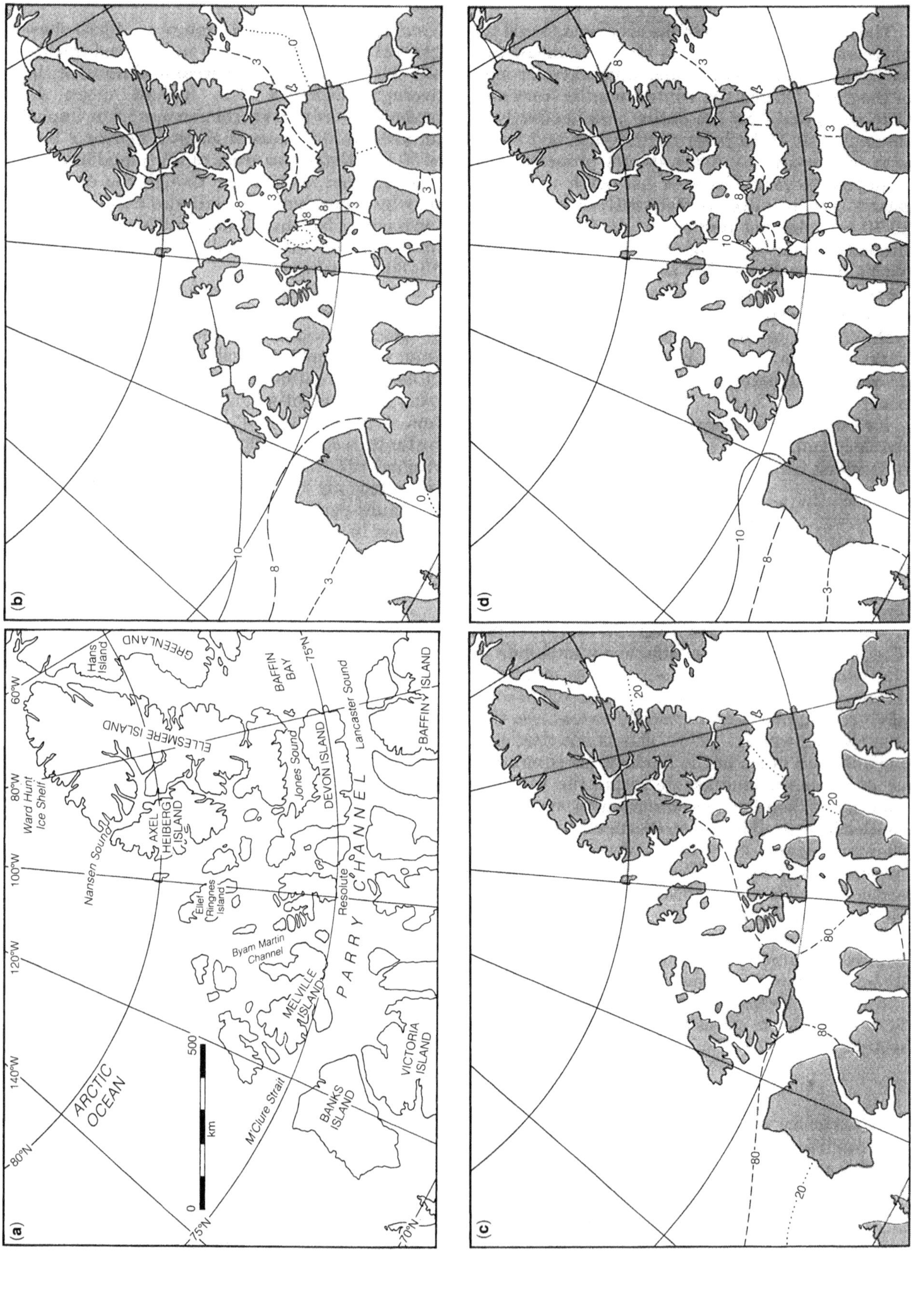
(a)
(b)
(c)
(d)
ARCTIC OCEAN
BANKS ISLAND
VICTORIA ISLAND
M'Clure Strait
MELVILLE ISLAND
PARRY CHANNEL
Byam Martin Channel
Eilef Ringnes Island
Resolute
DEVON ISLAND
Lancaster Sound
BAFFIN ISLAND
BAFFIN BAY
Jones Sound
AXEL HEIBERG ISLAND
Nansen Sound
Ward Hunt Ice Shelf
ELLESMERE ISLAND
Hans Island
GREENLAND
60°W
80°W
100°W
120°W
140°W
80°N
75°N
70°N
0
500
km

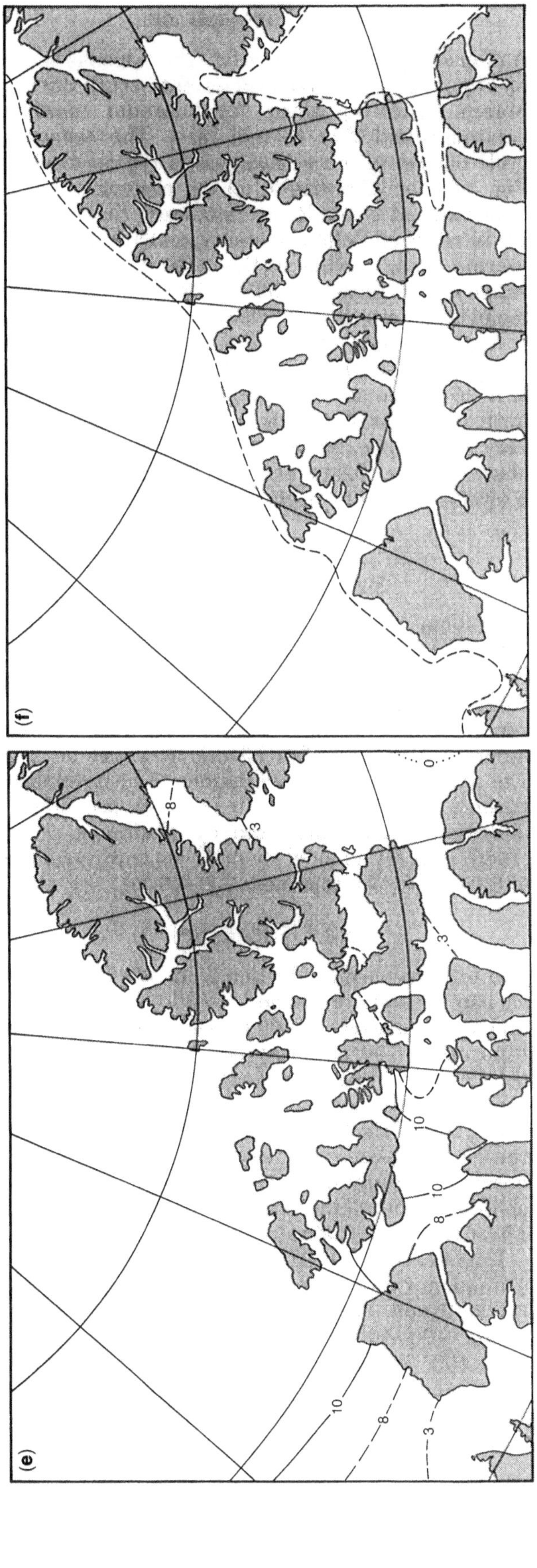

Fig. 3.18 Maps of average ice conditions in the Arctic Islands: (a) general map of region; (b) average ice concentration during August (minimum extent); (c) chance (%) of encountering ice of concentration greater than 3/10 during August; (d) average ice concentration during July (start of summer season); (e) average ice concentration during September (end of summer season); (f) edge of landfast ice during winter (February–March).

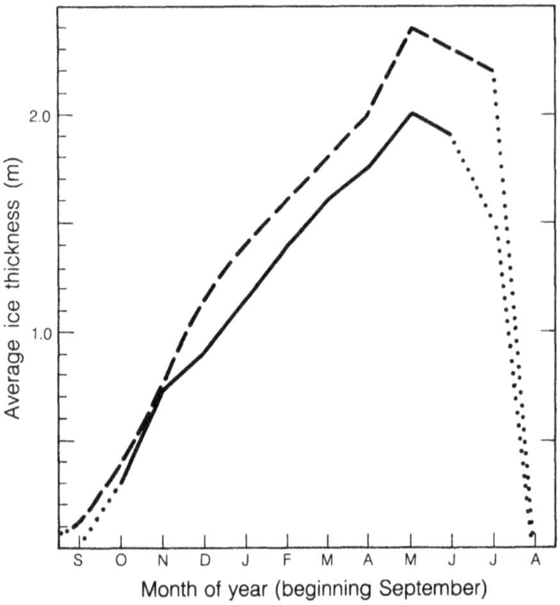

Fig. 3.19 Seasonal ice thickness for Resolute Bay in the Canadian Arctic Islands. Solid line shows average thickness and dashed line shows maximum thickness. (Data from Westermann, 1984, unpubl.)

through the Gustav Adolf Sea to McClintock Channel, and the ice island known as Station Charlie passed through M'Clure Strait in 1964 and is thought to have reached Lancaster Sound.

Numerous sightings have been reported along the north-west coast of the Queen Elizabeth Islands, as ice islands are transported in a southwesterly direction from Ellesmere Island towards the Beaufort Sea.

3.2.7 Labrador, Newfoundland and West Greenland

3.2.7.1 General

The waters between Canada and Greenland (Baffin Bay and Davis Strait) are notable chiefly for the huge numbers of icebergs drifting southwards in the Labrador current. The region is also characterised by seasonal first-year ice, which reaches as far as Newfoundland during February–April, and by multi-year ice drifting southwards from the Queen Elizabeth Islands and Robeson Channel. Multi-year ice seldom reaches further south than Baffin Island; ice of any type seldom survives long once it encounters the warm waters of the Gulf Stream south of Newfoundland.

3.2.7.2 First-year ice

A map of seasonal ice cover for the region is shown in Fig. 3.20; the maximum extent occurs during March. Break-up along the Labrador coast is generally mid-May to mid-June. The seasonal cycle of first-year ice thickness is represented in Fig. 3.21 for Cartwright, on the eastern coast of Labrador. At Clyde River, maximum thickness is 1.5 m while at Belle Isle it is typically 0.8 m. Submarine sonar profiles in Davis Strait indicated a range of thickness of 0.85–1.85 m, and an overall mean thickness of 1.18 m (Wadhams *et al.*, 1985).

Approximately 80% of the ice cover consists of level ice, but ridging is quite common, at the rate of 5–15 ridges per km. Since the ice is predominantly first-year, few keels are deeper than 5 m and very few deeper than 9 m. The deepest keel so far observed was 19.6 m, but it is likely that this was a multi-year ridge or even a bergy bit.

3.2.7.3 Multi-year ice

Multi-year ice concentrations in the region north of Cape Dyer, in Baffin Bay, are typically in the region of 1–10%, reaching up to 20% during August (Markham, 1981). Multi-year ice is very rare south of latitude 64°N, and can effectively be ignored south of Baffin Island. Statistics on floe size and thickness do not appear to be available, though secondary peaks in ice-draft probability density functions produced by Wadhams *et al.* (1985) suggest a modal ice draft for multi-year ice of 3.0–3.5 m (ice thickness 3.4–4.0 m).

3.2.7.4 Icebergs

The total iceberg production from the Greenland ice cap is estimated to be 150–200 km^3 per year, and approximately 40 000 icebergs calve annually from about 100 iceberg-producing glaciers along the Greenland coast. Most of these are on the west coast: the east coast of Greenland accounts for only about 10% of the total number of icebergs reaching the North Atlantic. An average of about 400 icebergs eventually drift as far south as 48°N (offshore Newfoundland) each year – approximately 1% of the total produced.

Icebergs offshore Labrador typically occur with a frequency of 50–300 per degree square of longitude/latitude. This is an areal coverage density of about 100–400 icebergs per 10 000 km^2 (a 100×100 km square). Maps of representative iceberg concentrations off the coast of Labrador are shown in Fig. 3.22, based on airborne observations over the 1963–77 period (Gustajtis, 1979).

The density of icebergs off West Greenland is typically rather less than this, and at latitudes 62–70°N is reported to be in the range 5–40 per 10 000 km^2 (Mangor and Zorn, 1983).

Farther south, offshore Newfoundland in the region of the Grand Banks, the density is very much lower, and is more easily expressed in terms of a flux of icebergs per year across a given latitude. An analysis of iceberg flux as a function of latitude is shown in Fig. 3.23, and the slope of this line illustrates that, between latitudes 67° and 48°N, icebergs are lost at the rate of approximately one per kilometre. Note, however, that these data were collected between 1963 and 1969, which was a period of unusually low iceberg flux. The long-term average may be two to three times higher than this.

The passage of icebergs is strongly seasonal, since during winter most icebergs in the region of Baffin Bay and Davis Strait are locked in landfast ice. This means that iceberg flux generally reaches a peak in the April–June period, as landfast ice melts and releases icebergs into the Labrador current. A picture of this seasonality is given by Fig. 3.24, which summarises the monthly average total flux of icebergs across various latitudes in the Labrador Sea for the 1963–9 period.

Iceberg flux is also highly variable from year to year, as shown in Fig. 3.25, which gives a long record, going back to 1900, of the annual iceberg count crossing 48°N. The number varies from 0 (in 1940) to 2000 (in 1984), with a mean of about 400. Note that the data used in Figs 3.23 and 3.24 were collected during a period in which the iceberg count was unusually low. Care should be exercised in using these data.

The size distribution of icebergs in the Labrador Sea is summarised in Table 3.1 as a function of latitude. Additional detailed statistics for the Grand Banks of Newfoundland are given by Hotzel and Miller (1983). In general, the proportion of large icebergs decreases with distance southwards, and the proportion of small icebergs and growlers increases. The course of melting of a large iceberg can be quite dramatic. During the three months of a journey from Baffin Bay southwards to the Grand Banks of Newfoundland a massive iceberg 80 m high above sea level and weighing 1 500 000 tonnes may shrink in height at the rate of 1–3 m per day and reduce in mass to only 150 000 tonnes. Once it encounters the Gulf Stream melting is still more rapid, at up to 10 m per day. In June 1926 an iceberg 127 m long was observed entirely to disappear in the course of 36 hours (Zubov, 1945).

Exceptionally, icebergs from West Greenland may be quite vast: the largest on record, encountered near Baffin Island in 1882, was 13 km long, 6 km wide and had a freeboard of about 20 m (thickness of about 120–150 m).

3.2.8 Greenland Sea and Svalbard

3.2.8.1 General

The Greenland Sea between East Greenland and Svalbard is dominated by the strong and massive southward-flowing sea currents passing through Fram Strait (Fig. 3.2). This is the only substantial deep ocean passage connecting the Arctic Ocean to the other seas of the world, and it plays a major role in the thermal and mass balance of the Arctic regions.

The current sweeps first-year ice, multi-year ice and icebergs southwards towards the Denmark Straits west of Greenland to meet the Gulf Stream east of Iceland. Much of the ice cover is an intimate mixture of pans of first-year ice and multi-year ice and, during summer in the Barents Sea, consists of about 90% first-year ice and 10% multi-year ice (Overgaard *et al.*, 1983).

3.2.8.2 First-year ice

The Gulf Stream, more properly known as the Atlantic Drift as it passes towards the North East Atlantic, keeps much of the Norwegian Sea and the Barents Sea free of seasonal first-year ice. Figure 3.26 shows typical seasonal ice cover in the region.

Ice typically grows to a thickness of up to 0.9–1.5 m during its annual cycle, depending on latitude. In the Greenland Sea at latitude 78–80°N a thickness of 0.91 m is typical, while at latitude 82°N the thickness is close to 1.5 m. North of this, the ice cover is progressively thicker and contains increasing proportions of multi-year ice. Ice thickness in the Barents Sea is generally less than in the Greenland Sea owing to the influence of the warm Atlantic Drift.

3.2.8.3 Multi-year ice

Multi-year ice is present for most of the year at latitudes north of about 80–82°N. Mean ice draft and thickness is variable with latitude, and this

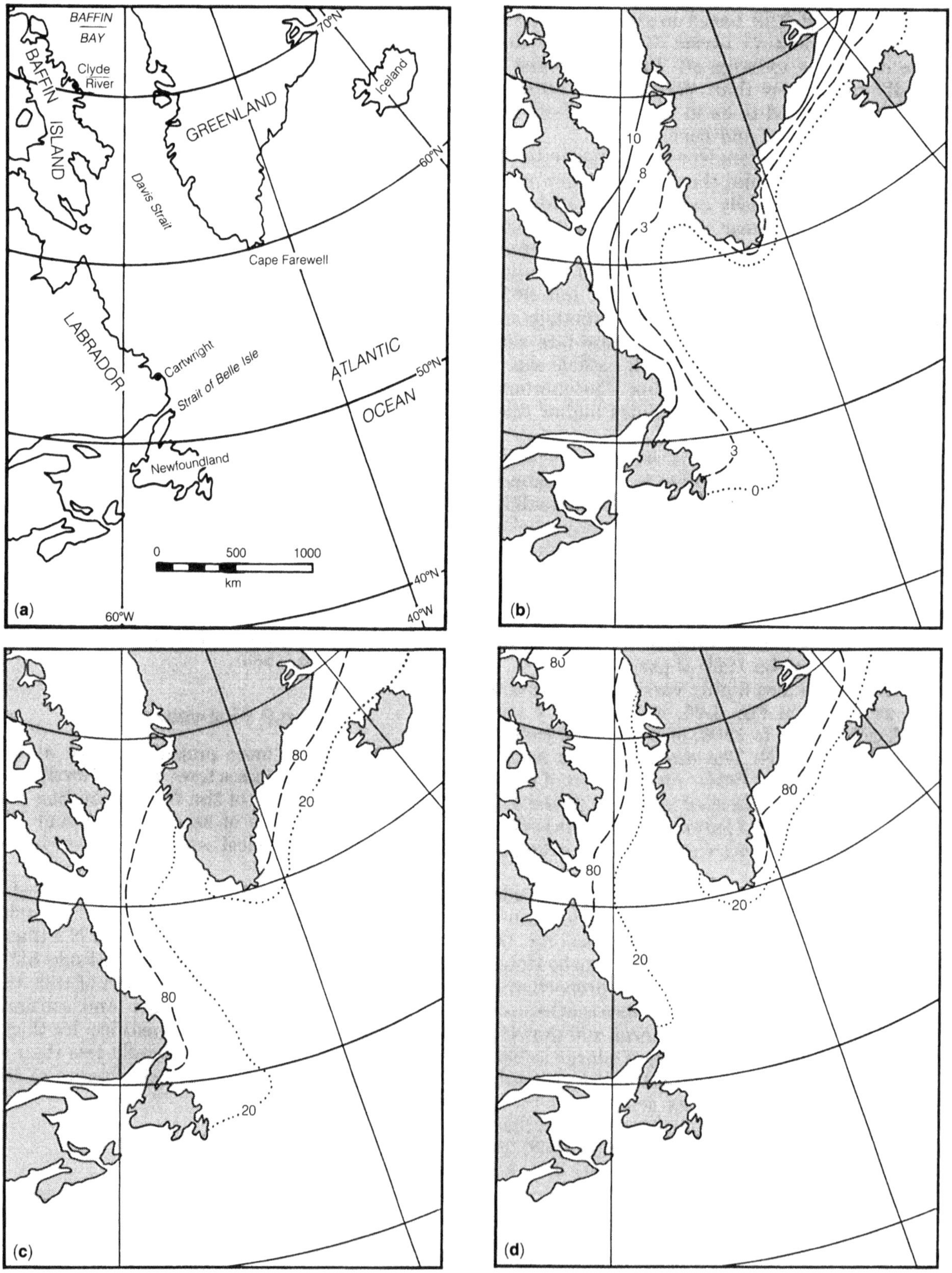
BAFFIN
BAY
BAFFIN ISLAND
Clyde
River
GREENLAND
Iceland
70°N
60°N
50°N
40°N
Davis Strait
Cape Farewell
LABRADOR
Cartwright
Strait of Belle Isle
ATLANTIC
OCEAN
Newfoundland
0 500 1000
km
(a)
60°W
40°W
10
8
3
3
0
(b)
80
20
80
20
(c)
80
80
80
20
20
20
(d)

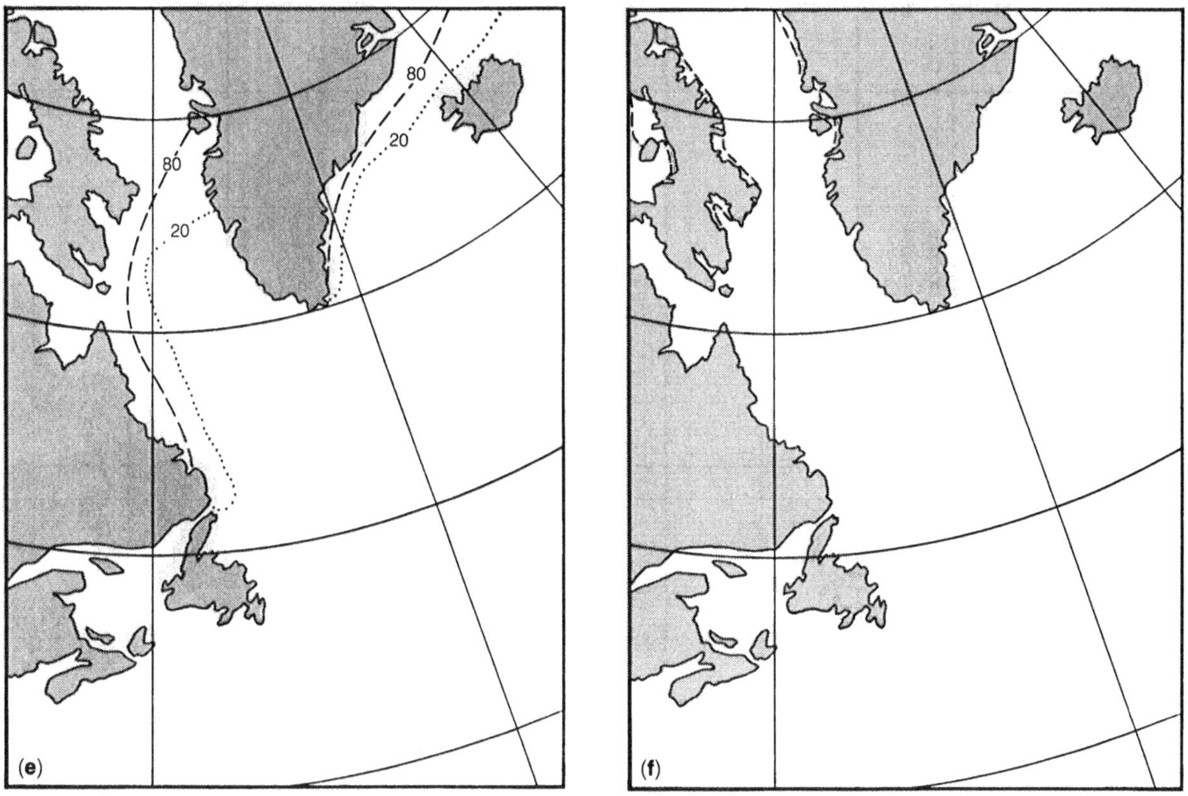

Fig. 3.20 Seasonal ice cover for Newfoundland, Labrador, and West Greenland: (a) general map of region; (b) average ice concentration during March (maximum extent); (c) chance (%) of encountering ice of concentration greater than 3/10 during March; (d) chance (%) of encountering ice of concentration greater than 3/10 during June; (e) chance (%) of encountering ice of concentration greater than 3/10 during December; (f) edge of landfast ice during winter (February–March). There is very little fast-ice in this region: it exists essentially from Baffin Island northwards only. During September, the month of minimum ice cover, the region is effectively ice-free.

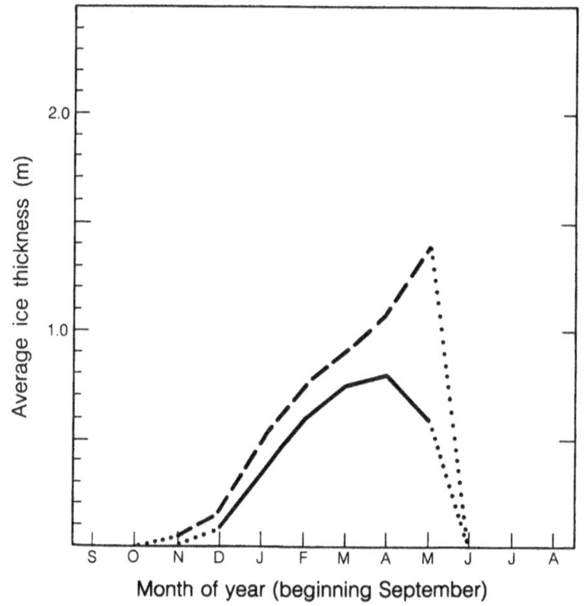

Fig. 3.21 Seasonal cycle of first-year ice thickness, for Cartwright on the Labrador coast. Solid line shows average ice thickness; dashed line shows maximum thickness. (Data from Westermann, 1984, unpubl.)

has already been summarised in Fig. 3.5. Mean thickness lies in the range 3–4 m.

The voyage of the *Ymer* during 1978 reported mean summer thicknesses of 3.1 m through smooth areas of selected multi-year floes. During the summer the average floe size in much of the region was very small: typically of diameter 30–120 m. Earlier in the season, when better protected from swell by surrounding ice cover, they are no doubt much larger.

Along the shores of Greenland, several observations have been made of exceptionally thick old fast ice, or "sikussak" (Koch, 1945), which remains

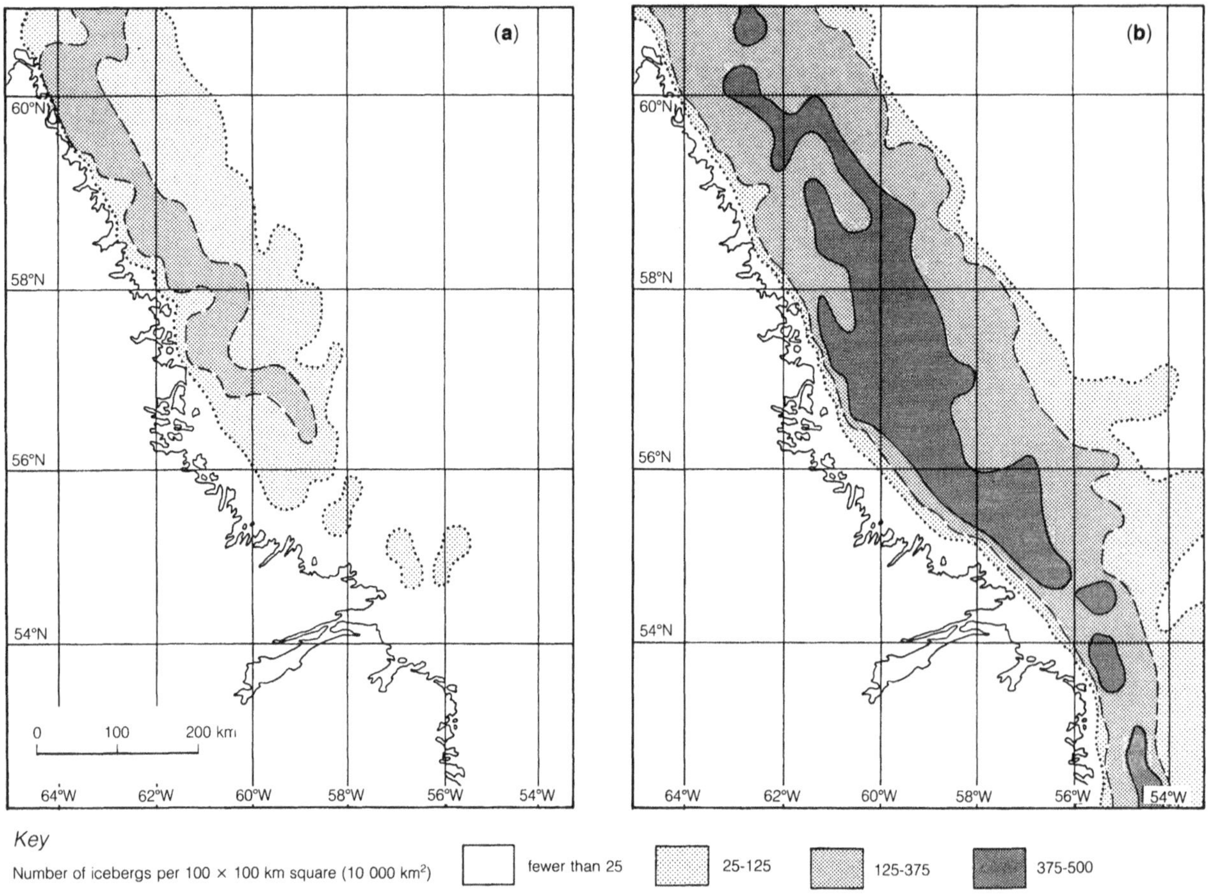

Fig. 3.22 Iceberg concentrations (number of bergs per 100 × 100 km square) in the region of Labrador: (a) winter (November, December, January, minimum period); (b) spring (February, March, April, maximum period). (From Gustajtis (1979). Data have been smoothed, and refer to the years 1963–77.)

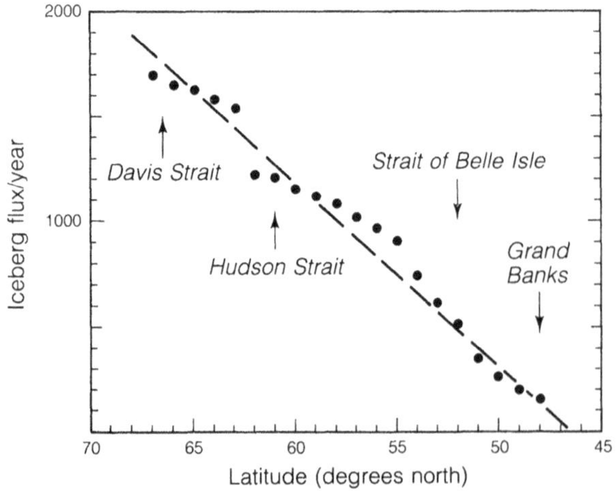

Fig. 3.23 Annual average number of icebergs crossing selected latitudes between Davis Strait and the Grand Banks of Newfoundland (data from Anderson, 1971). The dashed line represents the least squares fit between 67°N and 48°N, and the slope corresponds to a loss rate of about 1 iceberg km^{-1}. Note that these data were collected between 1963 and 1969, a period of rather low iceberg flux. The long-term average may be 2–3 times higher than this.

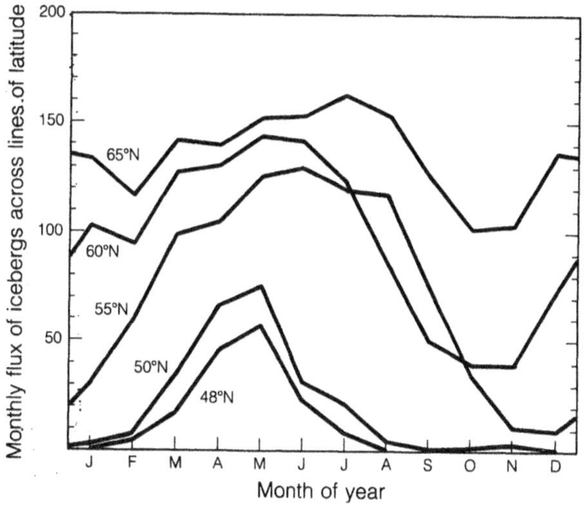

Fig. 3.24 Monthly flux of icebergs across various lines of latitude in the Labrador Sea, showing strong seasonal variation (data from Anderson, 1971). Note that these data were collected during a period of generally low iceberg flux (1963–9).

TABLE 3.1 Labrador Sea iceberg size distribution (% of total) as a function of latitude (after Gustajtis, 1979)

Latitude (°N)	*Percent distribution*			
	Growlers	*Small*	*Medium*	*Large*
60	6.4	43.3	33.5	16.8
59	4.2	49.9	29.4	15.5
58	4.2	48.0	36.6	11.2
57	4.3	41.9	42.2	11.7
56	5.1	51.6	34.1	9.3
55	3.9	61.4	27.8	7.0
54	16.6	52.3	24.5	6.7
53	10.2	52.4	26.5	10.9
52	16.7	52.2	24.6	6.5

in fjords from year to year (Zubov, 1945; Overgaard *et al.*, 1983). Its thickness is thought to be of the order 20 m. Such ice may periodically break away from the coast, and large thick floes, possibly from here or from the analogous features on Northern Ellesmere Island, have been observed drifting in the Arctic Ocean.

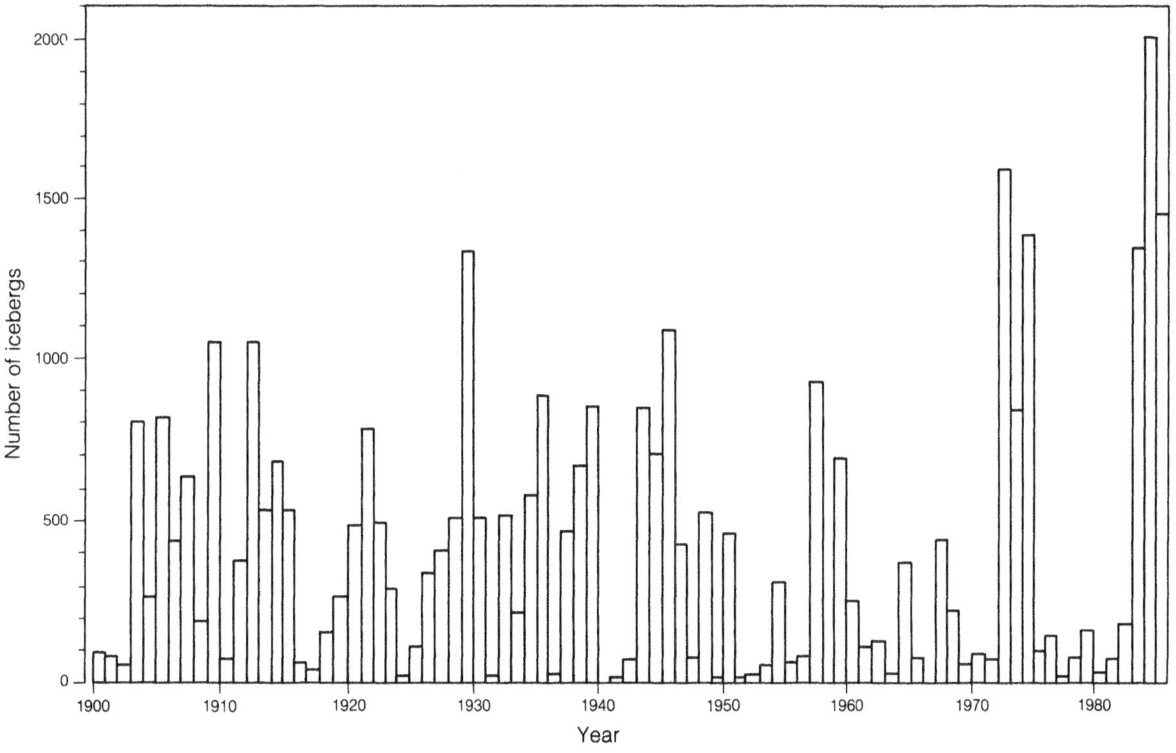

Fig. 3.25 Long-term annual variability in number of icebergs crossing latitude 48°N offshore Newfoundland (after Marko *et al.*, 1986).

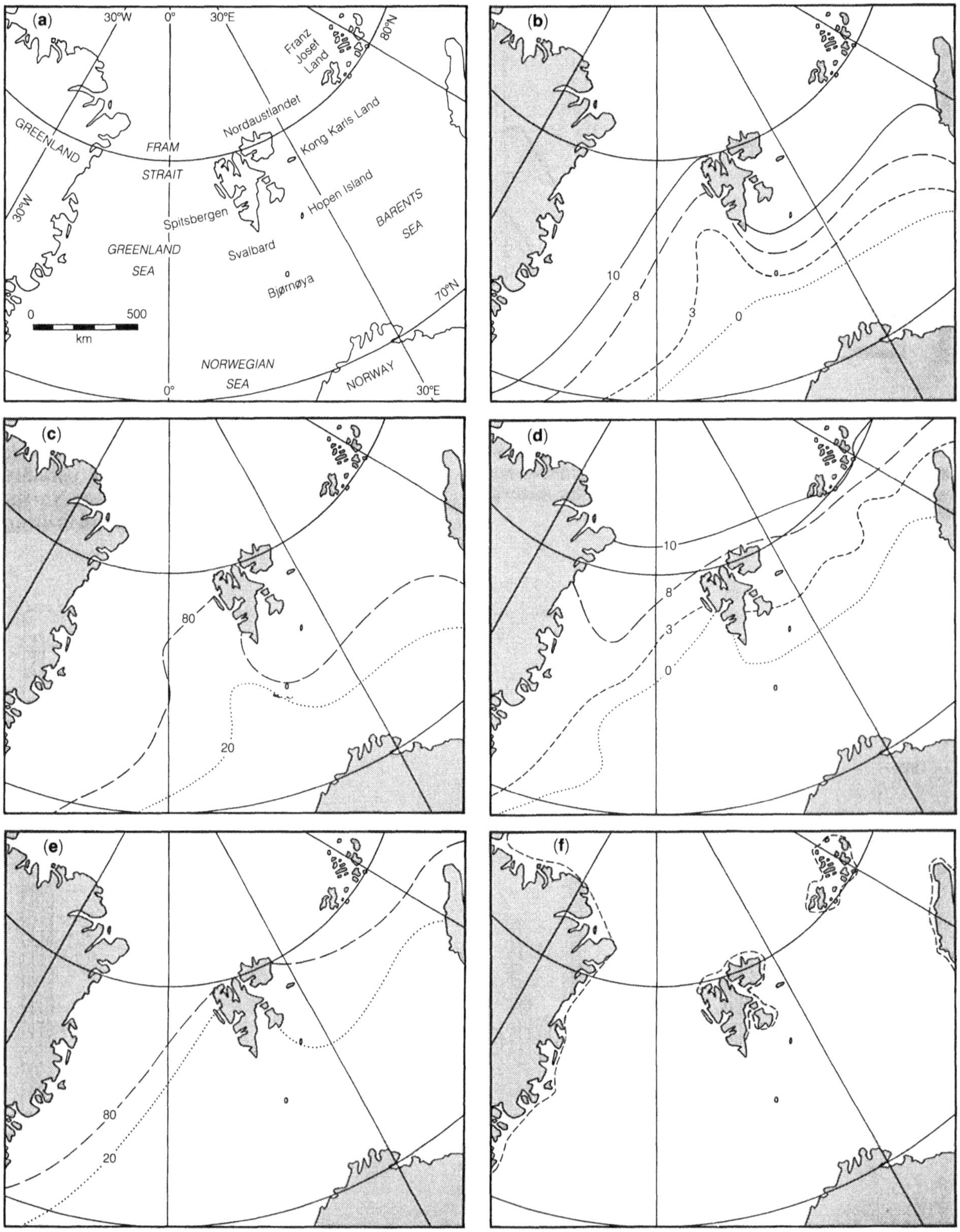

Fig. 3.26 Maps of ice conditions in the region of Svalbard and East Greenland: (a) general map of region; (b) average ice concentration during February (maximum extent); (c) chance (%) of encountering ice of concentration greater than 3/10 during February; (d) average ice concentration during August (minimum extent); (e) chance (%) of encountering ice of concentration greater than 3/10 during August; (f) edge of landfast ice during winter (February–March). During light ice years, negligible amounts of fast ice are observed around Svalbard.

3.2.8.4 Icebergs

Icebergs along the east coast of Greenland are not as common as might be expected from the conditions on the west coast. Along the east coast fewer icebergs are produced, and, more importantly, those that are produced frequently remain grounded or locked in fast ice in the fjords and never escape to drift freely out to sea. In general, they are smaller than those of the west coast, though the largest observed on record was 70 m high and about 1 km long (Zubov, 1945). Those that succeed in leaving the fjords tend to drift southwards close to the shore with the East Greenland current, drift around Cape Farewell at the southern tip of Greenland and then join the icebergs of West Greenland (Figs. 3.2 and 3.3).

Some icebergs are also produced by the glaciers of Svalbard, but they, too, often remain in deep fjords, many of which are separated from the open sea by a comparatively shallow shelf. Icebergs in the Barents Sea originate mainly from Franz Josef Land, which will be treated in the next section, on Soviet Arctic ice conditions.

3.2.9 Soviet Union

3.2.9.1 General

The Soviet Arctic encompasses a vast area of continental shelf, divided into six principal seas: the Kara Sea, the Laptev Sea and the East Siberian Sea, which are wholly within Soviet waters; the Barents Sea, which extends to Svalbard and Norway; and the Chukchi Sea, which extends into US Alaskan waters. The total area of USSR arctic continental shelf shallower than 100 m is approximately 3.5×10^6 km^2, about 10 times the total area of the shallow continental shelf of the US Chukchi Sea and Beaufort Sea.

Because this book is aimed principally at Western readers (and because literature on the Soviet Arctic is not readily accessible) the review here will be brief.

A general map of ice-coverage conditions is shown in Fig. 3.27 for the key periods of the annual cycle. The maximum extent occurs during March–April, and the whole central region east of Novaya Zemlya is effectively 100% ice-covered between November and May. The minimum extent occurs during August, when most of the northern sea route is navigable: the Barents Sea, the Kara Sea and the Laptev Sea are largely ice-free, but substantial concentrations of ice exist around Severnaya Zemlya, which divides the Kara Sea from the Laptev Sea, and around the coastal passage south of Wrangel island, which divides the East Siberian Sea from the Chukchi Sea.

3.2.9.2 First-year ice

First-year ice generally attains similar thickness to that of the Beaufort Sea. In the Laptev Sea, first-year ice reaches a thickness of about 1.8 m during May (Zubov, 1945), while towards the Chukchi Sea maximum thickness is typically some 1.6 m. In the Kara Sea it falls from about 1.7 m near Severnaya Zemlya to 1.1 m at the south tip of Novaya Zemlya.

Fast ice extends out to approximately the 25 m isobath in most regions, and an approximate map is shown in Fig. 3.24(e). The New Siberian Islands are particularly notable for large expanses of fast ice, associated with the shallow waters and the stabilising influence of the islands. The fast ice here extends for approximately 400 km northwards from the mainland coast. Fast ice towards the west of Novaya Zemlya (Barents Sea) is greatly inhibited by the presence of the warm Norwegian Current, and in the White Sea extends only to about the 10 m isobath, largely as a result of strongly fluctuating tidal currents (US Navy Hydrographic Office, 1958). The extent of fast ice is noticeably reduced in two other regions: the Khatangskiy Guba of the Laptev Sea, where tidal range is unusually high; and along the shores of the Soviet Chukchi Sea, where fast ice is broken by the Chukchi Current.

3.2.9.3 Multi-year ice

Few data appear to be readily available on frequency, size and thickness of multi-year ice in Soviet waters. The general map of Arctic Ocean ice thicknesses (Fig. 3.5) would suggest that thickness conditions may resemble those of the Beaufort Sea, and possibly be less severe. Concentration may be expected also to be less severe, since the flow of the Beaufort Gyre effectively by-passes the eastern Soviet Arctic, and ice drift in the western sector is predominantly away from the coast (Fig. 3.2(b)).

3.2.9.4 Icebergs

Icebergs in Soviet waters originate principally from Franz Josef Land, a group of islands which is practically 100% ice-covered. The other principal source is the Severnaya Zemlya group of islands,

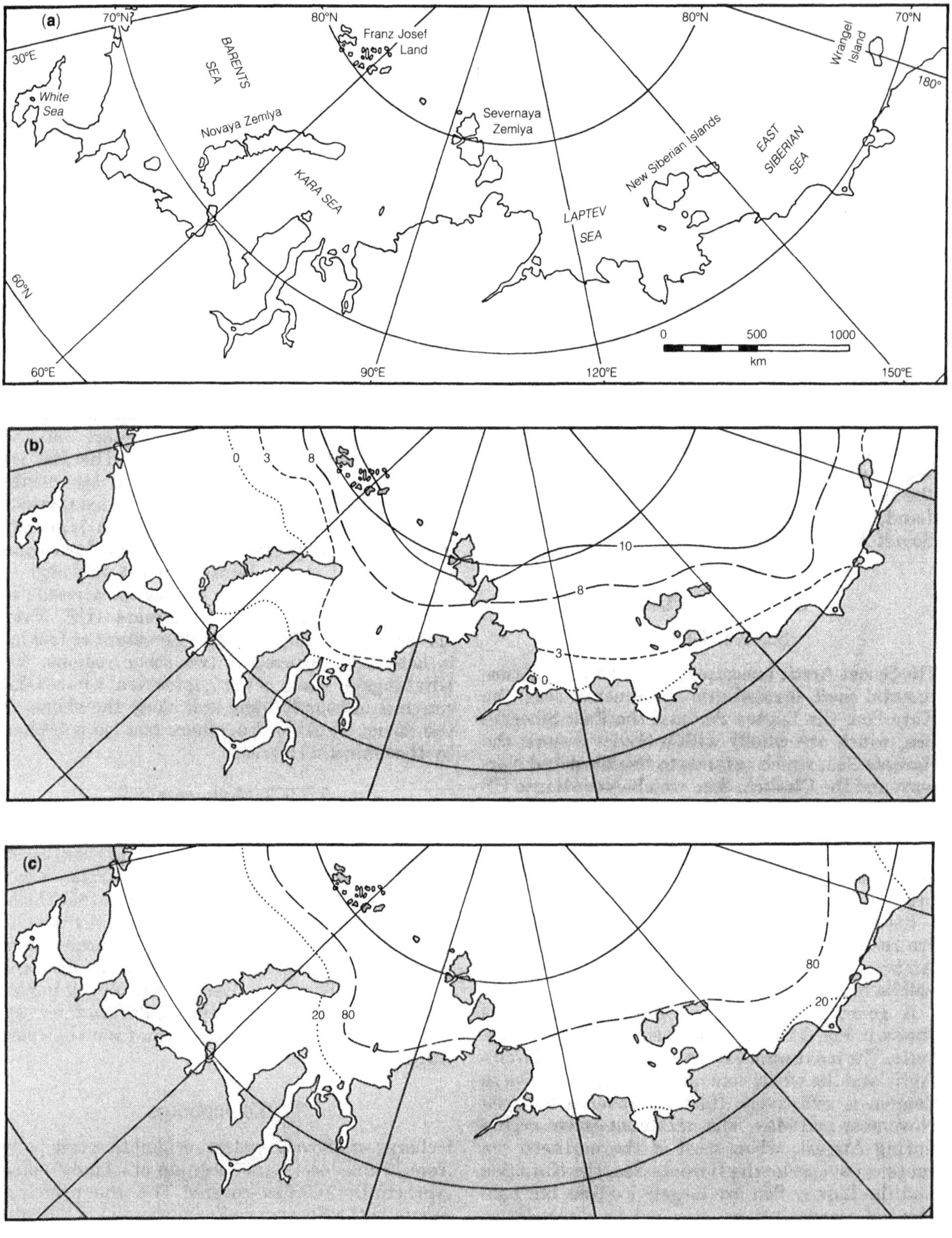
(a)
70°N
80°N
80°N
70°N
30°E
White Sea
BARENTS SEA
Franz Josef Land
Wrangel Island
180°
Novaya Zemlya
Severnaya Zemlya
New Siberian Islands
EAST SIBERIAN SEA
KARA SEA
LAPTEV SEA
60°N
0
500
1000
km
60°E
90°E
120°E
150°E
(b)
0
3
8
10
8
3
0
(c)
80
20
20
80

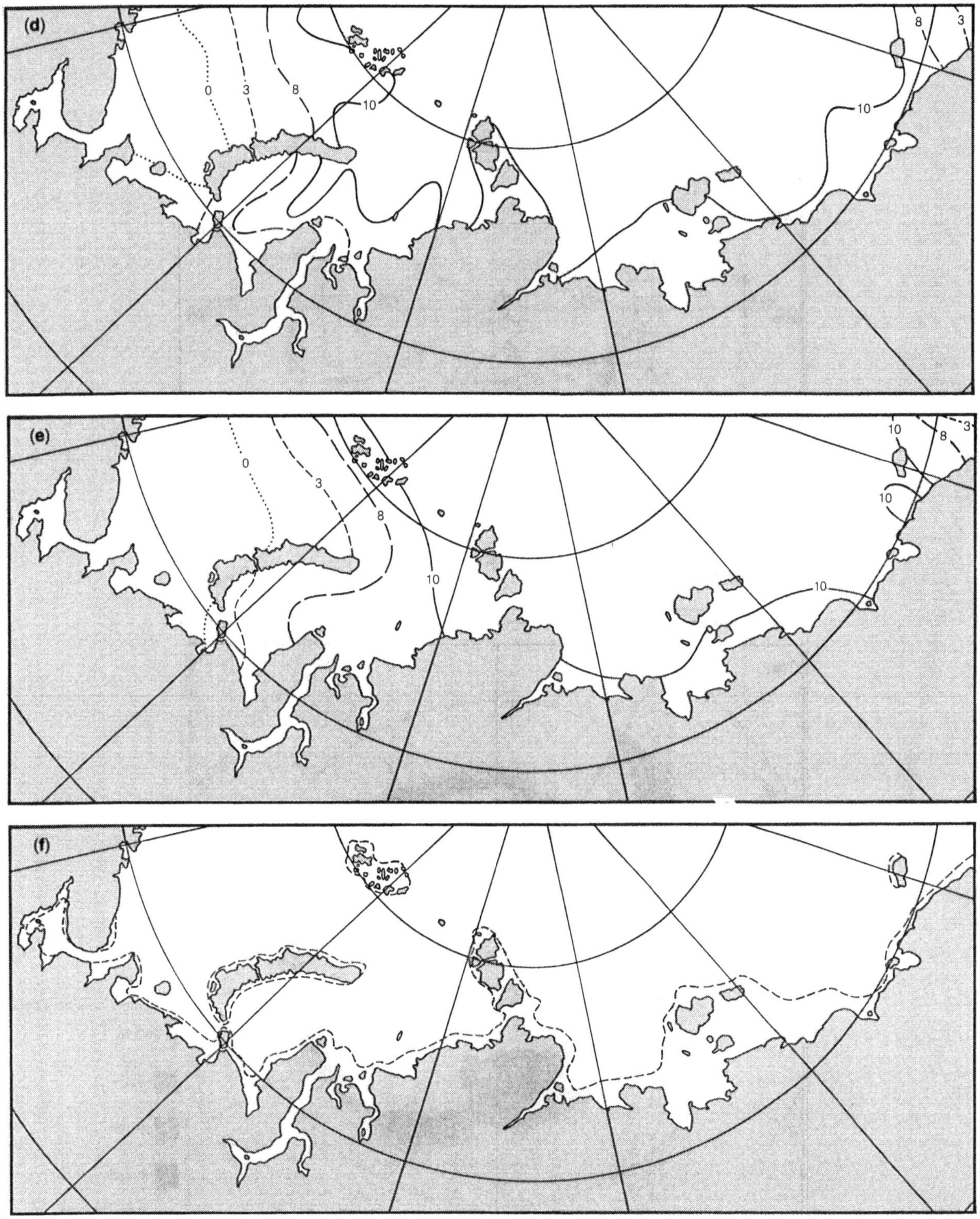

Fig. 3.27 General map of Soviet ice conditions: (a) general map of region; (b) average ice concentration during September (minimum extent); (c) chance (%) of encountering ice of concentration greater than 3/10 during September; (d) average ice concentration during June (beginning of summer season); (e) average ice concentration during October (end of summer season); (f) edge of landfast ice during winter (February–March).

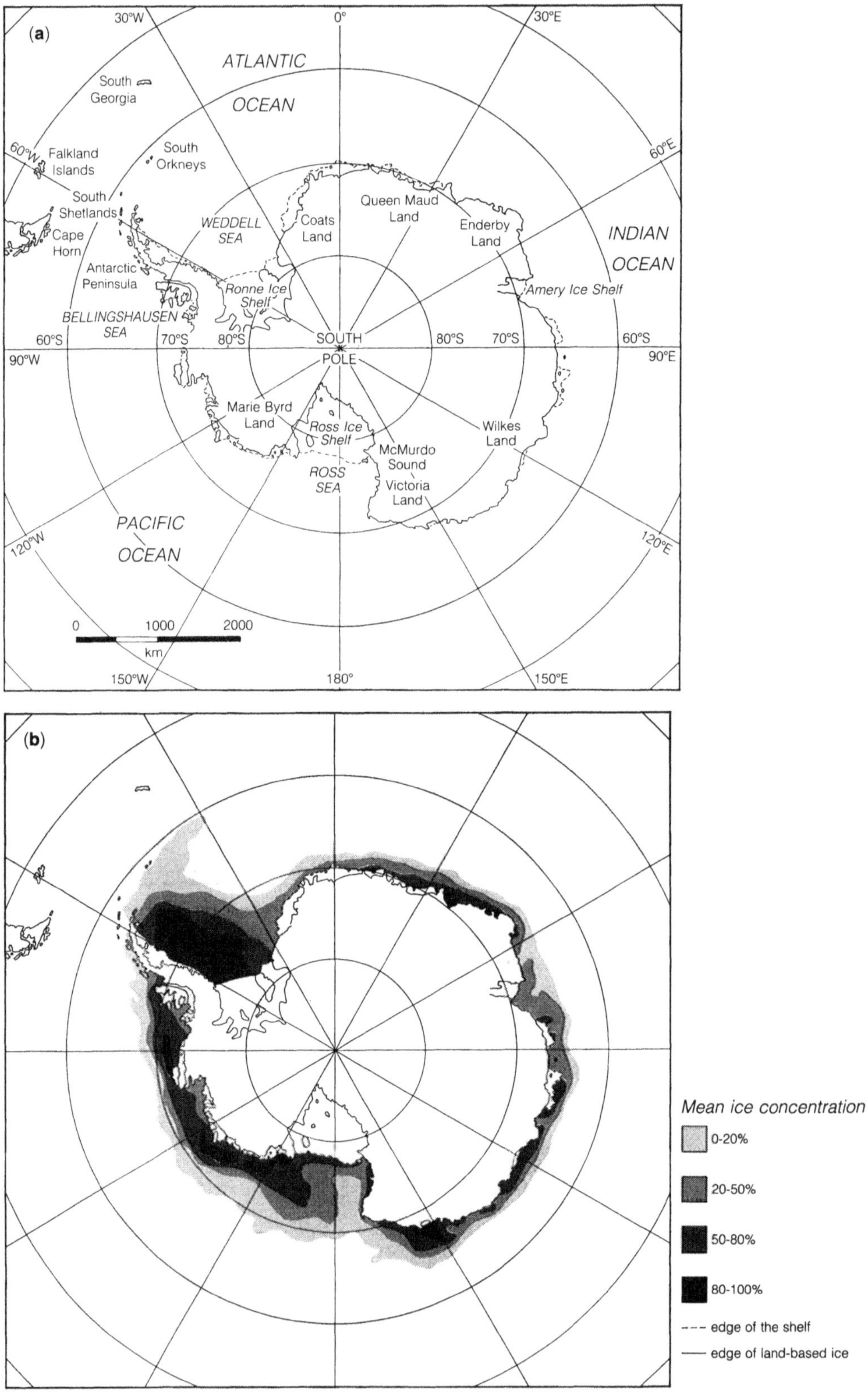
(a)
30°W
0°
30°E
ATLANTIC
OCEAN
South Georgia
Falkland Islands
South Orkneys
60°W
60°E
South Shetlands
Cape Horn
WEDDELL SEA
Coats Land
Queen Maud Land
Enderby Land
INDIAN
OCEAN
Antarctic Peninsula
Ronne Ice Shelf
Amery Ice Shelf
BELLINGSHAUSEN SEA
60°S
70°S
80°S
SOUTH POLE
80°S
70°S
60°S
90°W
90°E
Marie Byrd Land
Ross Ice Shelf
McMurdo Sound
Wilkes Land
ROSS SEA
Victoria Land
PACIFIC
OCEAN
120°W
120°E
0
1000
2000
km
150°W
180°
150°E
(b)
Mean ice concentration
0-20%
20-50%
50-80%
80-100%
edge of the shelf
edge of land-based ice

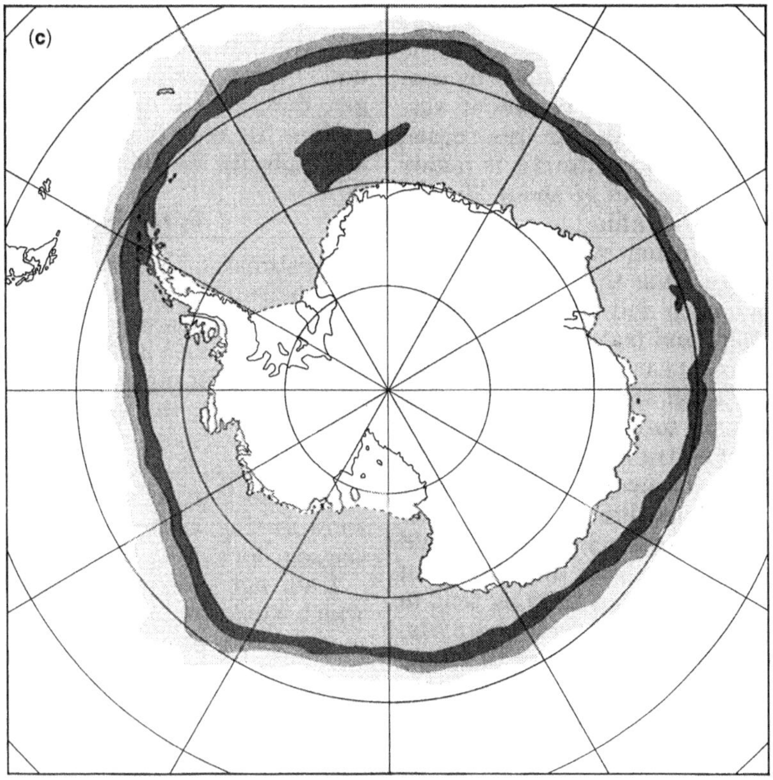

Fig. 3.28 Antarctica: extent of sea ice cover. (a) Map of principal locations; (b) mean ice concentration during March (minimum); (c) mean ice concentration during September (maximum) (after Naval Oceanography Command, 1985).

particularly the northernmost of the group, Komsomolets Island. Icebergs are also produced by the glaciers of Novaya Zemlya, but are seldom released into the open sea: they calve from land-based glaciers, but remain trapped in the fjords by prevailing sea currents and the shallow bathymetry.

Icebergs from Franz Josef Land and Severnaya Zemlya are regularly observed in the Barents Sea (mainly to the north-west), the Kara Sea and the northwestern parts of the Laptev Sea. Icebergs are not normally observed east of Severnaya Zemlya. A schematic map of occurrence of icebergs has already been shown in Fig. 3.3. Icebergs in the Barents Sea have been observed as far south as Murmansk.

The icebergs may be of very large size. An iceberg observed near Severnaya Zemlya in 1943 was 1.5 km long, 400 m wide and had a freeboard of about 10 m (implying a thickness of about 70 m). In the Barents Sea, icebergs are frequently observed to be grounded in waters of depth exceeding 100 m. Reliable statistics on frequency are apparently not available.

Ice islands have been observed in the East Siberian Sea and to the north of the New Siberian Islands, but are generally caught up in the Beaufort Gyre and are unlikely to drift farther west.

3.3 ANTARCTICA

3.3.1 General

This book is aimed principally at discussing ice hazards in Arctic regions. For completeness, however, it is worthwhile examining Antarctic ice conditions in the context of Arctic conditions.

Antarctica covers a vast area, including an area of about 4 million km^2 of continental shelf. The review I give here will be brief, however, since it seems to me highly unlikely that development of any offshore Antarctic mineral or hydrocarbon resources will take place in the forseeable future: ice conditions are extremely severe, continental shelf water depths excessive (at around 500 m), and political and ecological problems prohibitive.

The fundamental difference between the Arctic and the Antarctic is simple but important: the Arctic is essentially an ocean surrounded by continents while the Antarctic is a continent surrounded by oceans. It is mainly for this reason that the ice cover around the Antarctic is highly seasonal, reaching a maximum of about 20×10^6 km^2 during September and falling to a minimum of about 1.3×10^6 km^2 during March. This variability is much greater than that of the Arctic. Maps of typical maximum and minimum extent are shown in Fig. 3.28, and maximum ice extent varies about 10% from year to year. The high seasonal variability within any given year is due to exposure of the ice edge to open ocean conditions and to the lack of land restraint at the perimeter, allowing ice to drift freely away from the winter pack. In this respect the conditions are quite different from most of the Arctic. In addition, cold katabatic winds, originating from the interior of the Antarctic ice cap at elevation 2500 m, tend to stream off the continent under the force of gravity, being deflected to the left by the Coriolis effect. This results in a prevailing southeasterly direction for these winds, which tends to blow pack ice in a northwesterly direction.

The oceanic Antarctic convergence plays an important role in limiting sea-ice extent. This feature consists of a zone of sharp change in temperature of the surface waters of the oceans, often by as much as 5°C in the space of 250 km. The Antarctic convergence is located between latitude 47°S and 63°S, and its mean position corresponds closely to the maximum extent of sea ice shown in Fig. 3.28(c). Sea ice is not normally observed north of the Antarctic convergence, since it rapidly melts in the warmer waters.

3.3.2 First-year ice

First-year ice typically grows to an average thickness of about 1.5 m, but thicknesses of around 2.75 m are common around McMurdo Sound (Lewis and Weeks, 1971) and thicknesses up to 3.35 m have been observed. Much of this ice is probably infiltration ice, formed by the mechanism described in Section 2.3.2.1.

In general, Antarctic ice appears to be more even and to show fewer ridges and less evidence of dynamic action than Arctic ice. This is because of the lack of constraint by surrounding coasts, which in the Arctic often leads to convergent flow of floating ice. The Arctic pack may be viewed (loosely) as a rotating disc sending ice masses centrifugally out towards its restraining coastal boundaries. In the Antarctic, on the other hand, prevailing wind and current patterns lead to unrestrained divergent flow, and the opportunity for collision is rarer.

Near the edge of the pack, floes of first-year ice are typically 10–100 m in diameter.

3.3.3 Multi-year ice

It is estimated that roughly 15% of Antarctic sea ice consists of multi-year ice (Ahlnäs, 1983). Multi-year ice is particularly concentrated in the western Weddell Sea and in the Amundsen and Bellingshausen Seas. It may attain very great thicknesses, particularly in the vicinity of glacial ice shelves, where old sea ice of up to 15 m thickness has been observed attached to the ice cliffs. This ice probably attains such great thickness as a result of heavy snowdrifting in the lee of the ice shelf and subsequent formation of infiltration ice.

I am not aware of any submarine sonar data which would supply reliable statistics on mean and modal multi-year ice thicknesses.

3.3.4 Icebergs

The Antarctic ice cap and its peripheral ice shelves give rise to the largest icebergs in the world's oceans. The largest observed icebergs have exceeded 300×100 km in size, and icebergs are typically some 500 m in diameter, with thickness in the range 250–450 m. Table 3.2 shows statistics on iceberg size ranges derived from 2060 icebergs reported over the period 1960–78, mainly in East Antarctica (Neshyba, 1980). The modal size of icebergs is about 0.7 km.

TABLE 3.2 Size distribution for Antarctic icebergs (derived from data of Neshyba, 1980)

Iceberg width (km)	*Probability of exceedance (%)*
0.15	95
0.27	90
0.84	50
1.65	10
1.90	5
2.33	1

Icebergs typically drift at a rate of about 15–20 km per day, and their movement is dominated by sea currents. They are sometimes observed moving in a contrary direction to the pack ice, which is

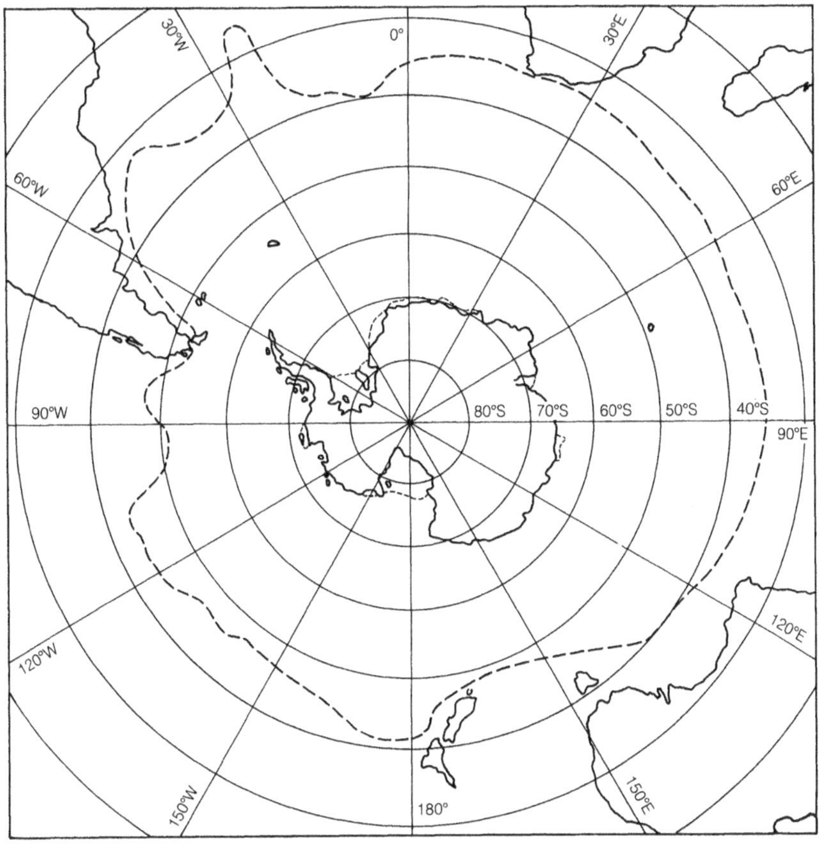

Fig. 3.29 Extreme limit of iceberg drift in the Southern Hemisphere (after US Naval Oceanographic Office, 1957).

largely wind-controlled. The extreme limit of iceberg drift corresponds to approximately latitude 45°S in the Pacific Ocean and 35°S in the Indian and Atlantic Oceans. It is represented in Fig. 3.29.

It is interesting to note that the draft of these icebergs is in many regions only just less than the water depth, largely because the sea bed has been scoured and eroded to this depth by the passage of ice shelves and icebergs. Ross Ice Shelf, for instance, is about 430 m thick, moving northwards at about 2–3 m per day and giving rise to tabular icebergs at its perimeter. The average bathymetry of the Ross Sea is generally in the region of 500 m.

3.4 LONG-TERM PERSPECTIVE

The ice conditions we have been examining are based on data collected over the past 30–50 years. As a postscript, it is worth putting these data into a longer-term perspective, since they are not necessarily representative of future ice conditions.

The longest record available of ice conditions comes from Iceland, where observations can be traced back to the year 1600 (Kelly *et al.*, 1982, unpubl.). Figure 3.30 shows the number of weeks of the year during which ice affected the coastline of Iceland. This shows that ice conditions, at least around Iceland, appear to have been anomalously light during the first half of this century, particularly in the interval 1920–50. Figure 3.31 shows mean annual Arctic temperature records since 1900. In broad terms, this set can be divided into three periods: cold in 1900–20; warm in 1920–50; and intermediate from 1950 onwards (Kelly *et al.*, 1982, unpubl.). At present we appear to be in a period of moderate temperature.

Indications are (Jones *et al.*, 1986) that global average temperatures are currently rising at a rate of approximately 0.4°C per century. This may largely be due to the celebrated CO_2 "greenhouse effect", the effect of which is also expected to be concentrated at the poles (see, for instance, Bolin *et al.* (in press) for an introduction to the

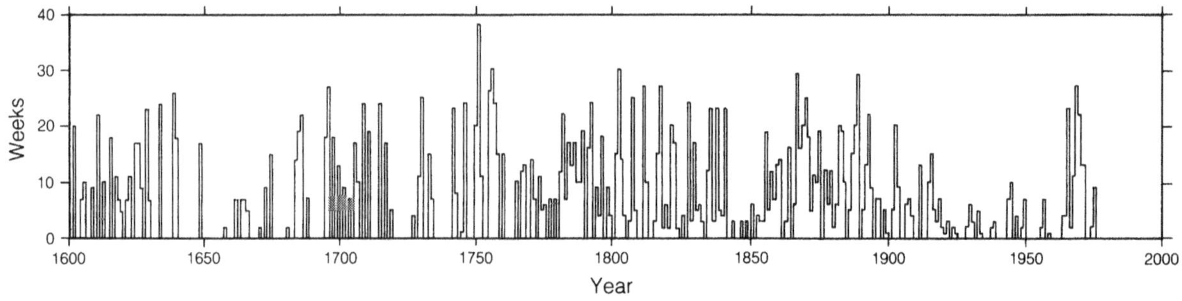

Fig. 3.30 The longest ice conditions record available: ice coverage around Iceland from year 1600 onwards (after Kelly *et al.*, 1982, unpubl.). The graph shows number of weeks per year during which ice affected the Icelandic coastline. A prolonged period when the ice hardly reached Iceland is apparent during the 20th century, and corresponds to the period of warmer temperatures identified in Fig. 3.31.

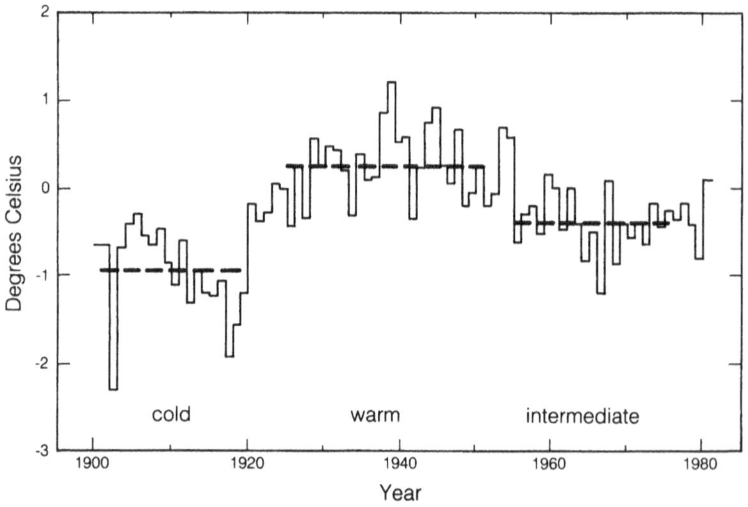

Fig. 3.31 Mean annual temperature records for the Arctic (after Kelly *et al.*, 1982, unpubl.). Temperatures are shown as departures from the mean over the period 1946–60, and fall into three natural intervals: cold, warm and intermediate.

greenhouse effect). Bearing this in mind, it would appear that ice conditions during the next 50 years may well be less severe than indicated by data collected over the past 30 years. However, if this turns out to be generally so, it will not necessarily be uniformly true for all regions. If climatic conditions change sufficiently that wind and current patterns alter, then dramatic local changes may well take place, of greater magnitude than any average global effect.

FURTHER READING

Principal data sources on ice coverage in polar regions

Arctic

US Navy Hydrographic Office 1958. *Oceanographic Atlas of the Polar Seas. Part 2: Arctic.* HO Publication No. 705. General atlas: monthly maps of average and extreme ice concentrations (twice-monthly during break-up and freeze-up). Not a coherent data set; mainly based on 1952–56 only. Good on currents, tide and physical oceanography.

Lindsay, D. G. 1976, 1977, 1981. *Sea Ice Atlas of Arctic Canada, 1961–68, 1969–74, 1975–78.* Three separate volumes giving detailed diagrammatic and photographic summaries of particular overflights by Polar Continental Shelf Project, 1961–78. Distinguishes new ice, first-year, multi-year ice and icebergs. No attempt at statistical synthesis.

Markham, W. E. 1981. *Ice Atlas, Canadian Arctic Waterways.* Gives unique analysis of median multi-year ice concentrations on monthly basis. Also provides median general coverage. Good on landfast ice boundaries. No mapping of icebergs.

LaBelle, J. C. *et al.* 1983. *Alaska Marine Ice Atlas.* Good probabilistic treatment of boundary of 5/10 ice coverage. Provides floe size indications and calculated ice thickness. Good on landfast ice. No treatment of multi-year ice concentrations.

Antarctic

US Naval Oceanographic Office 1957. *Oceanographic Atlas of the Polar Seas. Part 1: Antarctic.* HO Publication No. 705. General atlas. Monthly mean and maximum ice coverage. Limits of icebergs. Good on tides, currents and physical oceanography.

Naval Oceanography Command 1985. *Sea Ice Climatic Atlas: Volume 1, Antarctica.* Publication NAVAIR 50-1C-540, Ashville, USA. Atlas showing a wide variety of statistical information on sea-ice coverage: maximum, mean and minimum ice edge; mean ice concentration; maximum, mean and minimum extent of 5/10 coverage or greater. Based on passive microwave satellite imaging covering the period 1973–82.

Chapter 4

Mechanical Properties of Ice: Laboratory Studies

4.1 INTRODUCTION: THE CONCEPTS OF STRENGTH AND FAILURE

The mechanical properties of pure freshwater ice are well understood, at least at the scale normally encountered in the laboratory. The creep behaviour of ice, in particular, is extremely well understood, and has been the subject of numerous careful laboratory studies. In this chapter we shall look at the most important of these studies, and I shall also present the formalism which can be used to describe or explain the results.

Agreement on the subject of the fracture behaviour of ice is less unanimous: fracture behaviour is more difficult to examine under controlled laboratory conditions, and experiments are less repeatable. Also, the formalism for describing fracture is perhaps less familiar to the practising engineer or scientist than that for describing elastic, plastic or creep deformation, and this has led to some confusion in the literature about the appropriate way to represent fracture behaviour. However, there is now some measure of agreement on the groundwork of the subject. At the frontiers of the subject there is still room for argument (for instance, on the topic of crack nucleation) and this means that the discussion I present in this chapter may well be incomplete or partisan; however, the areas of uncertainty do not seriously affect our basic understanding.

The next question we shall look at concerns sea ice: given that pure ice is well understood, how do we treat sea ice, which for most practical applications is the more important form? This does introduce some difficulties, but it turns out that in most respects sea-ice behaviour can be described by the same equations as for pure ice behaviour, with appropriate modifications. Sea ice has brine inclusions which modify its behaviour, but if proper corrections are applied for the effect of brine then its properties can be satisfactorily reduced to those of pure ice. The description is by no means perfect, however: owing to its unpredictable formation process, naturally occurring sea ice shows a greater scatter in properties than carefully prepared laboratory ice. This is fact we have to live with.

Ice is not purely elastic, viscous or plastic and so does not readily lend itself to classical engineering analysis. When a stress is applied to a sample of ice, it displays a combination of responses: it shows an instantaneous elastic response but also immediately begins to creep, at a time-dependent rate. The deformation due to creep soon exceeds the initial elastic deformation: for stresses in the range 1–5 MPa, this occurs after about 5–200 s.

In addition, ice is an extremely brittle material: if the applied stress is high enough or is applied for long enough, then the ice may fail by brittle fracture rather than deform by ductile creep. It is most important to distinguish between brittle and ductile behaviour. Although both are a form of "yield", in the sense that the ice gives way under the stress, they are quite distinct processes and must be treated by quite different formalisms.

The words "failure" and "strength" are often applied too thoughtlessly to ice behaviour, and frequently cause confusion. When ice is creeping in response to stress the concept of strength does not properly apply to it – no more than it applies, for instance, to a viscous fluid such as treacle. At a given stress, both treacle and ice simply show a certain rate of deformation, and it is inappropriate to describe this stress as the "strength" of the material. If the stress is increased, the deformation rate increases, but it is misleading to say that the strength of the ice (or the treacle) increases. All that is happening is that the material is deforming in response to a load, and we should only describe the load as a strength once we reach the point where no further load can be applied. Nonetheless, having said this, the terminology of "strength" is commonly used even for creep deformation, and I have no doubt that I shall slip into it on more than one occasion in the course of this book.

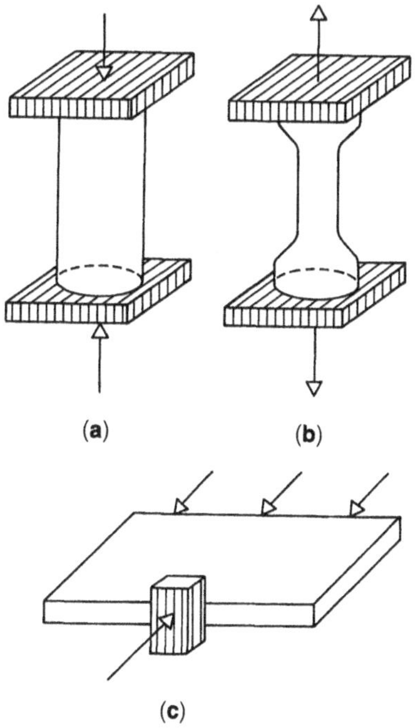

Fig. 4.1 Principal types of loading geometry: (a) uniaxial compression; (b) uniaxial tension (for practical reasons, the specimen is often dumb-bell shaped to begin with); (c) edge indentation.

Similarly, it is misleading to describe ice as failing or yielding under these conditions, since such terms are generally reserved for materials which have a fairly sharp transition between elastic and plastic yield behaviour. Many materials show this type of behaviour, but ice does not. It creeps at all stresses and shows no yield point.

If brittle fracture occurs the ice may, of course, be said to be failing. It ceases to be able to bear the applied load. However, there is unfortunately no such thing as a unique brittle failure strength: ice does have a certain well defined fracture toughness (which we shall describe in Section 4.3), but the stress at which a particular sample of ice fails depends also on the shape, size and number of flaws or imperfections in the sample. If the flaw-sizes in a number of very carefully prepared specimens of ice are perfectly uniform then the specimens should all have precisely the same brittle fracture strength. However, flaw sizes may be highly variable, with the consequence that failure strength in a brittle material is dependent on the distribution of flaws in the sample. Strength varies from sample to sample, and with sample size, and for this reason strength cannot properly be described as a material parameter.

It is convenient to divide the treatment of ice properties into two parts:

(1) *continuum behaviour* – elastic and ductile creep deformation without fracture or rupture;
(2) *fracture behaviour* – brittle and ductile-brittle processes in which cracks form.

Laboratory tests on ice over a large range of strain-rates show an apparently smooth transition between ductile and brittle behaviour. However, the two processes have quite different scaling properties, which means that they have to be treated separately.

The principal mechanical properties tests we shall be looking at are shown in Fig. 4.1. They are:

(a) *Uniaxial compression:* a simple compressive load is applied to each end of a sample, which is unrestrained laterally and free to deform. The sample is usually cylindrical, and the load is normally applied by means of end platens in contact with the ice which are selected to have similar Young's modulus to that of the ice. This helps avoid strain incompatibility and crack nucleation at the sample ends.
(b) *Uniaxial tension:* a simple tensile load is

applied to the sample. This is more difficult to perform, since it is hard to avoid applying a bending moment to the sample. Tests are normally carried out by freezing the ends of a dumb-bell shaped ice sample to the test machine platens. The dumb-bell shape helps avoid failure at the platen contact.

(c) *Indentation:* a rigid indentor, either flat or circular, is driven into a flat plate of ice. The stress conditions under this loading geometry are no longer uniaxial, since biaxial confining stresses are generated in the vicinity of the indentor. This type of loading is especially important to the problem of forces on structures.

All these geometries of test conditions may also be carried out in a variety of loading modes, principally:

(1) *constant strain-rate,* in which a load is applied and controlled in such a way that the sample deforms at a constant rate;
(2) *constant load,* in which an unchanging dead load is applied to the sample (for small deformations, this is effectively a constant stress test);
(3) *constant loading-rate,* in which a regular monotonically increasing load is applied to a sample.

Of the tests we shall be looking at, the vast majority were carried out under conditions of constant strain-rate (1), which are the most convenient to interpret. However, there is really no fundamental physical difference between these different methods of testing: a sample which has been deformed to 1% strain under constant load is not significantly different from a sample which has been deformed to 1% strain under constant strain-rate.

There are many other ways of testing ice (or any other material), among them:

- lateral strain constraint
- lateral stress constraint
- in bending (flexural strength)
- in shear
- in torsion
- with triaxial constraint

I will address these alternative test conditions as and when necessary, but for the most part we shall be looking at uniaxial and indentation conditions. First of all, we shall look at continuum behaviour (Section 4.2) and then at fracture behaviour (Section 4.3). Following this, and as a lead-in to subsequent discussion of large-scale indentation of ice with offshore structures, we shall look at laboratory and medium-scale indentation experiments (Section 4.4).

4.2 CONTINUUM BEHAVIOUR

4.2.1 Pure ice: general

We shall look first at the continuum properties of pure freshwater ice. They are the simplest to understand and provide a good basis for analysis of sea-ice properties.

Many different forms of ice can be identified, and they are generally classified according to their formation process, crystal type and orientation. For a full description of the diverse types of ice found in nature and in the laboratory, the reader should refer to Michel (1979).

For our purposes here there are really only two broad types of ice structure which are worth distinguishing:

(1) *Granular ice,* which is a conglomerate of randomly oriented crystals of ice, typically of grain size 1–3 mm. It is also known as random polycrystalline ice and equi-axed ice, and is the kind of ice normally found in glaciers, icebergs and the surface layers of lake ice. If the crystal orientations are truly random, then the mechanical properties of the ice are, of course, isotropic. Most laboratory studies on granular ice are carried out on ice formed by flooding fine snow crystals with cold water. This is known as T-1 ice.
(2) *Columnar ice,* which is an arrangement of crystals oriented so that their *c*-axes preferentially lie in-plane. Such crystals are generally in the size range 3–100 mm and under normal growth conditions the crystals are elongated in the vertical direction with their *c*-axes randomly distributed in the horizontal plane. This ice is generally referred to as S-2 ice and is found in the lower layers of lake and river ice. It is transversely isotropic (orthotropic) in properties; that is, its properties are the same in all directions in the horizontal plane. However, if *c*-axes also have a preferred orientation in their horizontal plane then it is wholly anisotropic.

The continuum behaviour of these two types of ice, granular and columnar, is markedly similar in form, with minor quantitative differences which I shall describe in this section.

Consider first, however, what happens as a load

is applied to a sample of ice. When ice is subjected to a stress it initially deforms in three distinct ways: it undergoes an immediate elastic strain ε_e, a transient time-dependent delayed elastic strain ε_d, and a time-dependent non-linear viscous creep strain ε_v. (The "delayed elastic" strain ε_d is referred to as primary creep, and is largely recoverable on unloading; the "viscous" strain ε_v is referred to as secondary creep, and is permanent.) Well verified constitutive laws exist to describe this stress–strain behaviour both for granular ice (Jacka, 1984) and for columnar ice (Sinha, 1978b). Once the stage of apparently stable secondary creep has been reached, and all transient creep effects have become negligible, the deformation process unfortunately becomes more complex again, and a stage of tertiary creep may be entered. As we shall see, the state of secondary creep can in fact be regarded merely as an intermediate stage between primary creep and tertiary creep, but the process of tertiary creep is not yet thoroughly understood.

Here we shall look briefly at each of the stages in turn, and then return (in Section 4.2.2) to a more detailed analysis of secondary creep, which for our purposes is the most important mode of continuum deformation.

A simple spring-and-dashpot model of the rheological description of ice behaviour is shown in Fig. 4.2(a), and a schematic illustration of the strain history during a constant-load test is shown in Fig. 4.2(b). The phases of deformation are as follows.

4.2.1.1 Elastic strain ε_e

A uniaxial stress σ is suddenly applied at time t_0. An instantaneous true elastic strain ε_e occurs (Fig. 4.2(b)), following Hooke's Law:

$$\varepsilon_e = \sigma/E$$

where E is Young's modulus. The value of E for pure ice is about 9.5 GPa (Hobbs, 1974; Sinha, 1983a) and is not strongly dependent on temperature or grain size (Sinha, 1979; Weeks and Mellor, 1984). Rather lower values, in the region of 5 MPa, are often quoted for Young's modulus E, and it is sometimes claimed that the Young's modulus of ice is a time-dependent material parameter, decreasing with time after the first application of load. This is not strictly true, but is a manifestation of subsequent delayed elastic strain effects, which should really be treated separately. The truly elastic properties of ice are best measured by applying high-frequency oscillating pulses or vibrations which allow no opportunity for significant creep deformation to take place, and it is under these conditions that pure ice displays a Young's modulus of 9.5 GPa. Young's modulus is, however, quite strongly dependent on porosity, whether due to inclusion of air or brine, and at 10% porosity (density about 0.83 Mg m^{-3}) E has a value of about 7 GPa (Weeks and Mellor, 1984).

Poisson's ratio ν for pure ice is approximately 0.33 ± 0.03 (Weeks and Assur, 1967) though again, different (higher) values are sometimes quoted from tests in which creep deformation played an important part. Ultimately, for a test in which creep is the dominant deformation process, an apparent value of 0.5 would be obtained for ν. This, however, is not a true Poisson's ratio.

Values of Poisson's ratio and Young's modulus measured under quasi-static conditions, rather than dynamic conditions, are sometimes referred to as "effective" Poisson's ratio and "effective" modulus, and they do, as expected, show quite complex dependence on time and loading-rate. I do not consider these concepts to be useful, and they should be avoided.

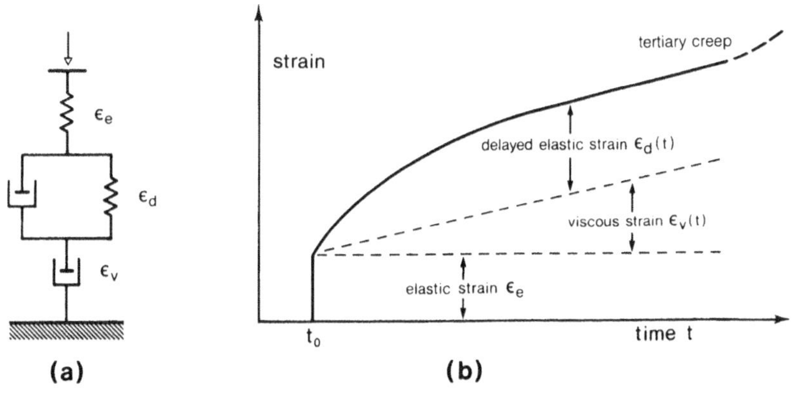

Fig. 4.2 (a) Simple spring-and-dashpot model of the rheology of ice. Deformation occurs initially in three ways: by simple elastic strain ε_e; by recoverable delayed elastic strain ε_d; and by permanent non-linear viscous creep ε_v. (b) Strain history for a constant stress suddenly applied at time $t = t_0$. Eventually, a stage of tertiary creep may be reached where strain-rate increases and simple constitutive laws no longer apply.

4.2.1.2 Delayed elastic strain $\varepsilon_d(t)$

From the moment the load is first applied, time-dependent delayed elastic strain begins to occur, at a variable decreasing rate (Fig. 4.2(b)). The deformation is not permanent, but reversible, and if the ice is suddenly unloaded it should under ideal conditions gradually recover all the deformation due to this delayed elastic strain process (Sinha, 1983a). This is why it is referred to as an "elastic" strain, even though it is manifested as a creep rate, and it is partly for this reason that it is often confused with true elastic strain. Delayed elastic strain is also referred to as primary creep, transient creep, recoverable creep and pseudo-elastic strain. Whatever it may be called, it has been thoroughly investigated by experiment and follows quite well defined laws (Sinha, 1978b, 1979, 1983a). Under conditions of constant applied stress σ applied at time $t_0 = 0$, the time-dependent delayed elastic strain ε_d at time t is given by:

$$\varepsilon_d = \frac{C\sigma}{Ed}\{1-\exp[-(a_T t)^q]\} \tag{4.1}$$

where d is grain diameter (m), E is Young's modulus, C is a constant, and a_T is a quantity dependent on temperature T. Experimentally determined values for these quantities for S-2 ice at temperature −10°C (263 K) are:

$C = 9\times10^{-3}$ m
$E = 9.5$ GPa
$a_T = 2.5\times10^{-4}\ \mathrm{s}^{-1}$
$q = 0.34$

Note that primary creep is grain-size dependent: the smaller the grain size, the greater the strain sustained for a given stress.

A similar equation to equation (4.1) can also be written on the basis of experiments with randomly oriented granular polycrystalline ice (Jacka, 1984; Ashby and Duval, 1985). There is still some room for argument about the most dimensionally satisfactory way of expressing equation (4.1); in particular, it appears that the constant q may simply be the reciprocal of the exponent n in Norton's law for secondary creep (equation (4.2)), but for our purposes equation (4.1) is a satisfactory description of experimental observation of delayed elastic strain.

4.2.1.3 Secondary creep strain $\varepsilon_v(t)$

After a certain length of time under load, the creep-rate of ice approaches an apparently constant minimum value (the dashed sloping line in Fig. 4.2(b)). This is known as secondary creep or steady-state creep. Secondary creep is present as a background strain-rate throughout the load history of the ice, but only becomes apparent when the delayed elastic strain-rate $\dot{\varepsilon}_d(t)$ approaches zero and allows the secondary creep strain-rate $\dot{\varepsilon}_v(t)$ to dominate.

Secondary creep is well understood both for isotropic polycrystalline ice and for columnar S-2 ice. For uniaxial loading under compressive or tensile stress σ_{11} the secondary creep strain-rate $\dot{\varepsilon}_v(t)$ is given by an equation of the form (Glen, 1955):

$$\dot{\varepsilon}_v(t) = B_i \sigma_{11}{}^n \tag{4.2}$$

where n is a constant in the range 3 to 4 (normally taken to be 3) and B_i is a parameter dependent principally on temperature and crystal type (the subscript i will be used to denote crystal type). This is an example of the flow law known as Norton's law. The law can be generalised for triaxial stress conditions and anisotropic deformation – and we shall examine this in detail in Section 4.2.2.

Secondary creep appears to be independent of grain size (Duval and Le Gac, 1980); reported apparent variations of creep-rate with grain size are probably due to contributions from delayed elastic strain effects, as described above, which *are* grain-size dependent. If the time taken to perform a test is insufficiently long, then secondary creep effects may not be properly isolated.

Although secondary creep is quite easily observed, in the sense that creep rate during experimental loading tests settles down to an apparently stable value, it is not a truly steady-state process which lasts indefinitely. Indeed, when viewed from one step back, in a longer timescale perspective, it emerges that secondary creep is really nothing more than a minimim creep-rate, marking a transition between primary creep and tertiary creep. This is shown strikingly in Fig. 4.3(a), in which strain-rate during constant-load tests is plotted against time on a logarithmic scale (Mellor and Cole, 1982). This shows that initially high strain-rates (primary creep) settle down to a minimum (secondary creep) and subsequently increase (tertiary creep). Figure 4.3(b) shows these data replotted as a function of axial strain instead of time, and it appears that the minimum creep rate occurs at an accumulated strain of $0.8–1.0\times10^{-2}$ (0.8–1.0%).

The creep rate at this minimum point follows some well established laws and is of importance for the following reason: because it represents the

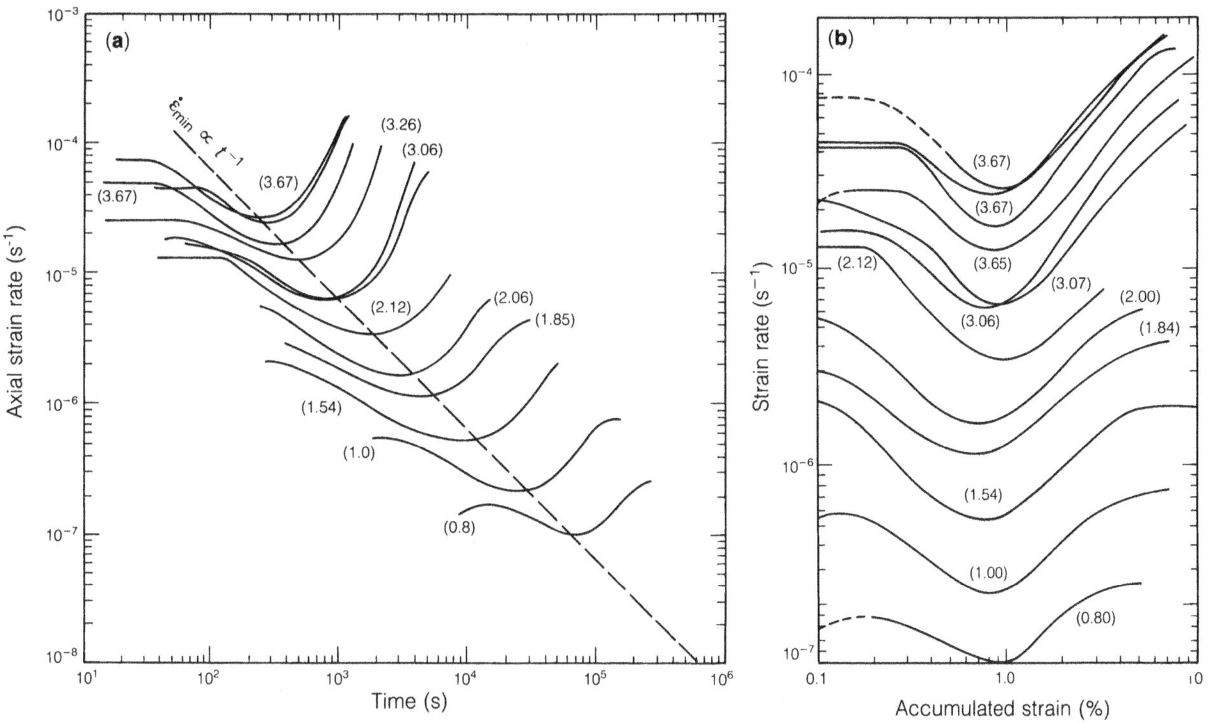

Fig. 4.3 Strain-rate of pure polycrystalline ice under constant load, plotted as a function of (a) elapsed time (s), and (b) accumulated strain (%). Note that the minimum strain-rate occurs at an accumulated strain of 0.8–1.0%. The numbers in brackets indicate the stress level (MPa) at which tests were carried out (after Mellor and Cole (1982), with some tests omitted for clarity).

minimum strain-rate associated with a constant stress, it also, conversely, represents the maximum stress associated with a given strain-rate. This means that if ice is tested at constant strain-rate, or if ice moves past a structure at a constant velocity, then the stress calculated on the assumption of secondary creep represents the maximum stress which will be sustained during the interaction process.

4.2.1.4 Tertiary creep

As is clear from Fig. 4.3, after a long time under compressive loading, strain-rates tend to increase, as the phase of tertiary creep is entered. This is frequently, though not always, associated with the formation of microcracks at grain boundaries in the ice fabric. These cracks may coalesce as strain continues, leading to accelerating deformation rates. Ultimately, complete failure may result. Tertiary creep is not well understood, and may also be observed without crack formation, as a result of dynamic recrystallisation at grain boundaries. If this occurs, then old deformed ice grains are effectively replaced by new undeformed ice, permitting a new wave of primary creep to occur, accelerating the strain-rate (Frost and Ashby, 1982).

Experimental evidence suggests that the onset of internal crack formation is related simply to total delayed elastic strain ε_d as discussed above, and described by equation (4.1). For S-2 ice of 4–5 mm grain size, cracks under any form of loading appear to begin to form when the axial delayed elastic strain ε_d exceeds 1.0×10^{-4}, or 0.01% (Sinha, 1982). This has interesting consequences. Equation (4.1) indicates that:

(a) If the applied stress is less than about 0.5 MPa then microcracks should never form, since the total delayed elastic strain will never exceed 0.01%, however long the sample is left under load. (The total strain $\varepsilon_e+\varepsilon_d+\varepsilon_v$ will of course continue increasing practically without limit, but the component ε_d will, according to equation (4.1) asymptotically approach a value just less than 0.01%.)

(b) If the applied stress exceeds 0.5 MPa then

> eventually cracks will form, as long as the sample is left long enough for the delayed elastic strain to exceed 0.01%. For a stress of 1 MPa the time required is 30 min; for a stress of 2 MPa it is only 2 min. Once cracks have begun to form, structural deterioration causes strain-rates rapidly to increase, and failure may follow soon afterwards.

I shall return to this later, in Chapter 6 (Section 6.3.2) and examine the influence of loading-rate on the stress at which cracks first form.

4.2.2 Microscopic deformation mechanisms

It is interesting to pause for a while and look at the microscopic mechanisms behind the various types of continuum deformation discussed in Section 4.2.1.

The subject is a vast one, and for a much fuller account the reader should refer to Frost and Ashby (1982), a work which discusses deformation mechanisms in the broad context of metals and ceramics in general. Ice is an interesting material, but not so very extraordinary when considered in this broader context: it is a solid at a high homologous temperature and it behaves, in fact, in a very similar way to most metals near their melting point.

We have already seen, in Chapter 2 (Section 2.2 and Fig. 2.3), how permanent deformation may occur in ice at the atomic level. The passage of dislocations along glide planes, perpendicular to the *c*-axis, allows a net permanent movement of atoms relative to each other. This process, known as dislocation glide, is principally responsible for the secondary creep behaviour of ice, and the rate at which proton rearrangement can occur appears

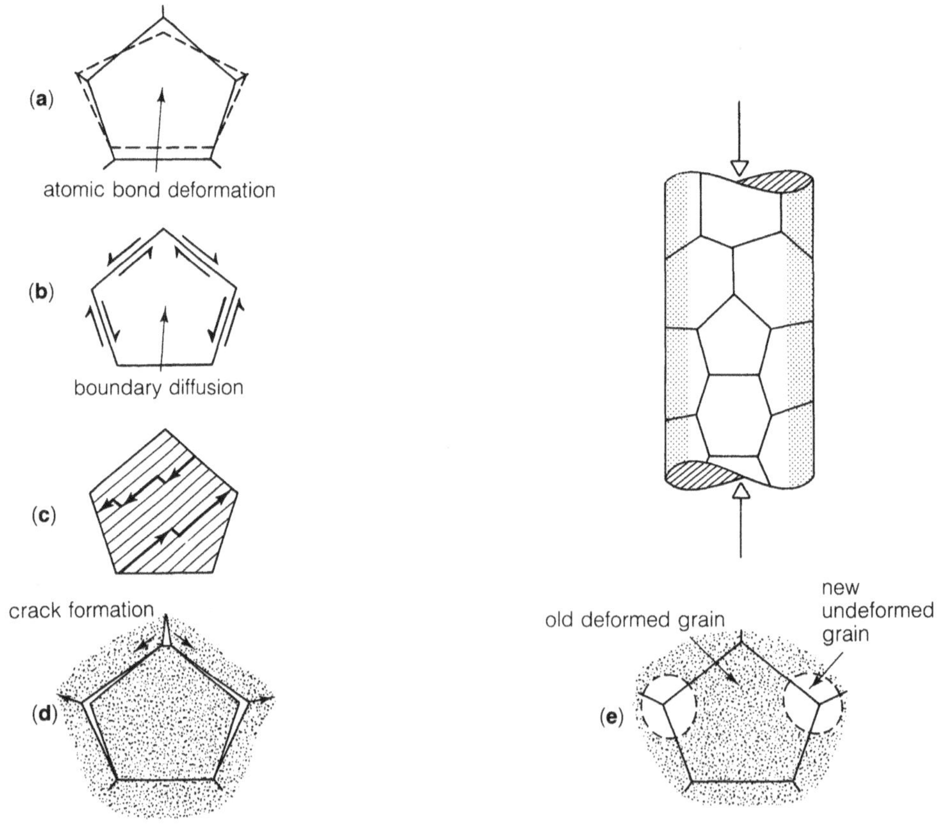

Fig. 4.4 Schematic diagram of processes within a grain during uniaxial loading: (a) elastic deformation, solely by atomic bond deformation; (b) delayed elastic strain, in which boundary diffusion allows further elastic deformation of a grain; (c) secondary creep occurring by dislocation glide (and climb) within a grain; (d) crack formation due to dislocation pile-up at grain boundaries; (e) dynamic recrystallisation, leading to formation of fresh grains which may begin the deformation cycle anew. See text for a more detailed commentary.

to be responsible for the rate at which creep occurs (Goodman *et al.*, 1981). But it is interesting also to look at the micromechanics of elastic strain, delayed elastic strain, tertiary creep and crack formation, and for this we need to consider the mechanics of individual crystal grains as they interact with each other in a polycrystalline solid.

A simplified picture of this is shown in Fig. 4.4, illustrating a cylindrical sample of polycrystalline ice placed under uniaxial compressive stress. Figures 4.4(a)–(e) represent, schematically, the processes occurring around a single grain which give rise to the various deformation modes we have considered:

(a) *Elastic strain:* as the sample is compressed, a grain initially deforms in a purely elastic and reversible manner, simply by atomic bonds lengthening and shortening as necessary to accommodate the strain. No dislocation motion occurs.

(b) *Delayed elastic strain:* as a result of shear stresses generated between grains, sliding occurs at the grain boundaries, storing up additional elastic energy within the grain itself. Sliding generally occurs by the mechanism of diffusional flow, at a rate which depends on the grain size, since diffusion of matter must take place over a distance of the order of a grain facet (Fig. 4.4(b)). The sliding occurs in conjunction with an elastic deformation of the grain, and in the case of pure delayed elastic strain, no internal permanent deformation of the grain takes place. All the atomic rearrangement processes occur at the boundary. This means that if the applied compressive stress is relaxed then the grain tries to recover its original undeformed shape again: but to do so, it must do work on the grain boundaries, and reverse all the sliding that has occurred. This again must occur by diffusional flow, and accounts for the reversibility of delayed elastic strain.

(c) *Secondary (power-law) creep:* within the grain, permanent deformation processes occur, leading to irreversible rearrangement of the material. In ice the principal mechanism responsible is dislocation glide, with dislocation climb occurring where necessary for compatibility. This is shown schematically in Fig. 4.4(c), where dislocation motion is represented by arrows.

(d) As a result of processes (b) and (c), dislocations tend to pile up at grain boundaries and migrate towards the junction of three grains. Eventually, this may lead to formation of cracks (Fig. 4.4(d)) and accelerating strain-rates. This is one cause of tertiary creep.

(e) Alternatively, especially at high temperatures, dynamic recrystallisation may occur in regions of high dislocation density (Fig. 4.4(e)). Instead of cracks forming, extensive rearrangement of dislocations occurs, resulting in the formation of a new, effectively undeformed grain. This may then deform again by primary creep, manifesting itself as an increasing strain-rate which may be identified as tertiary creep without rupture. Dynamic recrystallisation may occur repeatedly, resulting in waves of increasing deformation rate.

4.2.3 Pure ice: secondary creep

The simple power law creep relationship introduced above for secondary creep (equation (4.2)) is the special case for uniaxial stress states. The law can readily be generalised to cope with creep under biaxial and triaxial stress states, and it can also be generalised to cope with anisotropic material response to these multiaxial stress states. In this section we will look first at the basic data for uniaxial loading and then examine the formal treatment of more generalised stress states.

Consider a stress state described by the stress tensor σ_{ij}, giving rise to a strain-rate field described by the tensor $\dot{\varepsilon}_{ij}$, where $i, j = 1, 2, 3$ for three orthogonal spatial axes.

First of all, consider the data for uniaxial loading at stress σ_{11}, with $\sigma_{22} = \sigma_{33} = 0$. Along the axis of the applied stress the ice deforms at rate $\dot{\varepsilon}_{11}$, and for all types of ice the flow law for uniaxial stress can be written in the form

$$\dot{\varepsilon}_{11} = B_i \sigma_{11}{}^n \qquad (4.3)$$

where B_i depends on temperature and ice crystal type and orientation (the subscript i will be replaced by subscript g to denote granular ice and by c to denote columnar ice). A vast amount of experimental data exists to confirm this equation for pure ice. The most important studies are plotted in Fig. 4.5 (from Hallam, 1986), based on data of Tabata (1966), Barnes *et al.* (1971), Hawkes and Mellor (1972), Michel and Toussaint (1977), Carter and Michel (1978), and Michel (1978). These studies include data obtained in tensile loading as well as compressive loading: for pure creep (without crack formation) there is no difference between the properties of ice in tension and compression.

Over a wide range of strain-rates, from 10^{-10} s^{-1} to about 10^{-5} s^{-1}, the data of Fig. 4.5 show a clear and consistent slope of about 3, indicating that the power index n in the flow law equation (4.2) is approximately equal to 3. There are in fact good theoretical grounds for believing that it is precisely equal to 3, since this is the index which is expected to be associated with creep by the mechanism of dislocation glide (Goodman *et al.*, 1981).

At higher stresses, however, strain-rates tend to increase at a greater rate, and the flow regime enters a state known as power-law breakdown (shown after the change of slope in Fig. 4.5). This is probably due to an alternative process (such as kink nucleation) becoming the rate-controlling process. A fuller explanation of this is given in Frost and Ashby (1982) and Goodman *et al.* (1981). In the rest of this book I shall, for most purposes, set the power law index n equal to 3 when discussing secondary creep behaviour of ice.

The constant B_i in equation (4.3) is temperature-dependent, and follows the classical Arrhenius activation energy law (Glen, 1955), so that

$$B_i = A_i \exp(-Q_i/RT) \tag{4.4}$$

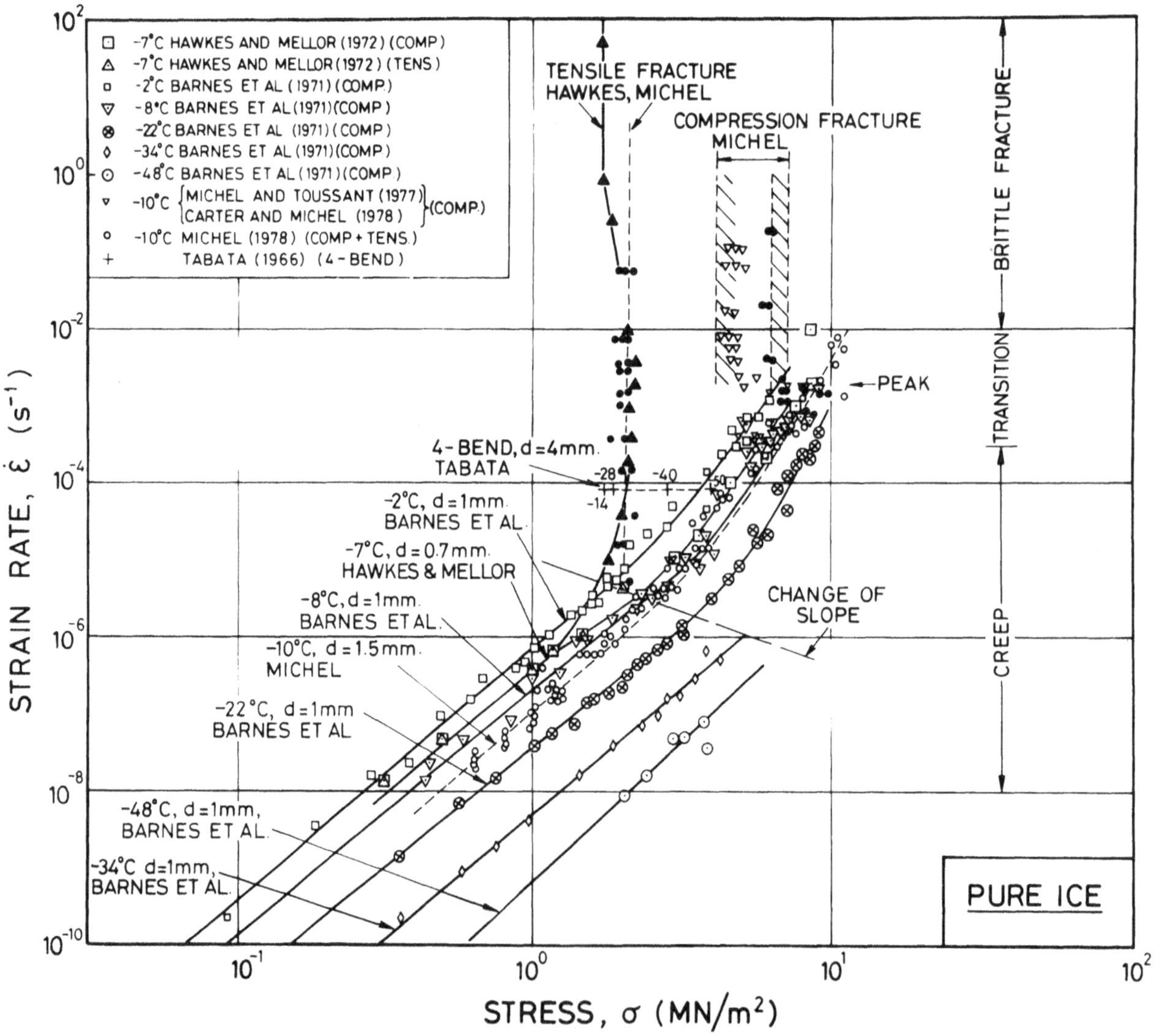

Fig. 4.5 Uniaxial loading of pure polycrystalline ice (after Hallam, 1986). Below strain-rate about 10^{-5} s^{-1} power-law creep occurs, with $n = 3$. At stresses exceeding 5–10 MPa, a transition to brittle behaviour occurs.

where $R = 8.314$ J mol^{-1} K^{-1} (the universal gas constant), T is temperature in kelvin (K) and Q_i is the activation energy. A_i is a constant which depends only on crystal type (i = g for granular ice and i = c for columnar ice). For engineering purposes there is negligible dependence of any of these constants upon overall hydrostatic pressure (Jones, 1982, and Section 4.3.4). Let us now look at these constants for the two principal crystal types.

4.2.3.1 Granular ice

Barnes *et al.* (1971) found for pure granular ice that the activation energy Q_g has two values, according to temperature T lying above or below about 265 K (−8°C). This is probably due to some grain boundary melting occurring above −8°C, which accelerates the creep-rate. Their data, and data of Michel (1978), are shown in Fig. 4.6, normalised to temperature 263 K (−10°C) using these values of Q_g. It can be seen that below a strain-rate of about 10^{-5} s^{-1} the data sit nicely on a straight line with slope of $n = 3$. At strain-rates above 10^{-5} s^{-1} the curve shows the power-law breakdown behaviour already discussed, but below this a general law can be written as

$$\dot{\varepsilon}_{11} = A_g \exp(-Q_g/RT)\sigma_{11}{}^3 \tag{4.5}$$

Where the constants A_g and Q_g take the following values:

above 265 K (−8°C): $Q_g = 120$ kJ mol^{-1}, $A_g = 7.8\times10^{16}$ (MPa)$^{-3}$ s^{-1}

below 265 K (−8°C): $Q_g = 78$ kJ mol^{-1}, $A_g = 4.1\times10^{8}$ (MPa)$^{-3}$ s^{-1}

The straight line in Fig. 4.6 represents equation (4.5) for the temperature 263 K (−10°C). The equation can, of course, be inverted to derive stress from strain-rate, when it becomes

$$\sigma_{11} = \left[\frac{1}{A_g}\exp(Q_g/RT)\dot{\varepsilon}_{11}\right]^{1/3} \tag{4.6}$$

4.2.3.2 Columnar ice

The same normalisation can be carried out for columnar ice. Gold (1973) found a value for Q_c of 65 kJ mol^{-1} for columnar ice, and Fig. 4.7 shows data from Gold (1973), Croasdale *et al.* (1977), Michel (1978) and Sinha (1983b) normalised to temperature 263 K (−10°C) using this value for Q_c. All the data refer to "horizontal" loading of pure S-2 columnar ice: that is, loading in the same plane as the c-axis. A general law of the same form

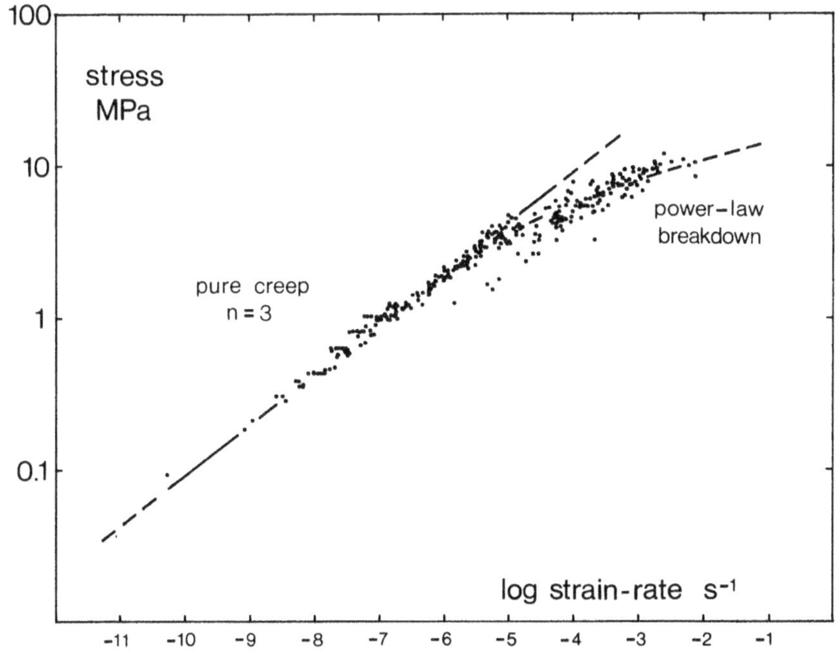

Fig. 4.6 Pure granular polycrystalline ice: stress against strain-rate under uniaxial compression normalised to temperature 263 K (−10°C). Data are from Barnes *et al.* (1971) and Michel (1978). (Note that axes are reversed with respect to Fig. 4.5.)

as equation (4.5) can be written for these data. It is

$$\dot{\varepsilon}_{11} = A_c \exp(-Q_c/RT)\sigma_{11}^{3} \qquad (4.7)$$

with constants:

$$Q_c = 65 \text{ kJ mol}^{-1}$$

and

$$A_c = 3.5\times10^{6} \text{ (MPa)}^{-3} \text{ s}^{-1}$$

It is possible that the activation energy Q_c for columnar ice should be two-valued, as for granular ice. Gold's test data, carried out between −5°C and −40°C, are consistent with either hypothesis, but for simplicity we shall just use one value.

Notice that equations (4.5) and (4.7) with their appropriate constants reflect the relative creep resistance, or "strength" of granular and columnar ice. At −10°C the stress required to make granular ice deform at a certain rate is about 1.5 times that for columnar ice.

4.2.4 Sea ice: secondary creep

Sea ice differs from pure ice in essentially only one respect: it contains brine. Brine pockets exist, as we have seen in Chapter 2, in the form of tiny, vertical, roughly cylindrical channels. They are fluid-filled, unable to support shear stress, and therefore allow the sea ice to deform more easily. They reduce its strength. In effect, any gross applied stress is transferred through the solid ice lattice alone and cannot pass through the brine. Additionally, the geometry of the brine pockets causes local stress concentrations which non-linearly increase the associated creep rate. The net result is that the solid ice lattice bears a greater local stress than the gross applied stress and deforms more rapidly in response to it. The behaviour of the medium as a whole may be described by reference not to the gross applied stress but instead to the local solid ice stress, or "net section stress". The same arguments apply also to the influence of air bubbles, but there is a shortage of measurements of this effect.

For brine pockets the effect has been investigated both theoretically and experimentally. Geometric considerations (Weeks and Assur, 1967) show that for horizontal loading of ice containing vertically arranged cylindrical brine pockets the net section stress σ' is given by

$$\sigma' = \sigma/(1-\sqrt{\nu_b/\nu_0}) \qquad (4.8)$$

where σ is the gross applied stress and ν_b is the brine volume. Brine volume is a function of temperature and salinity, and can be determined by using equation (2.1). The quantity ν_0 is a normalising constant which has to be determined. Physically, it is clear from equation (4.8) that ν_0

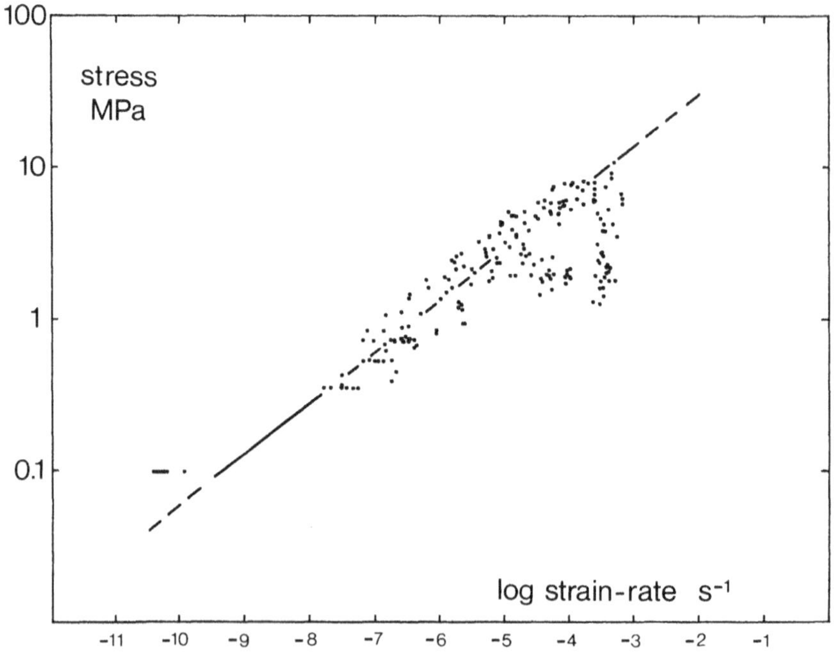

Fig. 4.7 Pure columnar S-2 ice: stress against strain-rate under "horizontal" uniaxial compression, normalised to temperature 263 K (−10°C). Data are from Gold (1973), Croasdale *et al.* (1977), Michel (1978) and Sinha (1983b).

corresponds to the brine volume at which the sea ice loses all strength: if $\nu_b = \nu_0$ the net section stress σ' tends to infinity, since it becomes transmitted by a vanishingly small section of the pure ice lattice. The ice must collapse. The concept of net section stress is not limited to secondary creep, but applies to all types of loading. A variety of tests on compressive strength, tensile strength and Young's modulus (Weeks and Assur, 1967; Vaudrey, 1977) show experimentally that ν_0 lies in the range 0.1–0.29.

The same is true for creep behaviour, and data for the creep of sea ice can be reconciled with data for pure ice simply by applying a net section stress correction of the form of equation (4.8). This is most easily demonstrated for columnar sea ice.

4.2.4.1 Columnar sea ice

Experimental data for horizontal loading of pure columnar ice have already been presented in Fig. 4.7. Comparable experimental data for sea ice, uncorrected for any effects of temperature or brine volume, are shown in Fig. 4.8, from a range of sources (Schwarz and Weeks, 1977; Saeki *et al.*, 1978; Wang, 1979a,b; Urabe *et al.,* 1980; Sinha, 1983b). These data were collected by performing horizontal compression tests on a variety of sea-ice samples. The tests cover a wide range of temperature, salinity and grain size and include sea-ice samples which have preferred *c*-axis orientation. Such samples are not truly axi-symmetric, so are not strictly comparable with samples of pure S-2 ice.

The data of Fig. 4.8 show a wide degree of scatter and the trend differs substantially from that of pure ice, represented by the dashed line. Practically all the data for sea ice lie below the line for pure ice, indicating, as expected, that the strength behaviour of pure ice represents an upper bound to that of sea ice.

Two corrections can be applied to these data which substantially reduce the scatter of the sea ice data and effectively normalise the behaviour of sea ice to that of pure ice. The corrections are:

(a) an Arrhenius temperature correction, as in equation (4.7), using the constants determined for S-2 ice;
(b) a brine volume correction using equations (2.1) and (4.8).

A value of ν_0 still has to be selected, and this can best be done by varying it until the fit of sea-ice data with S-2 ice data is optimised. It is found that a value of $\nu_0 = 0.16$ gives optimum agreement, and the result of correcting sea-ice data in this way is shown in Fig. 4.9: all the data of Fig. 4.8 are corrected for temperature and brine volume and are superimposed on the data of Fig. 4.7. The data still show a high degree of scatter even after correction,

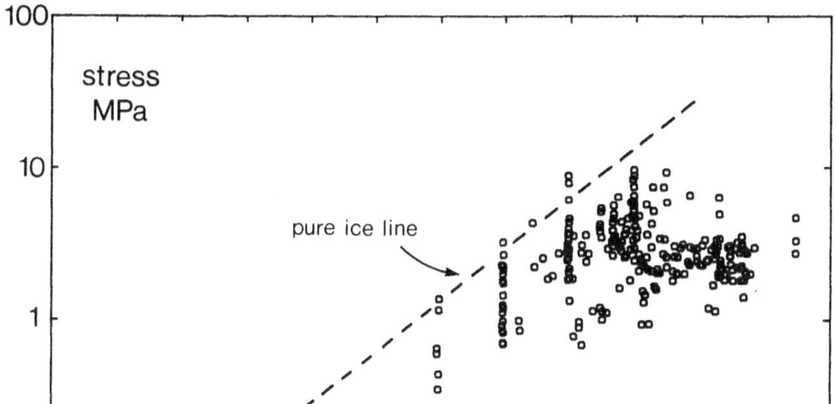

Fig. 4.8 Columnar sea ice: stress against strain-rate under uniaxial horizontal compression. Data were collected for a wide variety of temperature, salinity, grain size and crystal orientation, and no corrections have been applied. The dashed line is the best fit through data for pure S-2 ice (Fig. 4.7). Data are from Schwarz and Weeks (1977), Saeki *et al.* (1978), Wang (1979a,b), Urabe *et al.* (1980) and Sinha (1983b).

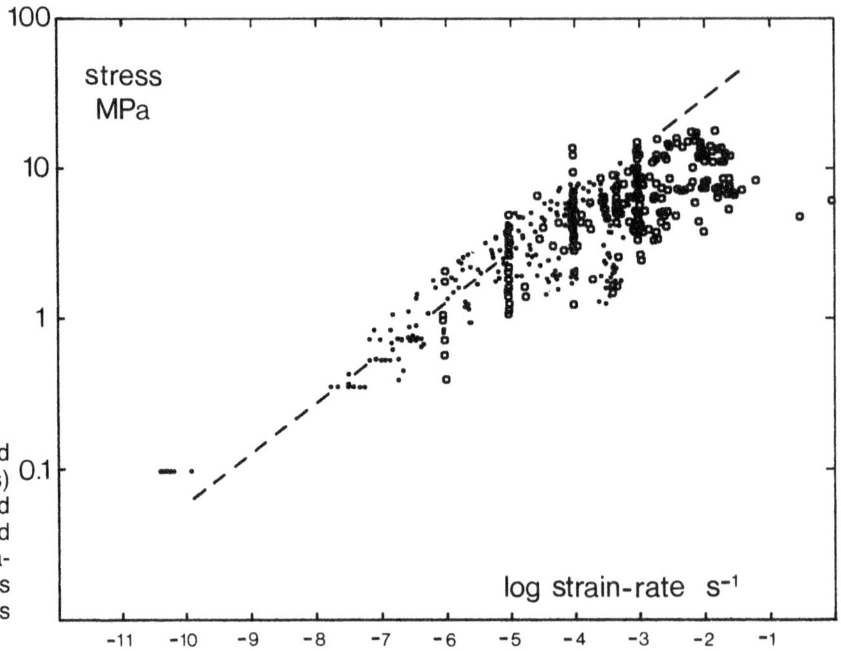

Fig. 4.9 Comparison of corrected columnar sea ice data (open circles) with columnar pure ice data (solid dots). All data have been normalised to zero brine volume and temperature 263 K (−10°C) using equations (2.1), (4.7) and (4.8). Data sources are as for Figs 4.7 and 4.8.

owing to variations inherent in a naturally formed material. Some of the variation is probably also due to preferential horizontal alignment of *c*-axes. On the basis of the data of Fig. 4.9 the law for horizontal uniaxial loading of unoriented columnar sea ice can be rationalised as

$$\dot{\varepsilon}_{11} = A_c \exp(-Q_c/RT)\left[\frac{\sigma_{11}}{(1-\sqrt{\nu_b/\nu_0})}\right]^3 \qquad (4.9)$$

where

$\nu_0 = 0.16$
$Q_c = 65$ kJ mol^{-1}
$A_c = 3.5\times10^6$ (MPa)$^{-3}$ s^{-1}

The behaviour of columnar sea ice is additionally complicated by the fact that sea currents cause *c*-axes to show preferential alignment in the horizontal plane (see Section 2.3.2). This results in a serious anisotropy in material properties with respect to azimuth of loading: the material is no longer axisymmetric and orthotropic but is fully anisotropic. Some data on this effect are shown in Fig. 4.10 (Wang, 1979b). It is clear that oriented columnar ice has minimum strength when *c*-axes are oriented at about 45° to the maximum compressive stress. This is because under such conditions the maximum shear stress acts along the basal planes of the ice crystals, which is the direction along which glide is easiest. Maximum strength occurs when loading axis is aligned with *c*-axes, and a slightly lower strength is measured when they are perpendicular. All this makes sense: imagine a pack of playing cards subjected to the stress states shown in Fig. 4.10(a). Deformation of the pack is easiest when it can occur by sliding of cards along their "basal planes". Additional data on anisotropic behaviour of columnar ice is also presented by Peyton (1966), but his experiments were carried out essentially on single crystals, which are not necessarily representative of polycrystalline behaviour.

Although anisotropic oriented ice is interesting, it typically only occurs in the warmer (softer and more briney) layers at the bottom surface of sea-ice cover. Because of this it is unlikely that it has a very serious influence on real structural interaction forces.

4.2.4.2 Granular sea ice

Not much work has been done on granular sea ice. The main public source is Wang (1979a), and a plot of these data is shown in Fig. 4.11, uncorrected for temperature or brine volume. The data are compared with the straight pure-ice line from Fig. 4.6.

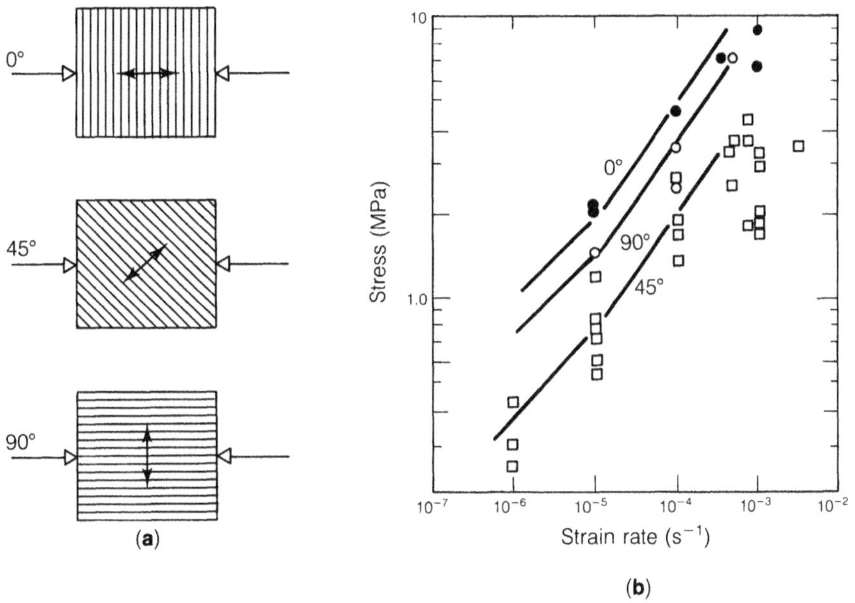

Fig. 4.10 Compressive loading of columnar sea ice with preferential *c*-axis orientation. (a) Three orientations of loading: stripes represent platelets and internal arrows represent *c*-axis orientation. (b) Stress against strain-rate for columnar sea ice at −10°C, at *c*-axis orientations of 0° (closed circles), 45° (open squares) and 90° (open circles). Minimum strength is measured when *c*-axes are at 45° to the direction of principal compressive stress. (After Wang, 1979b.)

The data again show a high degree of scatter. This is due largely to the fact that granular sea ice is generally formed in the early part of the ice season and is highly agitated during its formation. It contains many air bubbles, various flaws and a variety of crystal types, and this makes its mechanical behaviour highly variable. Most of the data of Fig. 4.11 are collected at strain-rates ex-

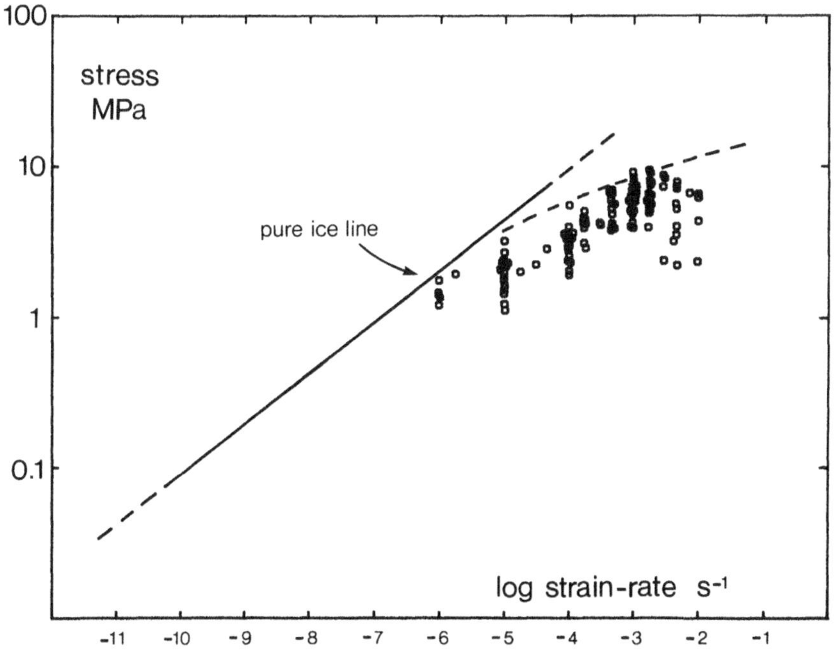

Fig. 4.11 Granular sea ice: stress against strain-rate under uniaxial compression. No corrections have been applied. The dashed line is the best fit through data for pure granular ice (Fig. 4.6). Data are from Wang (1979a).

ceeding 10^{-5} s^{-1}, when deviations from standard power-law creep would be expected, so there are not really sufficient data to justify a normalisation of the sea-ice data to the pure-ice data in the same way as was done for columnar ice. Also, it is not established whether brine pockets in granular ice are still cylindrical in form. If they are spherical instead, then theoretical considerations (Weeks and Assur, 1967) show that the net section stress correction should be of the form:

$$(1-\nu_b/\nu_0)$$

without the square-root dependence shown by equation (4.8) for columnar ice. A normalisation using this form has been attempted and it is found that the data are quite well described by the equation

$$\dot{\varepsilon}_{11} = A_g \exp(-Q_g/RT)\left[\frac{\sigma_{11}}{(1-\nu_b/\nu_0)}\right]^3 \qquad (4.10)$$

with $\nu_0 = 0.1$. The constants Q_g and A_g are as defined for pure granular ice in equation (4.5).

The result of applying the corrections of equation (4.10) to the granular sea-ice data is shown in Fig. 4.12, normalised to $-10°C$ and zero brine volume and compared with the best-fit line for pure granular ice. The fit is not very impressive, but nonetheless plausible. There are probably a number of alternative ways in which the data could be normalised, but the data are not good enough to permit any firm conclusions.

4.2.4.3 Multi-year ice

The studies presented above were all carried out on samples of granular or columnar first-year sea ice. There have also been a limited number of studies on mechanical properties of multi-year ice (Cox *et al.*, 1984). Multi-year ice is not fundamentally different from first-year ice: it is more chaotic in its crystallography, owing to its more agitated formation process, and generally has a lower salinity owing to brine migration processes (Section 2.3.2.3), but there is no evidence to suggest that it follows different laws from first-year ice. Data from Cox *et al.* (1984) are shown in Fig. 4.13, from experiments performed at strain-rates 10^{-5} s^{-1} and 10^{-3} s^{-1} and temperatures $-5°C$ and $-20°C$, on vertical multi-year ice specimens. The data show a high degree of scatter, owing to the great variability of grain and pore structures, but the data set is still broadly bounded by the pure-ice line from Fig. 4.6. The data points which exceed this line may refer to end-loaded vertical columnar ice specimens, which are expected to display exceptionally high strengths. With sufficiently detailed knowledge of each sample's crystal structure, orientation, salinity and temperature, it is likely that the data for multi-year sea ice could be nor-

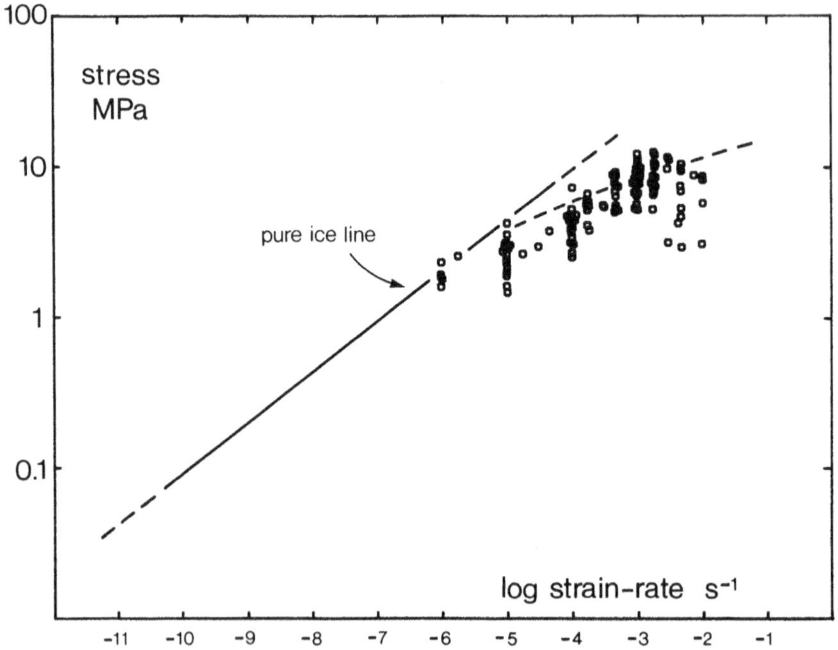

Fig. 4.12 Comparison of granular sea-ice data (open circles) with the power-law creep line for pure ice. All data have been normalised to $-10°C$ and zero brine volume using equation (4.10).

malised in the same style as equations (4.9) and (4.10) above, for first-year ice. However, this would be very difficult to perform, since many multi-year ice samples consist of a conglomerate of granular and columnar ice types in a variety of orientations. There are not yet enough data to be able to separate out all the factors which might affect the behaviour of multi-year sea ice, and I will not attempt it here.

4.2.5 General constitutive equations for secondary creep

The data we have looked at in the last two sections were for uniaxial loading of granular and columnar ice. For isotropic granular ice the loading direction was immaterial, but for columnar ice the data presented referred to loading in the same plane as the *c*-axes, which for natural columnar or S-2 ice generally corresponds to the horizontal plane. The simple formalism of equations (4.6) and (4.7) can readily be extended to cope with more general multiaxial stress states and with a variety of crystal orientations.

The analysis which follows is fairly involved and requires knowledge of tensor notation and scalar potential theory. It is not essential to the thesis of this book; it is interesting, but the disinclined reader may skip it. The analysis is also restricted to secondary creep (that is steady-state power-law creep behaviour) and is not easily applied to delayed elastic strain deformation. This is no major drawback for our purposes, since secondary creep is associated with the maximum stress experienced by ice subjected to a constant strain-rate, and thus it normally represents the design condition when creep is the controlling mechanism.

Denote the components of the three-dimensional stress tensor by σ_{ij} $(i, j, = 1, 2, 3)$. The strain-rate tensor is $\dot{\varepsilon}_{ij}$. We can in general now define an "effective stress", τ, which allows us to derive relationships between $\dot{\varepsilon}_{ij}$ and σ_{ij} for a general anisotropic power-law creeping material with flow-law index n.

For an isotropic material the effective stress is given by the second invariant of the stress deviator tensor (see, for example, Paterson, 1981) and may be written as

$$2\tau^2 = (\sigma_{11}-\sigma_{22})^2+(\sigma_{22}-\sigma_{33})^2+(\sigma_{33}-\sigma_{11})^2 +6({\sigma_{12}}^2+{\sigma_{23}}^2+{\sigma_{31}}^2) \qquad (4.11)$$

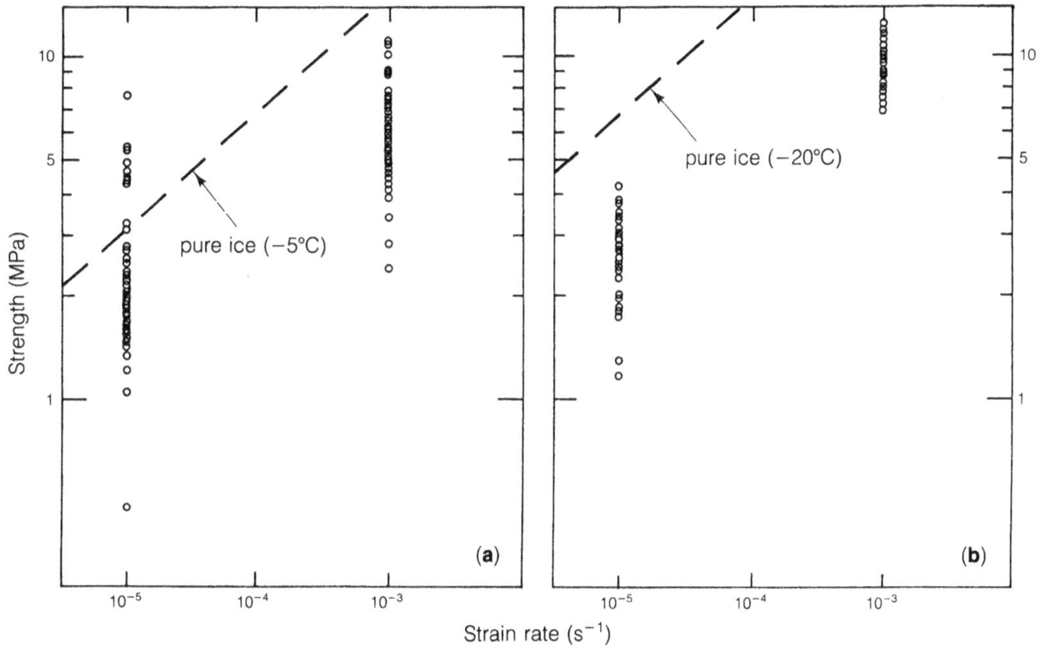

Fig. 4.13 Stress against strain-rate for vertical uniaxial loading of multi-year sea ice, at temperatures of (a) −5°C and (b) −20°C. The data are broadly bounded by the dashed "pure ice line" for granular ice derived from Fig. 4.6 and equation (4.6). Where data exceed this it is likely that crystal orientation is responsible (after Cox *et al*, 1984).

For an isotropic material this may be generalised to:

$$2\tau^2 = a_1(\sigma_{11}-\sigma_{22})^2+a_2(\sigma_{22}-\sigma_{33})^2+a_3(\sigma_{33}-\sigma_{11})^2 +6(a_4\sigma_{12}{}^2+a_5\sigma_{23}{}^2+a_6\sigma_{31}{}^2) \quad (4.12)$$

where a_1 to a_6 are constants. They are not all independent, and equation (4.12) may be written with just five constants instead of six. However, it is clearer with six, and we shall eliminate one constant later.

We can now define a scalar potential function Φ which obeys the associated flow rule and allows calculation of $\dot{\varepsilon}_{ij}$ from τ (Freudenthal and Geiringer, 1958). The scalar potential is

$$\Phi = b_i \frac{\tau^{n+1}}{(n+1)} \quad (4.13)$$

where b_i is a constant, depending on type of ice and temperature. The associated flow rule is:

$$\dot{\varepsilon}_{ij} = \frac{\partial\Phi}{\partial\sigma_{ij}} \quad (4.14)$$

If the constants a_1 to a_6 in equation (4.12) are known then equations (4.13) and (4.14) can be used to calculate any component of strain-rate under any state of stress.

Two special cases are of interest, corresponding respectively to randomly oriented columnar and granular ice. To simplify the treatment, we assume that σ_{11}, σ_{22} and σ_{33} are principal axes of stress. This can always be done by appropriate redefinition of reference axes, and makes all shear quantities σ_{ij} $(i \neq j)$ vanish. Further, we assume that one principal axis, σ_{33}, is vertical. The axes are shown in Fig. 4.14 for columnar ice.

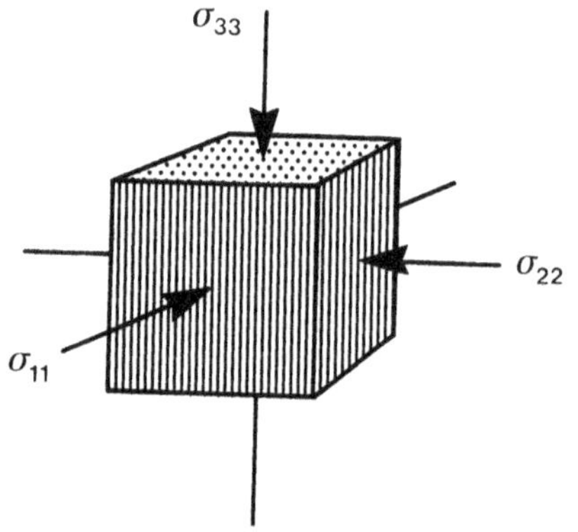

Fig. 4.14 Definition of principal axes of stress for columnar S-2 ice. The long axes of the columnar grains are vertical, and c-axes are randomly distributed in the horizontal plane.

4.2.5.1 Columnar ice

Assume we are dealing with S-2 columnar ice, which is orthotropic (axisymmetric) and has symmetry about the vertical axis. Equation (4.12) then becomes:

$$2\tau^2 = a_1(\sigma_{11}-\sigma_{22})^2+a_2[(\sigma_{22}-\sigma_{33})^2+(\sigma_{33}-\sigma_{11})^2] \quad (4.15)$$

since $a_2 = a_3$ by symmetry. The constants a_1 and a_2 are not independent, so we can choose $a_2 = 1$ without loss of generality. Equations (4.13) and (4.14) then supply the results:

$$\dot{\varepsilon}_{11} = [(a_1+1)\sigma_{11}-a_1\sigma_{22}-\sigma_{33}]\frac{b_c}{2}\tau^{n-1} \quad (4.16)$$

$$\dot{\varepsilon}_{22} = [(a_1+1)\sigma_{22}-a_1\sigma_{11}-\sigma_{33}]\frac{b_c}{2}\tau^{n-1} \quad (4.17)$$

and

$$\dot{\varepsilon}_{33} = [2\sigma_{33}-\sigma_{11}-\sigma_{22}]\frac{b_c}{2}\tau^{n-1} \quad (4.18)$$

where b_c is a constant for columnar ice. The constant a_1 expresses the anisotropy and can be derived by reference to

(a) comparison of vertical uniaxial tests with horizontal uniaxial tests, and
(b) analysis of plane strain compression tests.

Uniaxial tests. A horizontal uniaxial test (Fig. 4.15(a)) has $\sigma_{22} = \sigma_{33} = 0$ and equation (4.16) then reduces to

$$\dot{\varepsilon}_{11} = \left(\frac{a_1+1}{2}\right)^2 b_c\sigma_{11}{}^3 \quad (4.19)$$

where we have inserted $n = 3$. A vertical uniaxial test has $\sigma_{11} = \sigma_{22} = 0$, and equation (4.18) then reduces to

$$\dot{\varepsilon}_{33} = b_c\sigma_{33}{}^3 \quad (4.20)$$

Now consider two uniaxial compression tests at the same strain-rate: one vertical, one horizontal. The stress required for the vertical test (Fig. 4.15(b)) differs from that required for the horizontal test by

$$\sigma_{33}/\sigma_{11} = \left(\frac{a_1+1}{2}\right)^{2/3} \quad (4.21)$$

according to the analysis. By experiment (Butkovich, 1959; Peyton, 1966; Vaudrey, 1977; Sinha, 1983b) the ratio of vertical to horizontal strength lies in the range 2–5 (though the lower values are obtained at higher strain-rates and probably

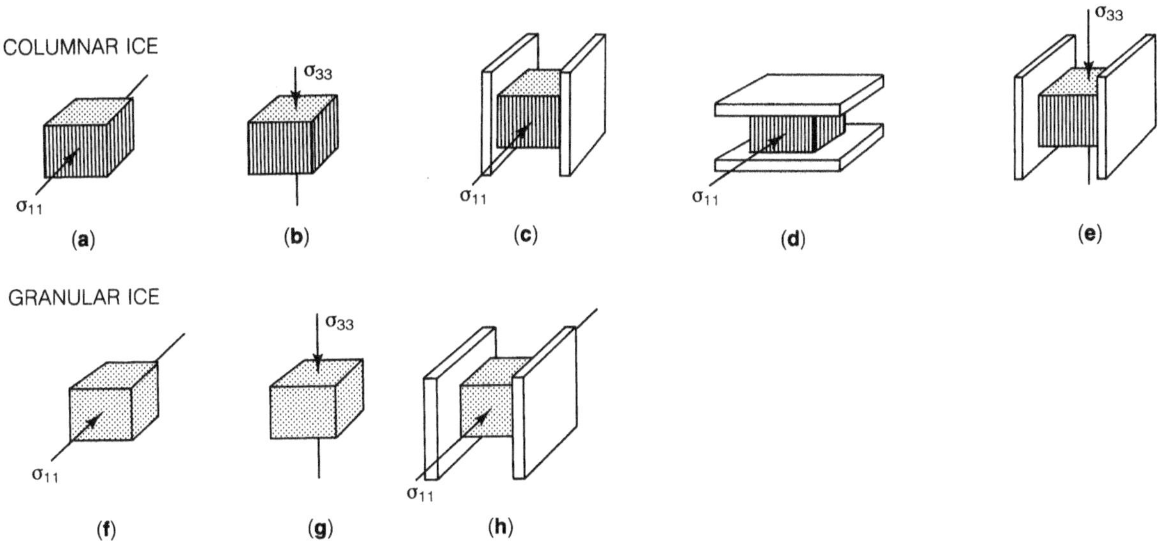

Fig. 4.15 Varieties of test geometry. For *columnar ice*: (a) horizontal uniaxial compression; (b) vertical uniaxial compression; (c) horizontal plane strain with horizontal constraint (Frederking's "Type A"); (d) horizontal plane strain with vertical constraint (Frederking's "Type B"); (e) vertical plane strain with horizontal constraint. For *granular ice*: (f) horizontal uniaxial compression; (g) vertical uniaxial compression (equivalent to (f)); (h) horizontal plane strain with horizontal constraint.

reflect stress reduction by fracture initiation). Equation (4.21) then shows that a_1 should lie in the range 5–20.

Plane strain tests. Consider a plane strain test in which a horizontal stress σ_{11} is applied, but the ice is prevented from straining in the transverse direction $\dot{\varepsilon}_{22}$ (Fig. 4.15(c)). It is compelled to strain vertically, with strain-rate $\dot{\varepsilon}_{33}$. This is known as a "Type A" plane strain test (Frederking, 1977).

In this case we have $\dot{\varepsilon}_{22} = 0$ and $\sigma_{33} = 0$, and we can solve equations (4.16) and (4.17) to obtain

$$\dot{\varepsilon}_{11} = \left[\frac{2a_1+1}{a_1+1}\right]^2 \frac{b_c}{4} {\sigma_{11}}^3 \tag{4.22}$$

for plane strain compression of columnar ice. Now compare this with equation (4.19) for unconfined uniaxial compression. It shows that the stress for plane strain compression should exceed the stress for unconfined compression by a factor of

$$\gamma = \left[\frac{(a_1+1)^2}{2a_1+1}\right]^{2/3} \tag{4.23}$$

For a_1 in the range 5–20 this factor lies in the range 2.2–4.9. Experimental results (Frederking, 1977, Fig. 5) measure this factor γ to be at least 2.5 and generally less than 4.4, with the larger values occurring at lower strain-rates. There is obviously good general agreement.

These considerations, on uniaxial and plane strain tests, indicate that for low strain-rate creep behaviour a suitable value for the constant a_1 in equations (4.16)–(4.18) is about

$$a_1 = 15$$

This implies a ratio of vertical to horizontal strength of 4.0 and a ratio of confined to unconfined stress of 4.1. These values are consistent with most data in the region of strain-rate 10^{-6} s^{-1}.

4.2.5.2 Granular ice

Granular isotropic polycrystalline ice is now trivial. Simply insert $a_i = 1$ ($i = 1$ to 6) in all previous equations. For uniaxial loading, equations (4.16)–(4.18) all become of the form:

$$\dot{\varepsilon}_{11} = b_g {\sigma_{11}}^3$$

where b_g is the constant for granular ice. The ratio of vertical to horizontal strength (Fig. 4.15(f) and (g)) is of course unity by equation (4.21). The ratio of confined to unconfined compressive strength is 1.21, by equation (4.23), for the test geometry shown in Fig. 4.15(h). Frederking (1977) found, by experiment, ratios of between 1.00 and 1.25, which is a satisfactory agreement.

A number of other geometries of plane strain compression may also readily be analysed by equations (4.16)–(4.18): for instance, horizontal compression with vertical constraint (Fig. 4.15(d)) or vertical compression with horizontal constraint (Fig. 4.15(e)). However, these are rare loading conditions in practice, and will not be analysed here.

Note that equations (4.16)–(4.18), with constants a_1 and b_c defined by experiment, allow the principal axes of strain-rate to be calculated for any multiaxial stress-rate.

If a sample of columnar ice shows preferred horizontal alignment of *c*-axes then its symmetry in the horizontal plane disappears and its behaviour is truly anisotropic. We have already examined, in Section 4.2.4 and Fig. 4.10, data on compressive strength of oriented columnar sea ice as a function of *c*-axis orientation, but it is beyond the scope of this book to analyse this in terms of the full anisotropic creep equation; it is doubtful, anyway, whether the data are sufficiently good to allow this.

This completes a general picture of the continuum properties of ice. It is important to note that all of these properties are scale-independent: it does not matter how large the sample being tested is, it will still obey the continuum mechanical laws derived from experiments at small scale in the laboratory.

4.3 FRACTURE BEHAVIOUR

4.3.1 Nucleation versus propagation

At a certain stage in the loading process, ice may display a sudden or gradual change of deformation mechanism and begin to exhibit brittle behaviour and crack formation instead of pure continuum creep and elasticity. This may occur for a variety of reasons and the onset of fracture behaviour may be interpreted as being due to

(a) The stress exceeding a certain level, (under uniaxial compression this appears from Fig. 4.5 to be about 5–10 MPa; under uniaxial tension it appears to be about 1–2 MPa).
(b) The strain-rate exceeding a certain level. From Figure 4.5 this appears to lie in the region of 10^{-3} s^{-1} for uniaxial compression.
(c) The strain in the material exceeding a certain critical level, generally 1% or more.

In this section we will look at the processes involved as cracks begin to form in ice and ultimately lead to its brittle failure.

The most important distinction to make at the outset is that between *nucleation* of a crack and *propagation* of a crack. A sample may contain pre-existing micro-cracks, or, alternatively, cracks may nucleate and develop during the loading process; however, if the stress field is insufficient to make these cracks propagate then the sample will not fail. It may undergo a phase of accelerating creep, but unless fractures extend and link then the sample will remain intact. Even if cracks do propagate under the prevailing stress field, they may do so stably, and in this case we have to examine how far they must propagate before failure occurs.

The fracture of ice may under different circumstances be either propagation-controlled or nucleation-controlled.

If an applied load induces the nucleation of very small micro-cracks they may be insufficiently large to propagate. In this case the load must be increased further until they do propagate, and the fracture process is then *propagation-controlled.* This is a common phenomenon with very fine-grained samples of ice. After nucleation the sample may contain many cracks, and show increased ductility and loss of stiffness due to their presence, but failure does not occur until they propagate.

If, however, an applied load results in the formation of larger microcracks we may find that the cracks are large enough to propagate the moment they nucleate. In this case the fracture process is *nucleation-controlled.* If the cracks had existed before the load was applied they would have propagated at a lower load. As it was, they were not there at the outset, and the load had to be increased until they nucleated.

The distinction between the two control mechanisms depends on the size of the cracks when they nucleate. This in turn depends principally on the grain size of the ice, and is best demonstrated by reference to tensile fracture of ice.

4.3.2 Tensile fracture

4.3.2.1 Nucleation

Consider an initially homogeneous crack-free specimen of polycrystalline ice subjected to a uniaxial tensile load. When subjected to load the ice deforms according to three distinct continuum processes, as discussed in Sections 4.2.1 and 4.2.2.

They are:

(a) immediate elastic strain ε_e,
(b) delayed elastic strain ε_d, and
(c) creep strain ε_v.

Crack nucleation (the appearance of first cracks) appears to be associated in some way with critical strain in the material. It is, however, not clear whether nucleation of cracks occurs at a critical level of total tensile strain

$$\varepsilon = \varepsilon_e + \varepsilon_d + \varepsilon_v$$

as proposed by Seng-Kiong and Shyam Sunder (1985), or at a critical level of delayed elastic strain ε_d alone, as proposed by Sinha (1982) and mentioned in Section 4.2.1.4. The delayed elastic strain criterion gives good agreement with data at low or moderate strain-rates, but may be inappropriate at very high strain-rates, where it apparently permits unlimited stresses to develop before crack nucleation occurs. In either case, the physical process of crack nucleation occurs in order to relieve stress concentrations which develop at grain boundaries during the early stages of deformation (Schulson *et al.*, 1984). As discussed above in Section 4.2.2, dislocations tend to glide within grains and pile up at grain boundaries until the stress concentrations are sufficient to nucleate cracks. The process of dislocation pile-up at grain boundaries has been physically observed by Sinha (1978a) using an elegant process of surface etching.

The process leads to nucleation of cracks which are of size of the same order as the grain diameter *d*, and under tension it is the largest of these cracks which generally leads to propagation and failure.

As to the stress – or strain, or delayed elastic strain – at which these cracks nucleate, I shall for the moment side-step the issue and concentrate on data obtained at quite high strain-rates (10^{-6} s^{-1}) by Schulson *et al.* (1984) and by Lee and Schulson (1986). For high strain-rate tests we can simply plot the data in terms of peak tensile stress, and ignore for the moment whether the peak stress corresponds to a critical strain or a critical delayed elastic strain.

These data are plotted in Fig. 4.16, in terms of tensile stress against inverse square-root of grain size. The data points are identified according to whether the stress plotted corresponds to crack nucleation (closed symbols) or crack propagation (open symbols). Note that, for tests on ice of grain size less than about 1 mm, nucleation of a crack may occur at about 1.0–1.2 MPa, while propagation only occurs once the stress has reached 1.2–2.0 MPa.

The stress for nucleation to occur (solid symbols) shows a clear dependence on grain size. The dependence can be well modelled by an equation of the form

$$\sigma_N = \sigma_0 + k_1/\sqrt{d} \qquad (4.24)$$

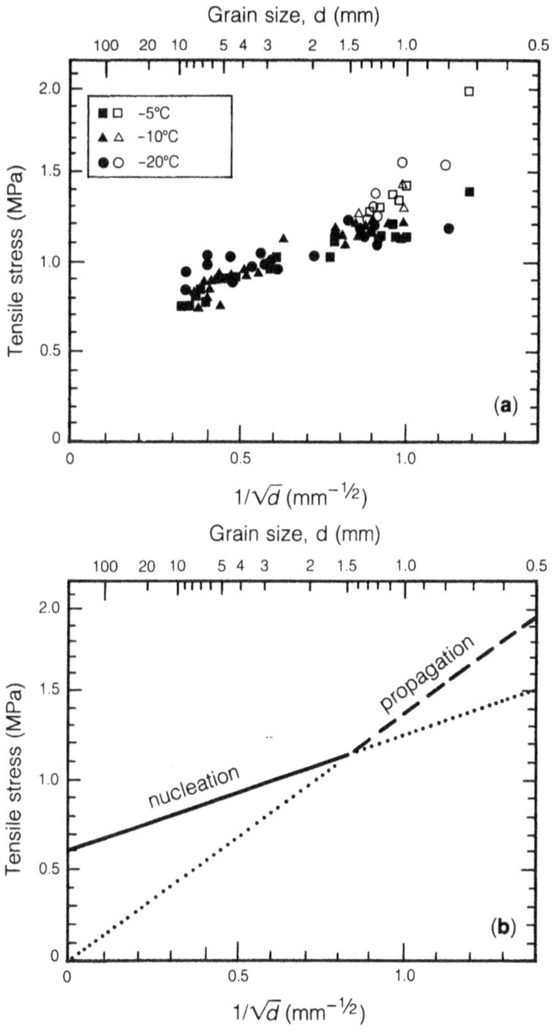

Fig. 4.16 (a) High strain-rate data for tensile loading of ice as a function of grain size (inverse square root scale). Closed symbols indicate crack nucleation while open symbols indicate crack propagation. (After Hallam (1986), from data of Schulson *et al.* (1984), and Lee and Schulson (1986). Note that data referring to different temperatures have all been plotted on the same graph.) (b) Nucleation and propagation criteria for ice: for medium to large grain size (*d* greater than 1.5 mm) failure is nucleation-controlled (solid line); for fine-grained ice (*d* less than 1.5 mm) failure is propagation-controlled (dashed line).

which supplies σ_N, the stress required to nucleate cracks in tension, as a function of the grain size d and two constants σ_0 and k_1. Equations of this form are also commonly found for crack nucleation processes in other materials such as metals. Appropriate constants for equation (4.24), in order to model the experimental data of Fig. 4.16, are

$$\sigma_0 = 0.6 \text{ MPa}, \qquad k_1 = 0.02 \text{ MPa m}^{1/2}$$

These constants show some temperature dependence (Schulson *et al.*, 1984) but here and in Fig. 4.16 I have amalgamated all the results onto one single plot. The temperature dependence leads to only minor variations (10–20%). The theoretical line based on equation (4.24) is plotted in Fig. 4.16(b) (solid and dotted line).

If the stress state is high enough that purely elastic deformation dominates the effects of transient and permanent creep strain, then the criterion of equation (4.24) can be re-written instead in terms of critical strain for nucleation ε_N, as:

$$\varepsilon_N = \sigma_N/E$$

where E is Young's modulus, or

$$\varepsilon_N = \varepsilon_0 + k_2/\sqrt{d} \qquad (4.25)$$

where ε_0 and k_2 are constants with values

$$\varepsilon_0 = 6.3\times10^{-5}, \qquad k_2 = 2.1\times10^{-6} \text{ m}^{1/2}$$

The formulation in terms of tensile strain rather than tensile stress will become useful in Section 4.3.3.1, when we come to consider nucleation of cracks under compressive stress states.

4.3.2.2 Propagation

Once a crack has nucleated (or, of course, if it existed already, prior to loading) we need to examine under what conditions it will propagate. For uniform tensile stress states the propagation process is well understood. The basic criterion for tensile propagation of a crack uses linear elastic fracture mechanics, which was developed originally for more conventional engineering materials such as steels (see, for instance, Knott, 1973).

The criterion is simple: a sharp crack or flaw of length $2a$ in a material (Fig. 4.17) will propagate unstably under a normal tensile stress σ if

$$\sigma > K_{IC}/\sqrt{\pi a} \qquad (4.26)$$

where K_{IC} is the fracture toughness of the material. K_{IC} is a material property, and is in general quite well behaved and easy to measure for truly elastic materials.

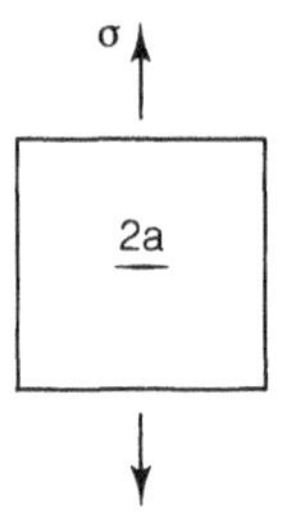

Fig. 4.17 Simple geometry for tensile propagation of a flaw of half-length *a*.

Ice is unfortunately rather more complex than, for instance, steels, owing to its creeping properties. Under rapid loading conditions ice may be treated by simple linear elastic fracture mechanics, but under slower loading conditions ductile processes at the advancing crack tip complicate its behaviour. At a slower rate of deformation the apparent fracture toughness of ice is higher – that is, it displays greater strength. Formal treatments of ductile crack-tip processes are available, using J-integrals (Hutchinson, 1979) or path-independent C* integrals (Riedel and Rice, 1980), but I shall not treat them here: there are many other more important uncertainties in the field of ice mechanics, and crack-tip ductility is the least of our problems. However, it probably does explain the apparent peak in the tensile strength of ice at strain-rates between 10^{-4} s^{-1} and 10^{-2} s^{-1} (see Fig. 4.5). At strain-rates higher than 10^{-2} s^{-1} the behaviour of ice becomes more truly brittle, and its strength is slightly lower than that when ductile crack-tip processes are occurring. The criterion for whether ductile processes may be important is as follows: if the creep zone around a crack tip exceeds the length of the crack, $2a$, then the J or C* integral approach should, strictly, be used; however, if the creep zone is small compared with the flaw size then the simple K_{IC} treatment can be used. This means that the linear elastic K_{IC} fracture treatment is especially applicable when large flaws or large grain sizes are involved. For most engineering purposes this holds true, and for the rest of this book I shall use only linear elastic fracture mechanics.

The fracture toughness K_{IC} of pure ice is approximately 0.115 MPa m$^{1/2}$ at high loading-rates or low temperatures (Goodman, 1979). It shows some variations with loading-rate and temperature, as a result of the non-linear ductile processes discussed above (Miller, 1979; Mellor, 1983). Sea ice has an apparent fracture toughness of the same order as or slightly lower than pure ice (Urabe and Yoshitake, 1981; Weeks and Mellor, 1984), but this may simply be due to effects of net section stress. For most purposes it is adequate to use a value of $K_{IC} = 0.1$ MPa m$^{1/2}$, and equation (4.26) then immediately supplies the tensile stress required to propagate a flaw of length $2a$.

Referring back to Fig. 4.16, look again now at propagation stresses shown by open symbols. They can be well described by a straight line through the origin, as shown by the dotted and dashed "propagation" line in Fig. 4.16(b). This line corresponds to an equation of the form

$$\sigma_P = K_P/\sqrt{d} \tag{4.27}$$

which relates the stress σ_P required to propagate a nucleated crack, to the grain size d. The quantity K_P is an empirical constant with the same dimensions as fracture toughness, and an experimentally determined value of about 0.044 MPa m$^{1/2}$. Note the similarity between equations (4.26) and (4.27), and that K_P has a value close to that of the fracture toughness K_{IC}. The identity of the equations does not appear to be quite perfect, however: if we try and force equation (4.27) into the same form as equation (4.26) it would lead to the conclusion that nucleated cracks in ice behave as if they have a crack length of about three grain diameters, that is

$$2a \sim 3d$$

This seems rather high, since we already know by observation that nucleated cracks actually have a length of order only d (see Section 4.3.2.1). The fact that ice behaves as if nucleated cracks are several grain diameters long indicates that it is weaker than expected: this may be due to cracks interacting with one another or even coalescing at high crack densities, but this is not thoroughly explained.

The stresses for nucleation and propagation as plotted in Fig. 4.16(b) indicate the following: to the left of the intersection of the two lines, that is for grain size d greater than 1.5 mm, fracture is controlled by nucleation, while to the right of the intersection, for d less than 1.5 mm, fracture is controlled by propagation. This is an interesting result, since most ice we deal with outside the laboratory in practice has grain size exceeding 1.5 mm. Once cracks have nucleated in such ice they propagate immediately. This has the following implications:

(1) If no cracks exist in a sample of ice to begin with (that is, it is initially flaw-free) then the stress for fracture under tension is equal to the stress required to nucleate cracks.

(2) If, however, cracks already exist in the sample and are of length greater than a few grain diameters then no further cracks need to be nucleated, and the pre-existing cracks will control ice failure through equation (4.26). Note that the stress required for failure drops markedly with size of flaw: the tensile stress required to propagate a flaw of length 5 mm is 1.3 MPa, while the stress required to propagate a flaw of length 50 mm is only 0.4 MPa.

It is because of this that the tensile strength of ice is expected to show a pronounced scale effect. Any sample of ice, however carefully prepared, contains a statistical distribution of grain sizes, and upon nucleation of cracks will contain a statistical distribution of crack lengths. Also, if cracks nucleate at adjacent grains there is a finite probability of a certain number coalescing to give rise to initial cracks of length exceeding the grain size. This is illustrated schematically in Fig. 4.18. Under applied tensile load the stress field will preferentially seek out and propagate the largest flaw (provided, of course, that it is appropriately oriented, approximately normal to the principal axis of tensile stress). Because under tension only one propagating crack is needed in order to fail the specimen, we find that the strength of the specimen is controlled by the largest, dominant flaw.

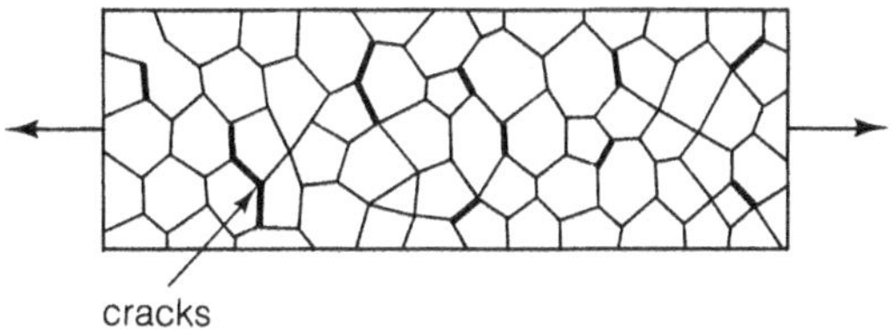

Fig. 4.18 Schematic diagram of ice under tension. Cracks nucleate at grain boundaries and may coalesce (bold lines). The largest nucleated flaw will propagate and lead to failure.

The statistical population of nucleated crack lengths likely to be found in a specimen leads to two simple observations:

(1) Samples of identical shape and size may not necessarily have identical strengths: their strength is determined by the largest nucleated flaw formed under tension, and this will show statistical variation from sample to sample.

(2) Larger samples should generally show lower strengths than smaller samples, since within their volume there is a greater chance of their containing larger nucleated flaws.

Unfortunately, there are very few data available to substantiate this expected scale behaviour. Data collected by Jellinek (1958) do indeed indicate an apparent scale effect, but the data were collected for tensile failure of very thin discs of diameters 5–20 mm. There is doubt about their reliability, since the scale of the experiments was small compared to the grain size.

4.3.3 Compressive fracture

4.3.3.1 Nucleation in compression

The nucleation of cracks under compressive stress occurs for the same reason as under tension. It is generally due to dislocation pile-up at grain boundaries and relief of stress concentration by parting along the grain boundary. However, cracks may also form across grains: studies by Hallam *et al.* (1986) indicate that as many as 50% of nucleated cracks under compression are transgranular in character.

The phenomenon of crack nucleation has been well investigated for low to moderate loading conditions (Sinha, 1982) and these studies indicate that crack nucleation is well described by application of the delayed elastic strain criterion (Section 4.2.1.4 and 6.3.2). However, there are problems with this criterion when applied to very high loading-rates: at high loading-rates Sinha's formulation would imply that the initial bearing capacity of ice is potentially infinite. This is because delayed elastic strain always takes a finite time to develop, and so fractures would not be expected to nucleate immediately, however high the applied stress might be. Unfortunately, few experiments have been carried out on crack nucleation at very high loading-rates. Nonetheless, it would be surprising if the bearing capacity of ice were potentially limitless.

If this apparent paradox is to be resolved it is necessary to introduce some alternative nucleation criterion, at least at high stresses (see for instance Seng-Kiong and Shyam Sunder, 1985). Hallam (1986) proposes that we use a similar criterion to that already developed for tensile loading. That is, assume that crack nucleation under compressive stress occurs simply when the lateral tensile strain induced by Poisson expansion reaches the level already defined as critical for tensile crack nucleation.

For tensile crack nucleation we have the strain criterion given by equation (4.25):

$$\varepsilon_N = \varepsilon_0 + k_2/\sqrt{d}$$

Now under compressive stress σ_{11} the lateral tensile strain ε_{22} is given by

$$\varepsilon_{22} = -\nu\sigma_{11}/E \qquad (4.28)$$

where ν is Poisson's ratio. The minus sign denotes compression. Taking $\nu = 0.33$ we can readily write a stress criterion for nucleation under compressive stress states: crack nucleation occurs when the lateral Poisson expansion strain ε exceeds the level ε_N defined by equation (4.25). This leads directly to the following expression for the compressive stress σ_N' at which cracks nucleate under compression:

$$\sigma_N' = -\frac{\sigma_N}{\nu} = -\frac{1}{\nu}\left(\sigma_0 + \frac{k_1}{\sqrt{d}}\right) \qquad (4.29)$$

with constants σ_0 and k_1 defined as above for equation (4.24). This implies that the stress required to nucleate a crack under compression is about three times that required in tension, since $1/\nu \sim 3$. This is consistent with data already presented in Fig. 4.5: for fine-grained pure polycrystalline ice, brittle tensile fracture occurs at about 1–2 MPa and brittle compressive fracture occurs at about 4–6 MPa.

4.3.3.2 Propagation in compression

The process of crack propagation under compressive stress is not as simple as that under tension, and does not so obviously lead to failure. This is because:

(a) propagation of cracks in compression is in general a stable process (while tensile propagation is generally unstable);
(b) final failure generally occurs by the linkage of a large number of cracks, not simply the catastrophic propagation of a single crack.

After nucleation of cracks at grain boundaries, a sample of ice under compression contains a wide variety of cracks of length of the order of the grain size, lying at various angles clustered around the axis of loading. Experimental studies by Cole (1986) provide quantitative data on length, density and orientation of cracks in polycrystalline ice. The

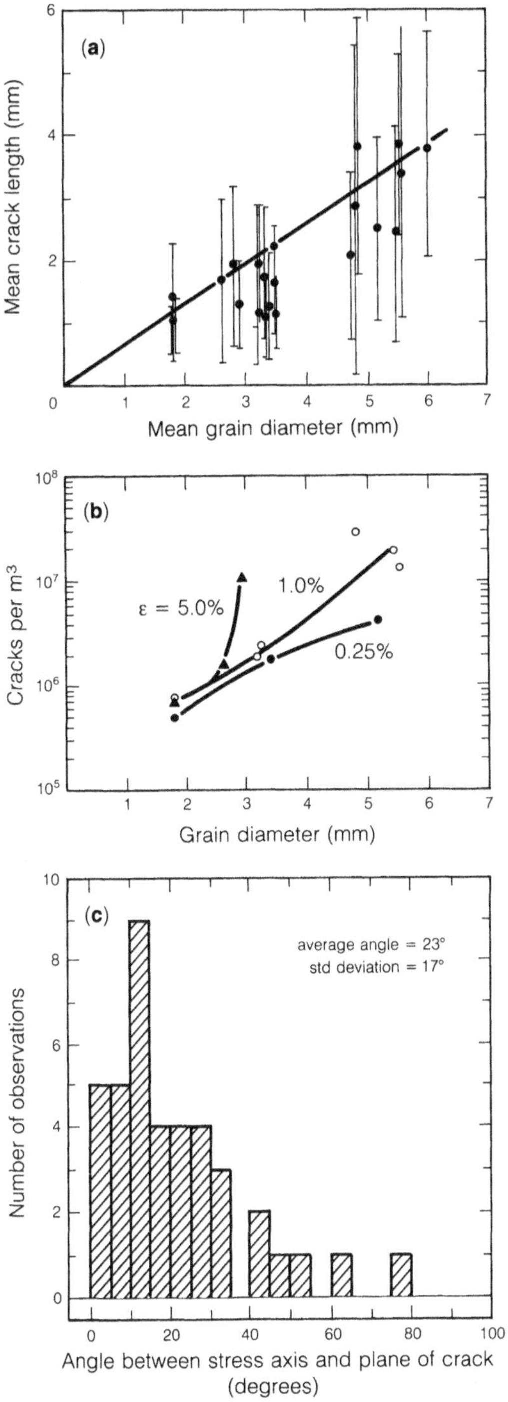

Fig. 4.19 Cracks nucleated in polycrystalline ice under compression: (a) crack length as a function of grain diameter; (b) crack density as a function of grain diameter (axial strain levels are indicated); (c) crack orientation (40 observations, after Cole, 1986).

results are summarised in Fig. 4.19 and provide the following broad conclusions:

(a) *Crack length:* the length of nucleated cracks, $2a$, is proportional to grain size d and obeys the approximate relationship

$$2a = 0.65d$$

This relationship can also be justified on a theoretical basis using concepts of elastic strain energy and surface energy (Cole, 1986) and it accords well with experimental data (Fig. 4.19(a)).

(b) *Crack density:* the density, or frequency of cracks appears also to be grain-size dependent (Fig. 4.19(b)), and for larger grain sizes (exceeding about 5 mm) it is reasonable to assume the formation of approximately one crack per grain.

(c) *Orientation of cracks:* cracks nucleate at orientations clustered around the principal axis of compressive stress (Fig. 4.19(c)). The average absolute angle is about 23°, with a standard deviation of 17°. About 90% of the observed cracks were oriented within 45° of the principal axis of compressive stress.

Consider now a single crack at an angle of 35–45° to the loading axis, and let us look at the manner in which it can propagate (Fig. 4.20). For generality, Fig. 4.20 represents the loading conditions by a principal compressive stress σ_{11} and a lateral confining stress σ_{33}. (The model is two-dimensional only, but an approximation to uniaxial conditions can be made by setting $\sigma_{33} = 0$.)

As the load σ_{11} is applied, the crack tries to slide (Fig. 4.20(a)) and gives rise to tensile zones at the two ends of the initial crack. At a certain critical stress level tensile "wing cracks" form and the process of propagation begins. The propagation initiates roughly perpendicularly to the crack ends, but as the wing cracks grow they tend to align themselves with the principal axis of compression (that is, perpendicular to principal axis of tensile strain). This problem has been studied numerically by Nemat-Nasser and Horii (1982) and an approximate analytic treatment has been presented by Ashby and Hallam (1986).

The analysis by Ashby and Hallam gives a two-dimensional model of the crack as idealised in Fig. 4.20(b): the crack is of length $2a$ and lies at an angle of ψ to the direction of the principal compressive stress σ_{11}. As the stress σ_{11} increases, wing cracks form and grow stably to a length l. The length l can be related to the stress conditions

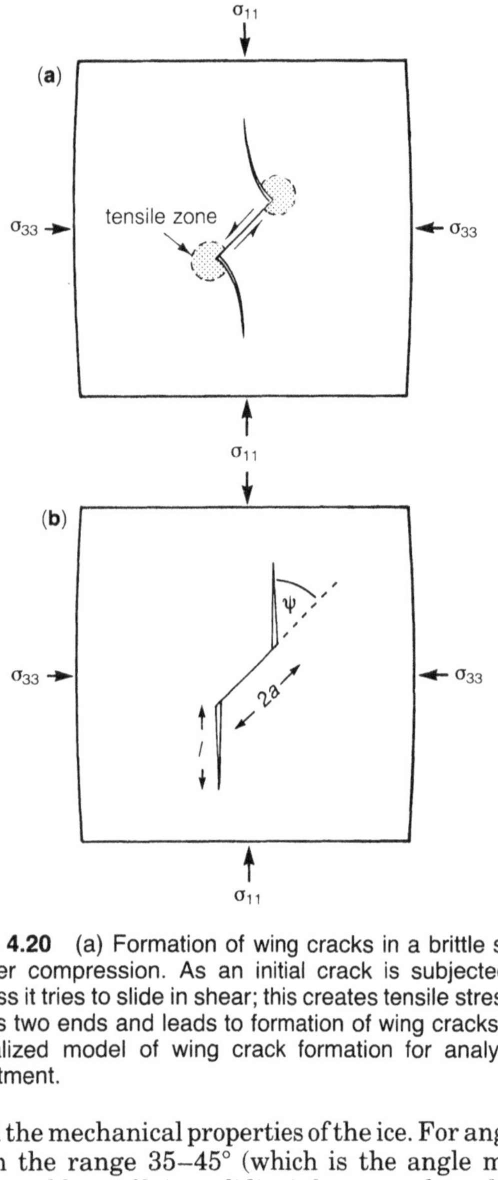

Fig. 4.20 (a) Formation of wing cracks in a brittle solid under compression. As an initial crack is subjected to stress it tries to slide in shear; this creates tensile stresses at its two ends and leads to formation of wing cracks. (b) Idealized model of wing crack formation for analytical treatment.

and the mechanical properties of the ice. For angles ψ in the range 35–45° (which is the angle most favourable to efficient sliding) the general result is

$$K_{IC} = \frac{\sigma_{11}\sqrt{\pi a}}{(1+\mathcal{L})^{3/2}}\left[1-\lambda-\mu(1+\lambda)-\frac{\sqrt{3}\lambda\mathcal{L}}{\beta}\right] \times\left[\frac{\beta\mathcal{L}}{\sqrt{3}}+\frac{1}{\sqrt{3}(1+\mathcal{L})^{1/2}}\right] \tag{4.30}$$

where K_{IC} is the fracture toughness, a is the half-length of the initial crack, l is the length of wing cracks and $\mathcal{L} = l/a$, λ is σ_{33}/σ_{11} (the ratio of confining stress to compressive stress), β is approximately 0.4 (a coefficient), and μ is the coefficient of friction across the crack.

Note that I am denoting compressive stress as positive in this instance. For an explanation of the detailed physics behind the equation, refer to Ashby and Hallam (1986). Here I shall simplify it, to illustrate its behaviour in simple conditions, and then attempt to apply it to the general problem of compressive failure of ice.

For purposes of illustration, take first simple uniaxial compression conditions, so that

$$\sigma_{33} = \sigma_{22} = 0$$

By definition this means that $\lambda = 0$ also, and equation (4.30) becomes

$$K_{IC} = \frac{\sigma_{11}\sqrt{\pi a}\,(1-\mu)}{(1+\mathcal{L})^{3/2}}\left[\frac{\beta\mathcal{L}}{\sqrt{3}}+\frac{1}{\sqrt{3}(1+\mathcal{L})^{1/2}}\right] \tag{4.31}$$

For additional simplicity, since equation (4.31) is still quite difficult to visualize, let us concentrate on the later stages of wing crack propagation, when wing cracks are longer than the initiating crack itself. In the limit, if we have $l \gg a$, then $\mathcal{L} \gg 1$ and we can simplify equation (4.31) to

$$K_{IC} \sim \frac{\sigma_{11}\sqrt{\pi a}\,(1-\mu)\beta}{\sqrt{3\mathcal{L}}} \tag{4.32}$$

Now, the coefficient of friction across the crack, μ, is not well known, and is likely to be very temperature dependent. It probably lies in the range 0–0.3. Adopting a value of $\mu = 0.3$ and setting $\beta = 0.4$, equation (4.32) then becomes simply

$$K_{IC} = 0.29\sigma_{11}\sqrt{a/\mathcal{L}}$$

or, rearranging:

$$\sigma_{11} \sim 3.5K_{IC}\sqrt{\mathcal{L}/a} = 3.5K_{IC}\sqrt{l/a^2} \tag{4.33}$$

For ice we can take $K_{IC} = 0.1$ MPa m$^{1/2}$. This means that if an initial crack of length 5 mm exists ($a = 2.5$ mm) then the stress required to propagate wing cracks of length, say, 10 mm is predicted to be

$$\sigma_{11} = 14 \text{ MPa}$$

This is rather high, but surprisingly close to reality, given that the model has been quite crude: typical S-2 ice containing grains of order 5 mm diameter might in fact have a failure stress of about 4–8 MPa in compression, or roughly half that predicted. But the model is indeed crude, and the choice of a wing crack length of 10 mm was arbitrary. We can now refine this and develop a more realistic model of compression failure.

Let us look in more detail at how "failure" under compression might occur. Essentially, we require

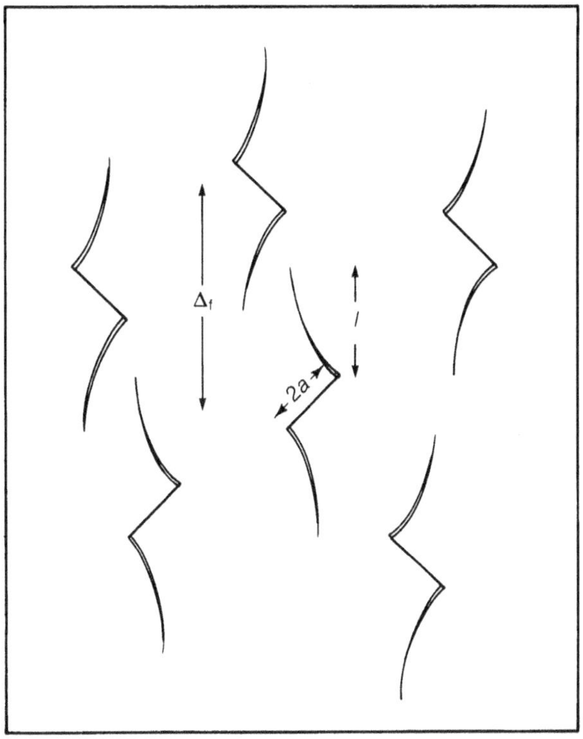

Fig. 4.21 Linkage of a population of propagating wing cracks. Failure will occur soon after $l \sim \Delta_f/2$.

linkage of a large number of propagating wing cracks with each other, until the material under compression loses all strength. A population of propagating wing cracks is shown schematically in Fig. 4.21. Two basic questions need to be addressed:

(a) How do wing cracks interact with one another and other flaws?
(b) What degree of inter-crack linkage is required before a specimen is sufficiently damaged that it collapses?

There are few firm data to answer these questions, but observations by Hallam (Cooksley, unpubl., 1984) indicate that the behaviour is much as might be expected: close proximity of other flaws allows a crack to initiate and grow wing cracks more easily than in a flaw-free solid. This has not been quantitatively formalised yet, but it means that if we assume cracks to be perfectly independent of one another we should be able to derive an upper limit to stress required for failure. We shall do this.

Now, it is not clear exactly what degree of crack propagation will necessarily lead to complete failure of a specimen. However, if initial flaws are separated by an average distance of order Δ_f (Fig. 4.21) it is likely that wing cracks will have to propagate a distance of no more than about $\Delta_f/2$ before the specimen is so damaged that failure occurs. This is especially true if cracks interact significantly with one another, and for a variety of rocks it is found that the propagation required is even less than this. A working criterion of $l \sim \Delta_f/4$ accords well with experiment (Cooksley, unpubl., 1984). Here, however, we shall assume that the propagation distance required is of the order $\Delta_f/2$, since we do not expect all cracks to be of the appropriate orientation to interact with one another. The separation of *active* cracks is always higher than the separation of cracks in general. If we adopt this criterion, it means that we can derive a nominal compressive failure stress σ simply from knowledge of initial flaw-size $2a$ and spacing Δ_f. The general simple result is, from equation (4.33), of the form:

$$\sigma \sim 3.5\frac{K_{IC}}{a}\sqrt{\frac{\Delta_f}{2}} \qquad (4.34)$$

Observation (Cole, 1986) suggests that initial crack density may be of the order of one nucleated crack per grain, so we may assume that $\Delta_f \sim d$, the grain diameter. Further, we can assume that $2a = 0.65d$, and we then find the approximate relationship

$$\sigma \sim 7.6\, K_{IC}/\sqrt{d} \qquad (4.35)$$

for the compressive strength of ice. For ice of grain size 5 mm this yields a value of about 11 MPa.

This is a sensible result, but not very rigorous: we have, for instance, violated the conditions that $\mathcal{L} \gg 1$, since here we have $\mathcal{L} \sim 1.5$. This can be rectified by calculating equation (4.31) in all its glory, but the result is not significantly different: the predicted strength is 11.3 MPa. Given the other gross uncertainties in the model this is trivial, and for most purposes equation (4.34) is an adequate approximation and much easier to interpret. Bear in mind, too, that we have perhaps not made adequate allowance for the fact that not all cracks are active. If they are oriented at 0° or 90° to the principal compressive stress they will not contribute significantly to failure. This may be one reason why the stress predicted by equation (4.35) is a little high. Another reason may lie with the choice of $\mu = 0.3$ for the coefficient of friction at internal crack faces: it may be much lower, especially at high temperatures.

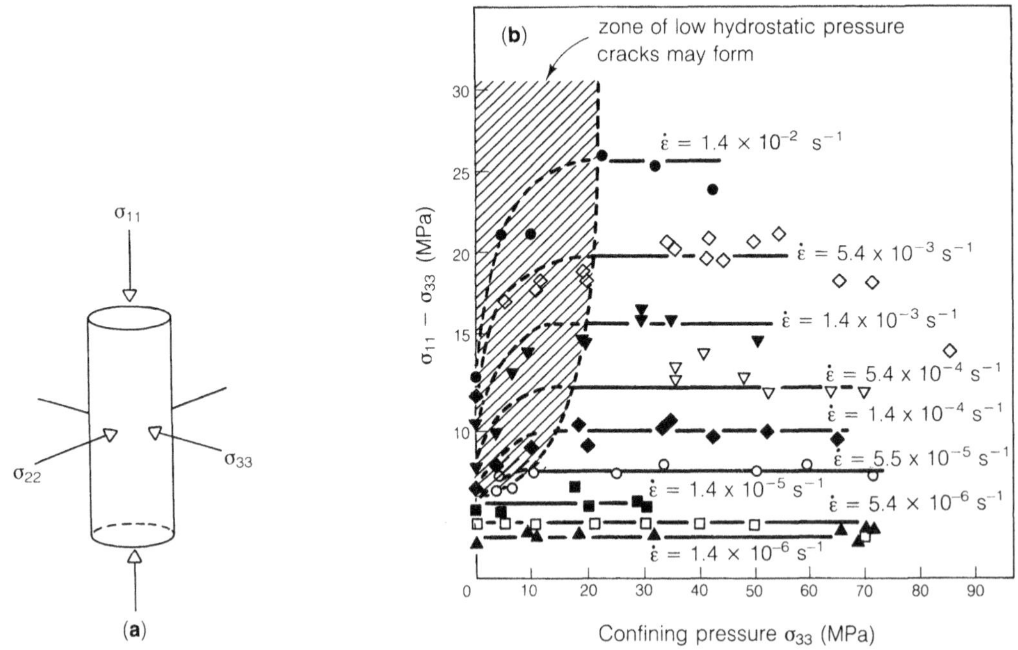

Fig. 4.22 Multi-axial tests. (a) Geometry: under most test conditions, confining pressure is hydrostatic, and $\sigma_{22} = \sigma_{33}$. (b) Plots of stress difference $\sigma_{11} - \sigma_{33}$ required to deform ice at various strain-rates ε_{11}, as a function of axial confining pressure σ_{33}. In the shaded zone shear damage and microcracking occur, but outside it, creep behaviour is independent of confining pressure (after Jordaan, 1986, from data of Jones, 1982).

4.3.4 Multiaxial behaviour

As already pointed out, the application of a multiaxial confining pressure to ice has negligible effect on its creep properties. Ice at the base of a glacier under enormous hydrostatic pressure creeps in very much the same way as ice under uniaxial loading in the laboratory: it responds only to the deviation of stress from hydrostatic pressure, and it is for this reason that universal creep laws can be written in terms of stress deviators alone. (The stress deviator σ_{ij}^{+} is defined in terms of direct stress σ_{ij} as

$$\sigma_{ij}^{+} = \sigma_{ij} - p = \sigma_{ij} - \frac{1}{3}(\sigma_{11} + \sigma_{22} + \sigma_{33}) \qquad (4.36)$$

that is, it is simply the direct stress minus the hydrostatic pressure p.)

The same is by no means true of fracture behaviour. Even quite a low lateral confining pressure can effectively inhibit fracture from occurring, and lead to dramatic increases in apparent strength.

The effect is well illustrated by experiments of Jones (1982), in which cylindrical specimens of ice were deformed at constant strain-rate in the axial direction ($\dot{\varepsilon}_{11}$) while subjected to an overall hydrostatic pressure p. Under these conditions, we have $\sigma_{22} = \sigma_{33}$, as illustrated in Fig. 4.22(a), and from equation (4.36) we find that the stress deviator in the axial direction, σ_{11}^{+}, is simply

$$\sigma_{11}^{+} = \sigma_{11} - \frac{1}{3}(\sigma_{11} + 2\sigma_{33}) = \frac{2}{3}(\sigma_{11} - \sigma_{33})$$

Figure 4.22(b) then shows $(\sigma_{11} - \sigma_{33})$, proportional to the stress deviator, plotted against lateral confining pressure σ_{33} for a range of tests at constant strain-rate $\dot{\varepsilon}_{11}$.

For "low" confining pressure, in the range up to about 25 MPa, it is seen that strength increases substantially as confining pressure is increased, as a result of internal cracking being suppressed.

Above 25 MPa, however, the family of curves becomes essentially flat. This is the region of pure creep. Pure creep can be induced, and fracture suppressed, for any rate of applied strain, however high, simply by applying a sufficiently high confining pressure. A strain-rate of $10^{-2}\ s^{-1}$ is, however, the limit of available data.

The behaviour of ice in the zone between pure uniaxial conditions and perfect constraint is represented by the shaded portion to the left of the plot in Fig. 4.22(b). This zone is not well understood in detail, but can probably be treated by application of the generalised crack-propagation equation given as equation (4.30). The introduction of even quite a small confining pressure σ_{33} (that is, $\lambda > 0$ in equation (4.30)) is very effective in inhibiting growth of wing cracks. This is intuitively obvious from Fig. 4.20, since a lateral confining stress σ_{33} actually tends directly to close any wing crack which has been initiated. The theory of this has been developed by Hallam (1986), and used to construct crack-propagation surfaces in stress space for a variety of confinement cases.

Experiments have also been performed using apparatus in which the three axes of stress can be varied independently (Häusler, 1981) using three orthogonal brush platens, but the results are subject to doubt because of problems of crack initiation at the platen contacts.

Triaxial experiments on sea ice have been presented by Richter-Menge *et al.* (1986). I do not propose to go into any further detail here on triaxial confinement behaviour, since it has yet to be shown how it can be properly applied to real loading geometries. It is difficult to quantify what confinement stresses are actually present during, say, impact of a thick multi-year floe with a structure, and for this reason it is not immediately useful to know precisely in what way the fracture strength varies with confinement.

For the moment it is enough to know that it increases, and that it is in all instances limited by the appropriate creep surface for the strain-rate in question.

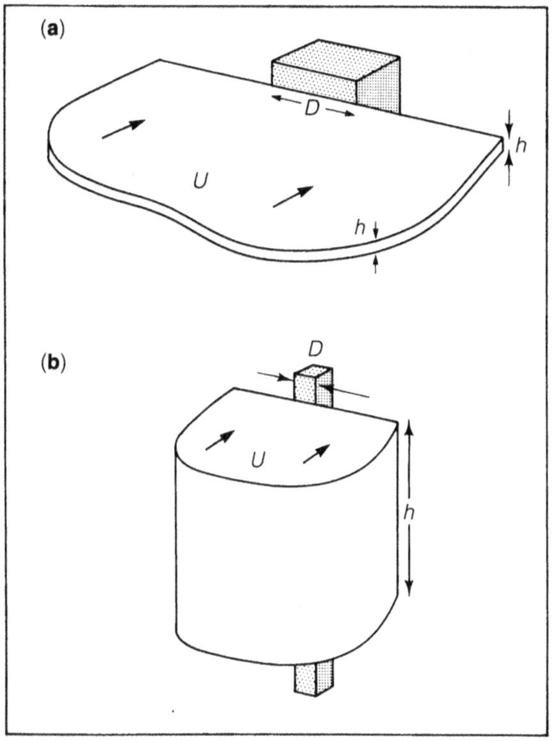

Fig. 4.23 Indentation geometries: a structure of width D indents ice of thickness h. (a) Plane stress conditions, where $D \gg h$; (b) plane strain conditions, where $D \ll h$.

4.4 INDENTATION TESTS

4.4.1 Test geometry

During most realistic ice interaction processes, for instance, impact of an ice floe with the leg of a structure or with an artificial island, the geometry of the interaction is far from uniaxial. The interaction conforms much more closely to the indentation geometry shown in Fig. 4.1(c) or Fig. 4.23.

In this section we shall look in some detail at a range of indentation experiments: most of them carried out in the laboratory at small scale, but also some carried out at medium scale through the full thickness of Arctic ice cover. In Chapter 5, we shall look at data from full-scale experiments on natural and artificial islands and structures.

As already foreshadowed in Chapter 1, these data will be used to build up a quantitative picture of the scale-dependent fracture behaviour of ice, which will then form the basis for the theoretical analysis proposed in Chapter 7.

The geometry of an identation test is shown in Fig. 4.23. Ice moves at a velocity U past an indentor of width D, and indentation occurs over the full thickness of the ice, h. Most identation tests are carried out with a flat punch indentor as illustrated, though experiments have also been carried out with a variety of semi-circular or wedge-shaped indentors. The indentor shape is of little importance when deformation is occurring by creep or by local crushing, though it can have significant influence on the formation of radial cracks.

The width of the indentor relative to the ice thickness does, however, have a significant effect on behaviour during indentation. Stresses and failure mode depend on the aspect ratio D/h, and the two extreme cases are illustrated in Fig. 4.23. They are:

(a) *Plane stress conditions:* under plane stress indentation conditions, the ice is thin and the indentor is very wide, and we have $D \gg h$. There is zero applied stress at the top and bottom surfaces of the ice and no internal vertical stress gradients are generated (since edge effects are negligible for such a wide indentor). This means that stresses are uniform through the thickness of the ice, and this is why it is known as the plane stress condition. Ice is free to deform in the vertical direction.
(b) *Plane strain conditions:* under plain strain conditions, the ice is thick and the indentor is narrow. We have $D \ll h$. Under these conditions, ice in the body of the material is not free to deform in the vertical direction, since it is constrained by the large surrounding thickness of undeformed material. In this case zero strain in the vertical direction occurs, and all deformation occurs in the plane of the indenting ice.

Clearly no real experiment will quite conform to these two theoretical extremes; they are merely paradigms.

4.4.2 Laboratory indentation tests

4.4.2.1 General

Experimental indentation tests have been carried out in the laboratory by various authors on a wide range of scales and geometries: indentor widths in the range 0.01–0.50 m, ice thicknesses in the range 0.01–0.1 m, and aspect ratios in the range 0.2–80.

I am not convinced of the value of small-scale indentation tests, at least from the point of view of the practising engineer. During a series of indentation tests a variety of failure models are observed: creep, spalling, buckling, radial cracking and crushing, and a bewildering number of empirical laws can be adduced to rationalize any particular set of experimental results. Stress and failure mode may be formulated as a function of ice thickness, indentor width, indentor shape, aspect ratio, indentation rate, temperature, grain size, etc. It is, however, my own experience that any such "general" laws are rarely of great value when applied outside the range of values of parameters from which they were initially derived. In Chapter 6, I shall present individual theoretical analyses for each of the principal deformation and failure modes of ice, which may then be applied quantitatively to a particular case in question. The problem with laying out general universal laws is that most of the important failure mechanisms (for instance, creep, elastic buckling and brittle fracture) show quite different scaling laws: what may be true at laboratory scale is not necessarily true for medium-scale full-thickness indentation, and is even less likely to be true for large-scale indentation across the full width of a structure. Here I shall restrict myself to a general survey of laboratory failure modes and the stresses associated with them.

4.4.2.2 Failure modes

Failure mode during indentation depends principally on two things: rate of indentation, defined as U/D, and aspect ratio D/h. The principal failure modes observed are illustrated schematically in Fig. 4.24. There are essentially four global failure modes:

(a) *Creep:* at low indentation rates, deformation is entirely in the continuum mode with no formation of cracks (Fig. 4.24(a)). The stress associated with creep deformation is dependent on indentation rate, in the same way as it is dependent on strain-rate for uniaxial loading conditions. It is also dependent on aspect ratio and on any anisotropy of the ice involved. This behaviour is well understood, but we shall delay a formal theoretical analysis of it until Chapter 6 (Section 6.2.3).
(b) *Radial cracking:* above a certain stress level, and especially at high aspect ratios, formation of radial cracks is commonly observed (Fig. 4.24(b)). Typically they radiate from the corners of the indentor or approximately along the median line of indentation, and are initially stable and of limited length. A theoretical analysis of their propagation is given in Section 6.4.2.
(c) *Buckling:* thin ice is prone to elastic buckle formation (Fig. 4.24(c)), and a theoretical analysis of this is given in Section 6.2.1. Such buckles are unstable and lead rapidly to the formation of circumferential cracks. At slow indentation rates, stably growing creep buckles may also be observed, but these are more common under full-scale field conditions, since laboratory experiments are seldom of long enough duration for creep buckles to develop noticeably. An analysis of creep buckling is given in Section 6.2.2.

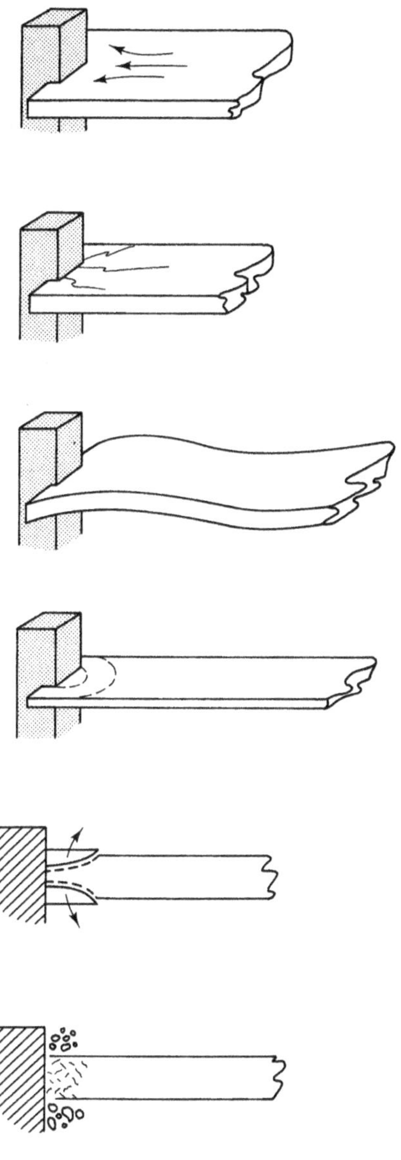

Fig. 4.24 Principal failure mechanisms observed during laboratory indentation experiments: (a) creep; (b) radial cracking; (c) buckling; (d) circumferential cracking; (e) spalling; (f) crushing.

(d) *Circumferential cracks:* as a result of elastic buckling, or an overall applied out-of-plane bending moment due to eccentric loading conditions, circumferential (or tangential) cracks may form (Fig. 4.24(c)). This mode is analysed theoretically in Section 6.4.4.

In addition to these global failure modes two local failure modes are of particular interest:

(e) *Spalling:* at high indentation rates and low aspect ratios local spalling failure is frequently observed, in which in-plane horizontal cracks grow away from the contact zone and fragments of ice break away upwards and downwards (Fig. 4.24(e)). This failure mode is also known as flaking, and, if the horizontal extent of cracks is very great, delamination. It is not thoroughly understood theoretically, but theoretical treatments have been proposed by Evans *et al.* (1984) and Wierzbicki (1985).

(f) *Crushing:* at higher rates still, cataclastic crushing is frequently observed, in which ice at the contact zone is pulverised to a powder and extruded upwards and downwards (Fig. 4.24(f)). This process can be tackled theoretically using the methodology given in Section 4.3.3.2; this theoretical analysis is in fact particularly applicable since, as presented in equations (4.30) and (4.31), the analysis is two-dimensional. This corresponds precisely to plane stress indentation conditions in which lateral strain close to the indentor is confined to zero but ice is free to deform in the vertical direction, where $\sigma_{33} = 0$.

It is unwise to make quantitative generalisations about indentation failure modes as a function of indentation rate and indentation geometry. However, as a general guideline, the deformation mode map of Fig. 4.25 provides a fair representation of the behaviour of ice under laboratory indentation conditions. I have deliberately not put scales onto the axes, since the graph is by no means universal, and not all the failure modes are simply dependent on aspect ratio and indentation rate alone. Elastic buckling, for instance, depends on absolute thickness as well.

The following useful generalisations can be made:

(1) for any given geometry or aspect ratio, there is an indentation rate below which deformation is entirely by creep;
(2) at low aspect ratios, especially if the ice is thick, increasing the indentation rate causes the ice to make a transition to spalling or crushing failure;
(3) at higher aspect ratios, radial cracks commonly form;
(4) at high aspect ratio but relatively low indentation rate, radial and circumferential cracking may occur without any crushing activity.

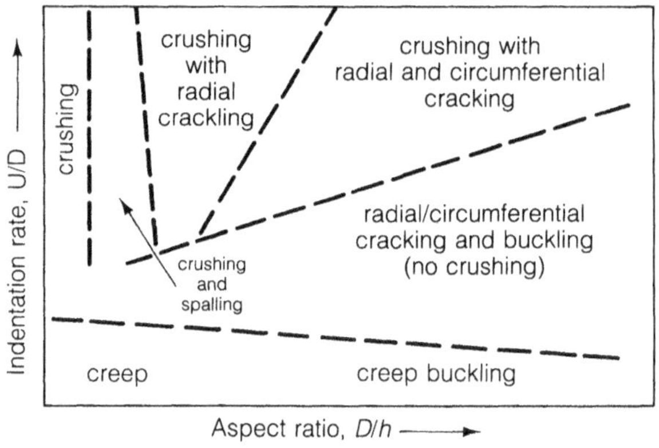

Fig. 4.25 Deformation mode map of ice failure as a function of indentation rate and aspect ratio (after Timco, 1986a, with modifications). The scale is deliberately omitted, since the map is not universal. However, it covers aspect ratios of about 0–25 and range of indentation rate about 10^{-3} s^{-1} to 1 s^{-1}.

4.4.2.3 Stresses during indentation

Peak stresses during indentation are higher than those during uniaxial compression, typically by a factor of up to about three, but depending on indentation rate and aspect ratio. It is simple to see intuitively why this should be so, at least in a qualitative sense. It is due to the lateral confinement in the x–y plane experienced by the ice sheet as it is indented: this biaxial confinement is applied to the deforming ice by the large body of undeformed ice surrounding it, and discourages the ice from deforming in the lateral direction. This has two effects:

(1) in the creep regime, it makes the stress conditions experienced by ice in the vicinity of the indentor approach more nearly to conditions of plane strain (Figs 4.15(c) and (h)). As we have seen, in Section 4.2.5, this is expected to result in stresses to be higher by a factor of up to about four for columnar ice or about 1.2 for granular ice. (This is a simplification, and we will look in more detail at modelling of creep indentation in Chapter 6 (Section 6.2.3).)

(2) in the brittle regime a similar effect occurs: biaxial stresses are developed in the same manner, and cracks are inhibited from forming and opening in the transverse direction, at least in the immediate vicinity of the indicator. This, however, is a complex phenomenon: although local crushing may be inhibited, since wing cracking should be suppressed, large-scale radial cracks may nonetheless form and effectively release the biaxial constraint. There is a further complication at very high aspect ratios (exceeding, say, 10–20). If failure is occurring by a brittle mechanism, then at high aspect ratio (very wide indentor) ice may begin to fail in discrete independent zones. This phenomenon is known as non-simultaneous failure and is of great importance since, although local stresses remain high, the average measured stress can be greatly reduced. We shall look at the theoretical analysis of non-simultaneous failure in Chapter 6 (Section 6.4.5).

An illustration of the effect of aspect ratio on brittle indentation strength is shown in Fig. 4.26 (after

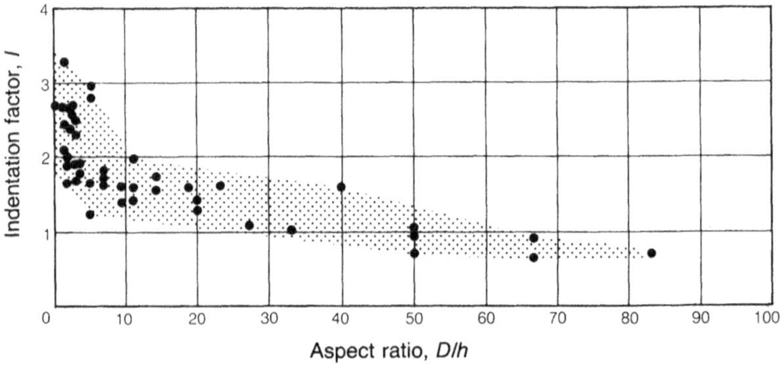

Fig. 4.26 Plot of indentation pressure P/Dh normalised by peak uniaxial strength, for a floating S-2 ice sheet interacting with an indentor of various widths. The normalised indentation pressure falls with increasing aspect ratio D/h, and the trend is highlighted by the shaded band (after Michel and Blanchet, 1983).

Michel and Blanchet, 1983). In this plot, average measured indentation stress of a floating S-2 ice sheet (that is, load P divided by contact area Dh) is normalised through dividing by the peak measured uniaxial strength of the ice. The ordinate then shows the factor by which indentation strength exceeds uniaxial strength for a wide variety of aspect ratios. The factor falls from about three at very low-aspect ratios of about unity ($D = h$), to around one at aspect ratios exceeding about 25. Note that the factor apparently falls below one at aspect ratios exceeding about 50. This is very likely due to effects of non-simultaneous failure.

In Fig. 4.27(a), I show peak indentation pressure as a function of indentation rate U/D for a number of small-scale measurements on S-2 ice (Hirayama *et al.*, 1974; Frederking and Gold, 1975; Michel and Toussaint, 1977; Timco, 1986a). The experiments by Frederking and Gold and by Michel and Toussaint were carried out at −10°C. Those of Timco were carried out at −3°C, and those of Hirayama *et al.* were carried out at close to 0°C.

Below an indentation rate of about $10^{-2}\ s^{-1}$, the data show very satisfactory grouping and consistency, and stress shows a dependency on strain-rate similar to that displayed in uniaxial conditions, that is:

$$\sigma = P/Dh \propto (U/D)^{1/n} \qquad (4.37)$$

where n is here in the range 3–4. Peak measured stress occurs at an indentation rate of $U/D \sim 10^{-2}\ s^{-1}$, where indentation pressures are approximately 20 MPa. This compares with peaks of about 10 MPa under uniaxial compression. At strain-rates exceeding this, peak measured stresses are lower, in the range 10–15 MPa. However, bear in mind that the higher strain-rate data (Timco, 1986) are measured with warmer, softer ice at −3°C, and may not be strictly comparable. In Fig. 4.27(c) all these data are replotted (open symbols) after correcting for temperature using the temperature normalisation formula given by equation (4.7). The data are normalised to −10°C. This considerably raises the normalised stress levels at high strain-rates, and removes the apparent peak at indentation rate U/D. It is dangerous to take this too seriously, though: the temperature normalisation of equation (4.7) is derived principally from non-brittle creep behaviour and may not be at all appropriate at high strain-rates. Unfortunately, the correction is not at all trivial, and amounts to a factor of three for the normalisation correction between 0°C and 10°C. Notice also that the stresses measured at indentation rates exceeding $10^{-2}\ s^{-1}$ show considerable scatter. This is due to the transition to brittle behaviour.

4.4.3 Medium-scale *in situ* experiments

A number of rather larger-scale experiments have also been carried out in order to determine the strength of ice when it is indented through the full thickness of natural floating ice cover. The ice cover tested in these experiments generally lies in the range about 0.5–1.5 m, and indentors of widths in the range 0.3–4.3 m have been used.

The first set of such tests, known as the "Nutcracker" tests because of the appearance of the two-armed hydraulic jacking system designed for the tests, were performed in 1969–70 by participants in the first projects carried out by the Arctic Petroleum Operators Association (APOA) of Canada.

The experiments are reported in Croasdale (1970, 1971), and involved a series of *in situ* tests on sea ice of 1.0–1.5 m thickness in Tuktoyaktuk Harbour and another series on lake ice of thickness about 0.9 m at Eagle Lake. The ice at Eagle Lake had predominantly vertical *c*-axes (not horizontal as is more normally the case). Quite low peak indentation pressures were measured during these tests, generally in the range 2.5–7.2 MPa, but the strain control of the apparatus was fairly rudimentary, and there is only a limited amount of data of good enough quality to plot on graphs of the form of Fig. 4.27. Data points from Eagle Lake tests on ice with horizontal *c*-axis are plotted in Fig. 4.27(b). A discussion of the tests is given by Croasdale *et al.* (1977). A further series of tests at Eagle Lake was carried out in 1972–73 using improved strain control, and this set of data, uncorrected for temperature, is also shown in Fig. 4.27(b).

A more ambitious series of tests on much thicker ice was carried out by Exxon Production Research Company (EPR) during the winter of 1980/81 on sea ice of thickness 1.5 m near Prudhoe Bay, Alaska (Petrie and Poplin, 1986; Lee and others, 1986). In addition to full-thickness indentation tests involving indentors of width 3.0–4.3 m, an extensive small-scale sampling and testing programme was carried out in an effort to relate small-scale strength to large-scale strength. The results are of outstandingly high quality, but unfortunately cover a range of indentation rate of only $4\times10^{-7}\ s^{-1}$ to $2\times10^{-5}\ s^{-1}$. This limits their value

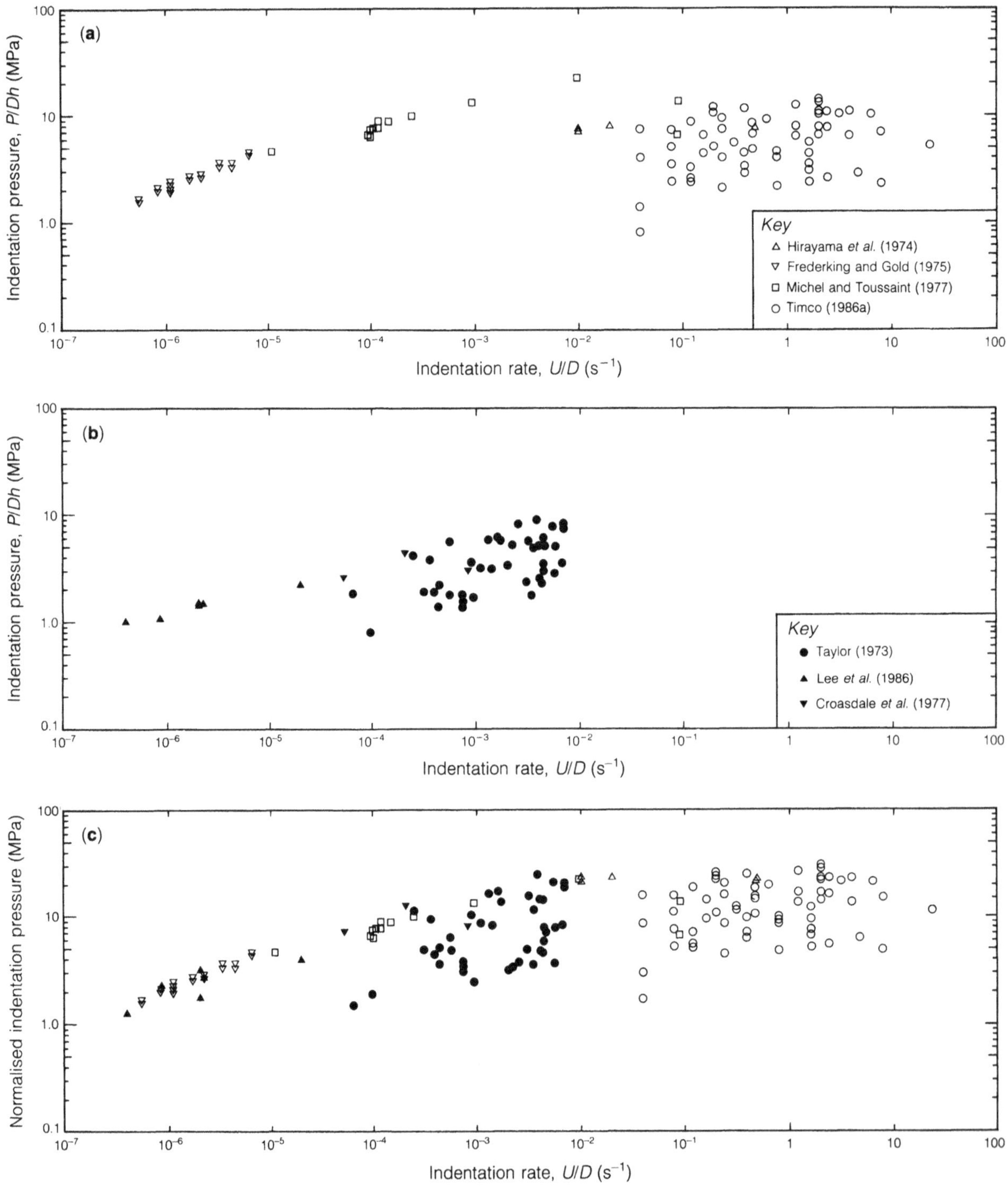

Fig. 4.27 Indentation data: indentation pressure *P/Dh* plotted as a function of indentation rate *U/D*. (a) Small-scale data measured in the laboratory (Hirayama *et al.*, 1974; Frederking and Gold, 1975; Michel and Toussaint, 1977; Timco, 1986a). No temperature correction has been applied. (b) Data from medium-scale *in situ* tests on lake ice (Croasdale *et al.*, 1977; Taylor, 1973) and on sea ice (Lee *et al.*, 1986). No corrections have been applied. (c) Data as in (a) and (b), normalised to zero salinity and temperature −10°C.

somewhat: no tests were carried out at brittle indentation rates of 10^{-3} s^{-1}, since equipment of sufficiently heavy duty to allow this is not readily available.

The results derived from full-thickness tests on ice with a pronounced temperature gradient and variable crystallography are difficult to compare with results obtained under controlled laboratory conditions at -10°C. Nonetheless, it was found that small-scale test results were compatible with the full-scale tests to an accuracy of better than $\pm 25\%$. This is well within the error bars of the technique, and demonstrates that, for the range of strain-rates employed, the strength of ice at full-scale *in situ* is essentially as expected, with no pronounced scale effect.

The principal results, uncorrected for temperature or salinity, are plotted in Fig. 4.27(b) as a function of indentation rate, along with data from Croasdale *et al.* (1977) and Taylor (1973). When approximately corrected for both temperature and salinity, using equations (2.1) and (4.9), these *in situ* data plot as shown in Fig. 4.27(c), superimposed on the temperature-corrected laboratory data of Section 4.4.2. The medium-scale experiments appear to be broadly consistent with small-scale experiments, at least at low indentation rates. At medium to high indentation rates, however, the agreement is considerably less satisfactory, and the medium-scale data show more scatter and generally lower values than the laboratory data. Looking at Fig. 4.27(c), the principal conclusions can be expressed as follows:

(1) The *upper bound* of ice pressures, as measured in the range of indentation rate 10^{-7} s^{-1} to 10^{-2} s^{-1}, shows no discernible dependence on size.

(2) At indentation rates below about 2×10^{-5} s^{-1}, data at both small-scale and large-scale show good consistency and reproducibility, and appear to follow a power law of the form given by equation (4.37). This is consistent with pure continuum (creep) behaviour.

(3) In small-scale laboratory experiments at indentation rates exceeding about 10^{-2} s^{-1}, the peak stresses (open symbols) show very much more dramatic scatter, and their upper bound appears to flatten off at a level of about 25 MPa. This corresponds to a transition to brittle behaviour, and the average stress (of about 10 MPa) is considerably lower than the peak.

(4) For medium-scale tests at indentation rates above about 10^{-4} s^{-1}, the same thing happens: peak stress results show considerably greater scatter than at lower strain-rates while the upper bound still conforms to the same upper bound as derived from small-scale tests.

This appears to indicate, within the limitations of the data set presented, that the upper bound of indentation pressures increases monotonically to about 25 MPa at 10^{-2} s^{-1} and then flattens off, but that a transition to irregular brittle behaviour may occur at a lower indentation rate than this. The transition indentation rate for small-scale tests appears to be around 10^{-2} s^{-1}, while for medium-scale tests it is lower, at around 10^{-4} s^{-1}. As we shall see in the next chapter, on full-scale measurements, it is likely that at really large scale the transition indentation rate is lower still.

Chapter 5

Full-scale Measurements

5.1 INTRODUCTION

Between the mid-1960s and the present day a large number of measurements of ice forces have been made at fullscale on real offshore structures and islands. Many of these measurement programmes were carried out by joint industry associations such as the Alaska Oil and Gas Association (AOGA) and the Arctic Petroleum Operators Association (APOA), and the results are generally subject to a five-year confidentiality agreement. A growing number of these programmes are, however, now in the public domain, and this fortunately includes almost all of the most significant projects carried out.

In this chapter we shall look at these projects in turn and examine the most important results from each one. At the end of the chapter, in Section 5.6, I shall draw together all the data on ice indentation from this chapter and the last and plot it all onto a single set of graphs (Figs 5.36–5.38) which will form the basis for load calculation methods expounded in Chapter 7.

The full-scale projects available fall into three broad categories:

(a) Measurements on narrow structures:
- Cook Inlet, Alaska
- lighthouses (e.g. Baltic Sea)
- bridge piers (river ice)

(b) Measurements on Arctic structures and islands:
- Canadian Beaufort Sea
- Hans Island

(c) Measurements from surface vessels:
- icebreakers
- floating drilling units (e.g. *Kulluk*)

I shall concentrate here on measurements at Cook Inlet and on measurements on Arctic structures and islands, as these are the most important. My treatment of lighthouses and river bridge piers will be very brief: the data are included in the general data set to be considered in Section 5.6, but I do not propose to give a detailed explanation of the results. Similarly, I do not propose to look in great detail at icebreaker measurements (Section 5.5). The reasons for this are twofold: firstly, much of the data collected from lighthouses and surface vessels is highly dependent on compliant structural response characteristics of the structure or vessel, and it is beyond the scope of this book to consider these in great depth; secondly, most of the lighthouse and bridge pier data refers to very warm ice at or near its melting point, and it is difficult to apply the data to much more severe conditions in the high Arctic.

Before looking at projects in detail, it is worth looking first at the methods used for collecting the data: the data are not easy to collect, and it is much easier to appreciate their value and limitations if the measurement methods are made familiar.

5.2 MEASUREMENT TECHNIQUES

A wide variety of techniques have been used for measuring ice forces on structures. Conceptually, there are four types of method (Fig. 5.1), which we shall look at in turn.

5.2.1 Interfacial method

Instruments for measuring stresses at the ice/structure interface are typically designed in the form of flat panels (often called "flatjacks") affixed to the structure at the water line (Fig. 5.1(a)). A panel is generally of dimensions 1 or 2 m by 1 or 2 m and is a few centimetres thick. It contains some form of well calibrated structural stiffening sandwiched between two steel plates, and the panel operates by measuring the compression of the plates during application of stress. Stiffening is generally provided by an internal complex of steel reinforcing members (for instance, the "Hexpack" sensor, Graham *et al.*, 1973) or by internal buttons of elastomeric material (for instance, the "Medof" panel, Metge *et al.*, 1983, Pilkington *et al.*, 1983). Although the panel contains stiffening, it is in effect a "soft" load sensor, in the sense that it measures load by deforming in response to the applied load. Most of these sensors were originally designed as *in situ* ice sensors, to be installed in slots in the ice cover, and are designed to have a stiffness of about 1–5 GPa, comparable with that of ice.

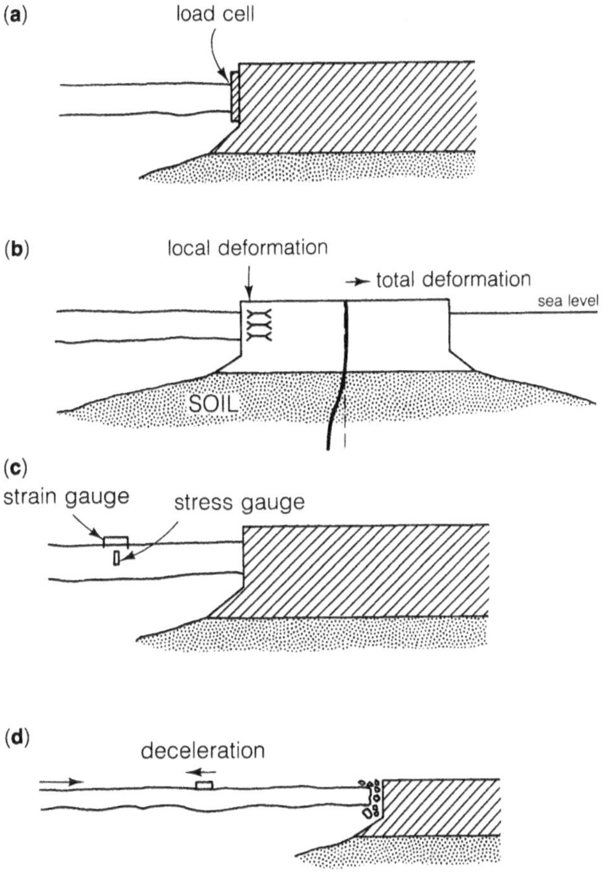

Fig. 5.1 Methods for measuring ice forces on structures: (a) interfacial load cells; (b) structural response; (c) stress or strain in ice; (d) floe deceleration.

The deformation of the panel can be measured by a variety of methods. The *Medof panel* (Pilkington *et al.*, 1983) uses fluid displacement. The panel is filled with calcium chloride solution, and as it is squeezed by the applied stress the internal volume of the panel decreases, displacing the contained fluid into a narrow measuring standpipe. The level of fluid in the standpipe is then monitored, either visually or by means of a sensitive pressure transducer. Because gross volume change only is being measured, this panel supplies only an average response to applied stress; it cannot distinguish between a locally applied high stress and an overall lower stress. There is also some doubt about its ability to average an areally variable stress in a truly linear fashion, and its response to stress varying at a frequency exceeding about 1 Hz is not satisfactory. Nonetheless, the Medof panel has yielded a large amount of extremely valuable data.

An earlier sensor, the *Esso sensor*, was constructed along similar lines, but used the changing electrical capacitance of the panel as an index of its overall deformation. Its non-linear behaviour and susceptibility to drift gave rise first to the Medof panel and later to the *EPR sensor*.

The *EPR sensor* is a thin metal panel 0.5 m wide and 2 m long, and it uses internal strain-gauged metal elements to measure load. Its construction and performance is described by Templeton (1979) and Chen (1981).

The *Hexpack sensor* (Graham *et al.*, 1983) again uses strain-gauging to measure deformation of internal stiffening members. The interior of the panel may be fitted with a large number of strain sensors, which in principle allow the panel to measure areal distribution of stress over the panel.

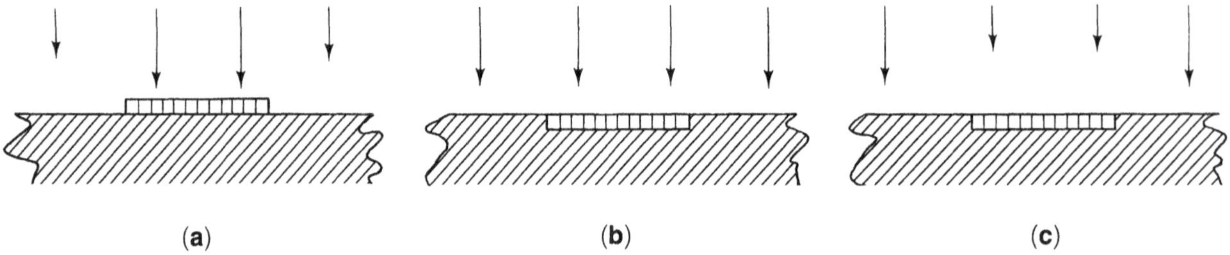

Fig. 5.2 Problems of mounting a sensor. (a) If mounted proud on the surface it may initially bear more than its share of load. (b) If flush-mounted and of the same stiffness as the structure it should provide a uniform stress distribution. (c) If flush-mounted and of lower stiffness than the surrounding structure it may be "bridged" and bear less than its share of load. Creep deformation helps relax these stress concentrations.

Since all of these sensors are effectively soft they present some problems when mounted on the outside face of a structure (Fig. 5.2):

(a) they may project significantly;
(b) even if flush-mounted they have a stiffness which probably does not match that of the structure itself.

If the sensor projects significantly (Fig. 5.2(a)) then it may, at least during the initial stages of contact, bear more load than the surrounding structure.

On the other hand, if the sensor is flush-mounted then the difference in stiffness between the sensor and the surrounding structure leads to a "bridging" effect, by which the stress distribution at the interface is not uniform, and load may be shed preferentially onto the stiffer surrounding material (Fig. 5.2(c)). Because ice has complex elastic/viscoelastic/brittle properties it is not appropriate simply to analyse this stiffness contrast as a problem in elasticity theory (this is a pity, since such an analysis is not too difficult). The effects of creep, for instance, tend rapidly to relax any stress imbalance caused by mismatch of elastic stiffness.

It is my own feeling that during a real ice impact event, when ice may be undergoing highly non-uniform crushing over a random distribution of independent failure zones, these stiffness contrast problems may pale into insignificance compared with the more general problem of knowing over what area the ice is failing, and to what extent the stress over the limited area of a panel may be representative of the average stress on the entire structure. Nonetheless, in cases where comparison has been possible between panel measurements and inferences of load from total structural deformation, agreement between the two methods has been apparently satisfactory (for instance, the Molikpaq structure: see Section 5.4.1.12).

5.2.2 Structural response

The most reliable method of measuring loads by monitoring structural response involves strain gauging of internal structural members or measurement of relative movement of components of the structure (Fig. 5.1(b)). In principle this is much the same method as that of monitoring the internal deformation of a load panel, but the structure itself is here being used as the load cell. Calculation of stresses from structural response naturally requires some form of calibration between strain and stress. This may be performed either analytically or empirically. Analytical calibration is often fairly readily and reliably available by application of finite-element analysis to the structural design: often, such an analysis will have been carried out anyway during the design phase of the structure. However, it is desirable to carry out physical calibration as well, if possible. This may be achieved by application of known static loads and measurement of resulting strains and deflections, but such a procedure is difficult and expensive to perform.

The major difficulty with structural response methods lies with structural resonance problems. These may lead to "measurement" of dynamic loads, which apparently greatly exceed simple static loads, but which are not always easily distinguishable from genuine peak static loads. This question of the relevance of dynamic loads is a difficult one, and will arise again when we come to look at observed loads at Cook Inlet (Section 5.3.1) and on the Molikpaq structure (Section 5.4.1.12). In both instances the structures appear to show pronounced resonant excitation at a frequency of around 1 Hz. A dynamic load felt by a structure, and measured by structural response such as an internal strain gauge, is not necessarily equal to the applied force. This is because of the effect of energy stored dynamically in the resonating system. The simplest everyday example of this, which makes the problem very obvious, is that of a child's swing being pushed by a parent: by repeatedly, and gently, pushing the swing at its resonant frequency, a dynamic deflection of the swing can be achieved which far exceeds the static deflection which could be achieved by one single static application of the same, gentle pressure. A strain-gauge attached to the swing would indicate, from the swing's deflection response, that a force much greater than the actual force was being applied.

Now, from the point of view of the structural designer, this amplified "load" at resonance is indeed important, since real stresses and strains corresponding to it are being generated within the structure. The stresses are, however, an artefact of the structure design and response characteristics, and they are not necessarily related directly, even approximately, to the magnitude of driving force being applied. For the purposes of describing ice loading applied to a structure it is preferable to rely on interfacial measurements or *in situ* ice measurement methods, which bypass the problem of structural response.

Total deflection of the structure (as opposed to relative movement of internal members) may also be measured (Fig. 5.1(b)) and under certain circumstances provides an estimate of applied load. Deflection may, for instance, be monitored by use of geotechnical inclinometers, which provide a measure of elastic and permanent soil deformation. However, soil properties are rarely constant enough or well enough known to permit reliable calculation of total applied loads. At Adgo P-25 artificial island in 1974–5, the observation that the island had permanently deformed during an ice loading event was used to back-calculate a lower bound to total applied load (see Section 5.4.1.2). Other geotechnical stress sensing devices may also be employed in the soil, such as embedded total pressure cells or piezometers, but these too suffer from uncertainties in physical description of the medium under load.

5.2.3 *In situ* ice measurements

Measurements of stress in an ice cover may be performed by two principal types of sensor (Fig. 5.3): the thin flat sensor and the stiff cylindrical sensor.

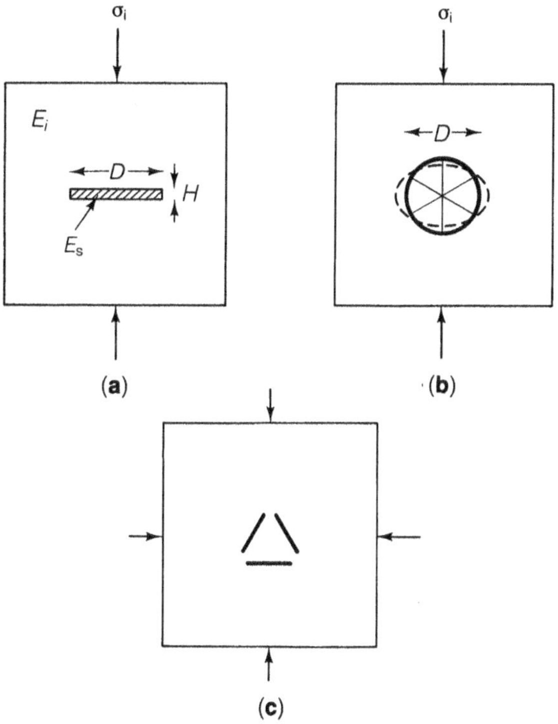

Fig. 5.3 Two types of *in situ* ice stress sensor (vertical view of horizontal sections): (a) thin flat sensor, $D \gg H$ – the sensor has elastic modulus E_s and the ice has elastic modulus E_i; (b) stiff cylindrical sensor. (c) Three flat sensors deployed in a rosette formation to measure biaxial stresses.

5.2.3.1 Thin flat sensors

The panel-type sensors already described in Section 5.1.1 are thin flat sensors in the sense that their width D greatly exceeds their thickness H (Fig. 5.3(a)). When used as *in situ* ice stress sensors these are normally deployed through the full ice thickness (of about 1.5 m) by cutting slots with a chain-saw and allowing the panel to freeze in. As already mentioned, their stiffness E_s is designed to be close to the stiffness E_i of the ice.

An alternative type of thin flat sensor is the much smaller disc-shaped sensor, typically of diameter about 10 cm. Such sensors are hydraulically filled and function by measuring fluid pressure under applied load. They are generally designed to have stiffness as high as possible and, in any event, greater than that of ice. Sensors of this type have been deployed by ARCO Oil and Gas Company (filled with hydraulic fluid; McBride, 1981) and by BP Petroleum Development Ltd (filled with mercury, to achieve greater fluid incompressibility), though the details of construction of these two sensors differ in important respects.

For any *in situ* sensor, an "inclusion factor" needs to be calculated to take account of any stiffness contrast between the sensor and the ice in which it is embedded. Since ice is a creeping material, it does not have an unambiguous static Young's modulus (see Section 4.2.1.1), so a stiffness contrast will invariably be present however carefully the sensor is designed. The problem has been tackled through two-dimensional elastic analysis by Metge *et al.* (1975), Chen (1981) and Templeton (1981). The result is as follows: let the ice sensor be of width D and thickness H, with an elastic modulus of E_s, and suppose that it is embedded in ice with an elastic modulus E_i. Now, if σ_i is the stress applied to the ice then the sensor will feel an average stress σ_s given by the expression:

$$\frac{\sigma_i}{\sigma_s} = \frac{(E_s/E_i)+(H/2D)}{(E_s/E_i)(1+H/2D)} \tag{5.1}$$

Notice that this tends to unity if $D \gg H$, irrespective of the relative stiffness E_s/E_i: that is, for a wide thin sensor it does not matter how stiff or soft the sensor is. However, the stress ratio σ_i/σ_s (the inclusion factor) is plotted in Fig. 5.4 for three different aspect ratios D/H and it seems that:

(a) the inclusion factor is best-behaved at large values of E_s/E_i – that is, the stiffer the sensor the better;
(b) for panel-type sensors which are designed to have a stiffness equal to that of ice ($E_s/E_i = 1$), the inclusion factor is theoretically equal to 1, independently of aspect ratio.

Ice, of course, is not a purely elastic material; only over very rapid rate loading cycles can its behaviour be adequately approximated by elasticity analysis. For longer periods of loading, stress relaxation occurs by creep. The general effect of this is to reduce inclusion effects and make the inclusion factor approach more closely to unity. In practice, if the panel has an aspect ratio of 10 or greater then it does not matter precisely what its stiffness is, so long as it is comparable with or greater than that of the ice. In this case the inclusion factor may be assumed to be equal to 1.

A single sensor of this type placed in a biaxial stress field measures only the normal component

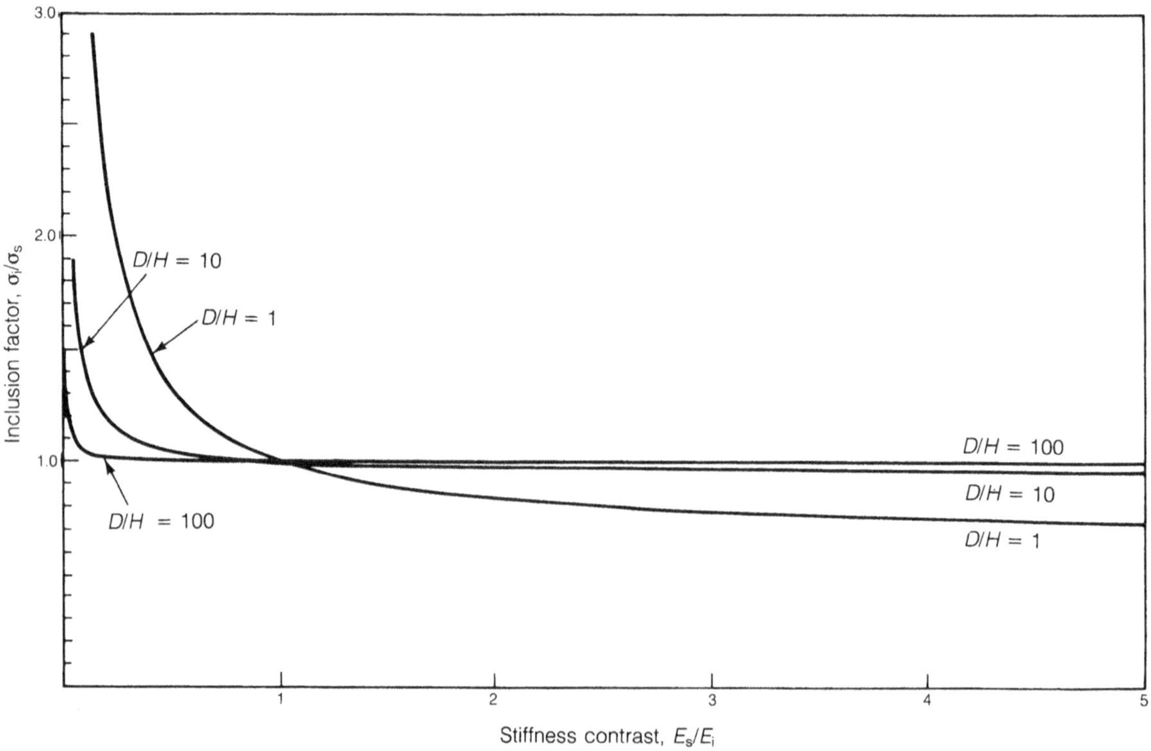

Fig. 5.4 Inclusion factor σ_i/σ_s as a function of stiffness contrast E_s/E_i for a flat panel with aspect ratio D/H of 1, 10 and 100.

of stress. However, an array of three or more sensors can be deployed in a rosette formation to yield principal axes of stress (Fig. 5.3(c)). Sensors of the smaller, disc-shaped variety may be deployed at several depths within the ice cover, and are best installed by augering or coring an access hole and then cutting a fine chain-saw slot at the base of the hole, into which the sensor is inserted. The sensor is then frozen in by addition of a small amount of fresh water to ensure good contact. All types of sensor which are frozen in by such methods experience substantial "freeze-in stresses" as a result of the expansion of ice as it forms from water. These stresses may amount to several hundred kiloPascals, but are typically relieved by creep over a period of several hours.

5.2.3.2 Cylindrical stress sensors

An alternative approach to the problem is to use a stiff cylindrical sensor installed in a hole in the ice. Sensors of this type measure deformation of a thick-walled cylinder in response to a biaxial load, either by internal strain-gauging (Nelson, 1975) or by measurement of vibration frequency of internal wires under tension (Cox and Johnson, 1983). The theory of biaxial cylinders embedded in ice has been developed by Cox and Johnson, and it shows that for a very stiff sensor we have an inclusion factor of

$$\sigma_i/\sigma_s = 1/1.5 = 0.67$$

(this is equal, as it happens, to the inclusion factor for a square-shaped sensor with $D = H$ in equation (5.1)).

This means that apparent measured stresses are high by a factor of 1.5 and need to be corrected by this factor to yield true ice stresses. Note that this depends on the ice being adequately modelled as a simple elastic material. It seems to me that stress relaxation effects should mean that during application of a constant applied load the inclusion factor should in time approach unity. The effective inclusion factor thus lies in the range 0.67–1.0.

5.2.3.3 Strainmeters

Ice interaction with a structure may also be monitored by directly measuring the strain behaviour of the ice sheet. A strainmeter is installed on the surface of the ice sheet, and consists of two anchored end-points joined by a stable gauge material. Movement of the end-points relative to the stable gauge length is then measured by a sensitive Linear Voltage Differential Transformer (LVDT). Original designs of ice strainmeter (Goodman, 1980) were based around an invar wire under tension over a gauge length of about 1 m and could resolve strain of about 5×10^{-8}. Later developments have used a rod as gauge material, which is more robust and portable but gives a lower resolution of around 5×10^{-7}. The strainmeters used to measure strains around Tarsiut Island (Sanderson *et al.*, 1983; and see Section 5.4.1.7) were a further development still, in which three strainmeters, each using a 30 cm gauge length of carbon fibre rod, were integrated into a single highly portable delta-shaped unit capable of providing complete information about principal axes of surface strain.

Measurement of strain, and hence strain-rate, gives useful and important quantitative information about ice deformation during the full-scale indentation process. Strainmeters also have the attractive characteristic that they cause negligible disturbance to the ice cover (no analogue of the inclusion factor correction is necessary). They can also in some circumstances be used to make estimates of the accompanying ice stress. Such an inference can be made only if the following conditions hold:

(a) the loading event is slowly varying and of long duration;
(b) negligible bending of the ice cover takes place;
(c) the ice in the vicinity of the strainmeter is well characterised for temperature, salinity and crystal structure.

If these conditions hold (and they are rather restrictive conditions) then it is reasonable to make the assumption that the ice is deforming by secondary creep, and that it is doing so uniformly throughout its thickness without bending. In this case the characteristics of the ice can be used to back-calculate the stress by applying equations (4.9) and (4.10). These are the equations for uniaxial stress, but the full generalisation to biaxial stress and strain-rate can also be made using equations (4.16)–(4.18). The inversion of the equations is lengthy and tedious and I will not provide it here. It is available in Sanderson (1982, unpublished). The results supplied by this method are in general an upper bound to the actual stress level existing in the ice, since it makes the assumption that ice is deforming by secondary creep. If any degree of primary creep is also occurring the inferred stress would be higher. Unfortunately, it is not possible to incorporate the effects of primary creep since it is a history-dependent process.

It will be obvious that the technique of strain measurement is not in general a satisfactory method for measuring stress: it provides a good scientific picture of the ice deformation process, but as an operational method for measuring stresses it is fraught with difficulties.

A comparative study of various available varieties of stress sensor and strainmeter was made in the winter of 1984 under controlled conditions in Calgary, and is reported in Croasdale *et al.* (1986).

5.2.4 Floe deceleration

Measurement of floe deceleration is perhaps the most elegant method of measuring ice forces. The technique has been used with great success at Hans Island (see Section 5.4.1.11), where it was the principal method of measurement of total force exerted by vast multi-year floes colliding with a natural island.

The method is appealingly simple. We measure the deceleration of an ice floe during the impact, estimate its mass and then calculate total force by applying Newton's second law of motion:

force = mass×acceleration

The deceleration is in general measured by installing a biaxial or triaxial accelerometer close to the centre of mass of the floe, but can also be estimated from time-lapse photography or measured by high-resolution electronic distance meters (EDM) in fast-tracking mode. Mass is estimated volumetrically by estimating area, thickness and density of the floe. There seems little that can go wrong with such a simple and fundamental method, but the following aspects require some consideration:

(a) *Added or entrained water mass.* A moving floe undergoing deceleration does not behave simply as a decelerating rigid body in a vacuum: it interacts hydrodynamically with the water flowing with it and behaves as if it had a mass greater than its actual mass. This is well known to any pilot berthing a supertanker. However, calculations by Dunwoody (1981) demonstrate that the effect is typically insignificant, of the order of about 1–4%, for realistically shaped floes in deep water. The effect would be more pronounced in water depth comparable to floe thickness (see Section 6.4.5.4).

(b) *Dynamic elastic effects.* The floe does not behave as a perfectly rigid body during impact and thus the deceleration measured at the centre of mass of the floe is not necessarily the same as the deceleration at the leading edge of the floe. What is ideally needed is the volume-averaged deceleration over the whole flow. Of course, this is not generally available, and I know of no calculations of the effect it might have. Measurements from accelerometers placed at different positions on an impacting floe do indeed show some differences, especially in the magnitude (and even the phase) of peaks. It seems to me that the use of results from a single accelerometer, rather than an average of many, is likely to lead to an overestimate of the total force during impact.

5.3 MEASUREMENTS ON NARROW STRUCTURES

5.3.1 Cook Inlet, Alaska

Cook Inlet, on the south coast of Alaska (Fig. 3.15), has been the site of concentrated offshore oil development since the first discovery wells were drilled in 1962. In global Arctic terms the ice conditions there are far from severe – indeed it is not truly Arctic at all – but the developments were the first of their type, the structures were well instrumented, and a wealth of ice force data is available. The principal published data source is Blenkarn (1970).

The ice cover in Cook Inlet is thin seasonal first-year ice, and is typically present between late November and early March. It is generally of thickness about 0.4 m and reaches a maximum thickness of between 0.5–0.65 m during the months of January and February. Salinity is generally in the range 4–6 ppt. The ice is not landfast, but highly mobile, in the form of medium to large floes several hundred metres across. This is largely due to vigorous tidal action: the tidal range is more than 6 m, with accompanying tidal currents of 3 m s^{-1}. In addition, prevailing winds tend to drive ice towards the south-west, and influx of river water leads to a general outflow from Cook Inlet towards the open ocean. As a result it is estimated that few ice floes remain in Cook Inlet for more than about 30 days. This limits the ice thickness to which a floe can normally grow. However, the constant dynamic motion of the ice cover and the interaction of floes with the coastline leads to the formation of numerous first-year pressure ridges: these normally have poorly consolidated keels projecting some 3–5 m below the waterline and quite well consolidated sails about 1 m above sea level.

During the mid-1960s, 14 offshore structures were built in Cook Inlet. The structures are fairly conventional, though ice-strengthened, piled steel jacket structures, providing a platform for drilling and oil production operations. The platforms are supported by four cylindrical steel legs of typical diameter 4.3 m, and cross-bracing members are in general confined to levels well below the waterline, to avoid problems with build-up of ice rubble. The structures are oriented in such a direction, with respect to the prevailing ice motion in the channel, that in general only two legs are subjected to significant ice interaction. The most common form of interaction consists of continuous crushing, and the two upstream legs interact with the ice independently of each other.

Several of the structures in Cook Inlet were instrumented for ice forces during the 1963–9 period, including a specially installed instrumented test pile. Instrumentation was generally in the form of strain gauges on piles, which measured bending strain under load. These were physically calibrated to yield ice force measurements to an accuracy of around 10%, or, in certain cases, calibrated by theoretical calculations with accuracy estimated to be 25%.

The force records provided from these instrumented structures are complicated by their dynamic response characteristics: the structures in general have resonant frequencies of the order of 1 Hz, and show excitation at this frequency as a filtered response to the randomly varying ice crushing force being applied. This means that force records during level ice interaction typically show quite widely oscillatory dynamic loads scattered around more or less steady loads associated directly with the ice crushing. In addition, genuine peak loads were frequently observed, and appeared to be associated with impact of pressure ridges.

Dynamic loads were often observed to be 50–100% higher than the apparent static loads. A representative summary of loads on a single leg of a structure is shown in Fig. 5.5, for midwinter 1964–5. The total loads on a leg of diameter 4.3 m lie in the following ranges:

(a) "Steady loads", associated with level ice, of 0.3–1.4 MN. For ice of thickness about 0.4 m this corresponds to an average indentation pressure of 0.2–0.8 MPa.
(b) "Static peak loads", associated with ridged ice, of 0.9–2.2 MN. For ridges of consolidated thickness, say 2 m, this would correspond to an average indentation pressure of 0.1–0.3 MPa.
(c) "Dynamic peak loads", associated with structural response to the impact, of 1.8–3.8 MN. As already discussed above (Section 5.2.2), these dynamic peaks are characteristic of the structural resonant response of the structure, and do not necessarily correspond to pressures applied by the ice.

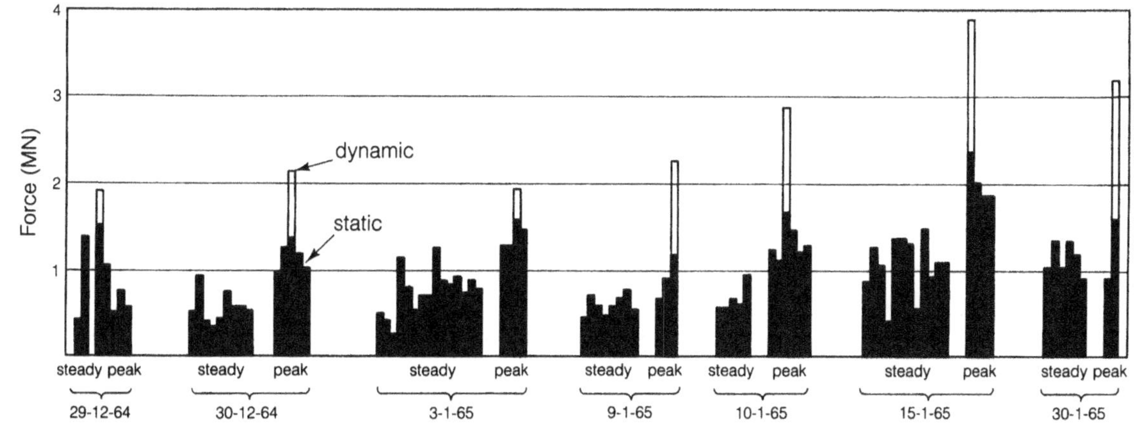

Fig. 5.5 Summary of peak loads at Cook Inlet 1964–5, on a structure leg of diameter 4.3 m. Two groups of data are shown for each day: steady loads and peak loads. Dynamic peak loads are identified by the white histogram bars (after Blenkarn, 1970).

An analysis of steady loads for the entire 1964–5 season is shown in Fig. 5.6, where load is plotted against ice thickness. It is seen that the maximum steady load observed due to level ice during 1964–5 was 1.6 MN, associated with ice of thickness 0.35–0.4 m. This corresponds to an average indentation pressure of about 1.0 MPa. A further analysis of the data in terms of ice temperature (Fig. 5.6) yields the interesting, and surprising, result that the highest loads appeared to be associated with the warmest ice: the peak loads were associated with ice of surface temperature −3°C, compared with ice of surface temperatures more generally around −10°C. The reason for this is not clear.

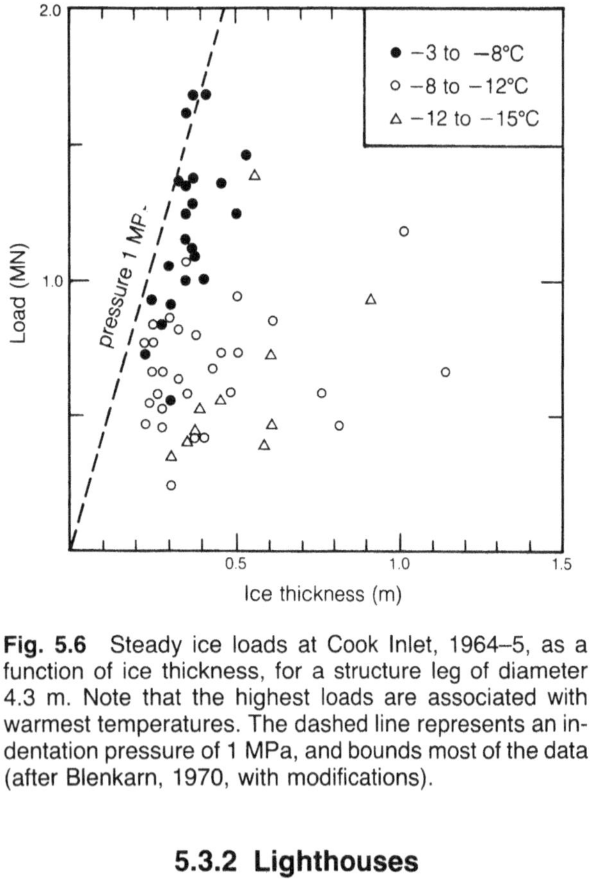

Fig. 5.6 Steady ice loads at Cook Inlet, 1964–5, as a function of ice thickness, for a structure leg of diameter 4.3 m. Note that the highest loads are associated with warmest temperatures. The dashed line represents an indentation pressure of 1 MPa, and bounds most of the data (after Blenkarn, 1970, with modifications).

5.3.2 Lighthouses

Offshore lighthouses have now been in use in the Baltic Sea for some 30 years. Although not truly Arctic, these Scandinavian waters are subject to drifting winter first-year ice of thickness up to about 1 m, often rafted, with ridges up to 10 m thickness. However, ridges are seldom consolidated, and the Baltic Sea contains no multi-year ice or icebergs.

In the Gulf of Bothnia and adjacent regions of the northern Baltic Sea there are now some 20 offshore Swedish lighthouses which have been exposed to ice action for an average of about 10 years (Engelbrektson, 1983). These lighthouses are not in general instrumented for ice force measurements. However, as a result of limited field data at the design stage several of them were built with insufficiently high design load specifications, and have since collapsed or been damaged. Two lighthouses were destroyed by ice after two years' service, while two other structures were laterally displaced by several centimetres. An approximate hindcast calculation of peak ice loads can be made for these lighthouses.

We do this by making the assumption that collapse or displacement occurred when the ultimate design load of the lighthouses was exceeded. The four lighthouses concerned had water-level diameters of 2.5–3.0 m, and according to data of Engelbrektson (1983) their ultimate horizontal load capacities were in the range 6.0–8.0 MN. Assuming that the ice involved in the interaction was of the order of 1 m in thickness, it appears that the ice must have been subjecting the lighthouses to average indentation pressures in the range 2.0–3.0 MPa over gross contact areas in the range 2.5–3.0 m^2.

The response of most lighthouses to ice action is highly dynamic, since they are in general compliant, non-rigid structures. This has the result, as with Cook Inlet structures, that measurements based on structural response or failure may give an overestimate of the static equivalent applied load. Experiments by Määttänen (1983) on model lighthouse structures indicate that the standard deviation of dynamic peaks about the average load level may typically be about 50% of the average load level. An estimate of the maximum peak load can then generally be made by calculating the mean load plus three standard deviations – that is, peak loads may be roughly 2.5 times the average load. This represents a very high level of structural response, and failure of the Swedish lighthouses described above may largely be due to excessive dynamic response rather than excessive applied loads.

Some direct measurements of ice forces have also been made in the Gulf of Bothnia, on two Finnish test lighthouses which were specially designed to minimize effects of dynamic excitation (Määttänen, 1981). These two vibration-isolated lighthouses, the *Kokkola* and the *Kemi II*, have been in operation since 1977/8, and ice forces have been inferred from calibration of (elastic) structural displacements. Total ice loads measured in this way were found to lie in the range about 2–3 MN, applied to structures of water-line diameter 0.64 m (*Kokkola*) and 1.1 m (*Kemi II*).

5.3.3 Bridge piers

Bridge designers for high-latitude inland waterways have long been aware of the need to design bridge piers to cope with impact of seasonal river-ice, especially during the highly mobile spring break-up season. Until recently, the design of such bridge piers has proceeded on the basis of fairly broad design code recommendations: prior to 1974, for instance, the Canadian Highway Bridge Design Code recommended simple application of a single effective ice failure pressure of 2.8 MPa, irrespective of ice type, thickness or temperature. This effective pressure, multiplied by the pier width and the maximum expected ice thickness, supplied the ice design load. While this procedure appeared to work satisfactorily, in the sense that it did not apparently lead to any appreciable risk of structural collapse, it became increasingly clear in the late 1960s that the design level might be over conservative. Many older bridges, for instance, were designed to ice pressures much less than 2.8 MPa, yet still survived severe seasons of ice impact and they, too, appeared to be in no danger of collapse.

As a result, three bridge piers in Alberta, Canada, were constructed with built-in instrumentation in the period 1966–9. Two of these have supplied many years of good quality data: the Hondo Bridge over the Athabasca River (1966), and the Pemridge Bridge over the Pembina River (1969). These bridges are instrumented using a specially designed nose section mounted on the upstream face of the bridge pier. In both cases this consists of a simply-supported steel-encased concrete beam connected through a load cell to the solid bridge pier support behind. At Hondo the nose section is of diameter 2.32 m, inclined at 23° to the vertical. At Pembridge the nose section is vertical and of diameter 0.86 m.

Measurements of ice forces during the seasonal "ice runs" at break-up have been taken each spring since 1967 at Hondo, and since 1969 at Pemridge. The best review account of the series of measurements is that by Lipsett and Gerard (1980). Their account gives details of data collected during the 1973–9 period, and provides full documentation of ice thickness, ice velocity, ice failure mode and peak forces generated. Lipsett and Gerard also provide a summary of maximum ice forces over the entire 1967–9 period.

At Hondo, peak forces in the range 0.3–2.3 MN were measured during ice interaction events involving ice of thickness in the range 0.33–1.07 m. These events represent effective ice pressures of 0.3–1.0 MPa applied over gross contact areas in the range 0.7–2.5 m^2. Ice velocities of up to 2.9 m s^{-1} were recorded, and ice failed in a variety of modes: principally crushing, bending and splitting. During the break-up season the ice temperature was close to 0°C throughout the ice thickness.

At Pemridge, peak forces in the range 0.2–1.1 MN were measured, with ice thickness generally of the order 0.5 m and ice velocities in the range 1.0–1.9 m s^{-1}. These events represent effective ice pressures in the range 0.5–2.7 MPa applied over gross contact areas of the order just 0.4 m^2. Note that these effective ice pressures at Pemridge, measured over relatively low contact areas, are higher than those measured on the Hondo Bridge Pier: Hondo, a larger bridge pier, experienced slightly higher forces, but lower effective ice pressures.

In the case of both Hondo and Pemridge dynamic excitation of the piers was highly damped, with the result that the effects of structural resonance were negligible.

Measurements on bridge piers are of some interest to the offshore designer, since they give a large-scale model of conditions to be encountered in the High Arctic. However, quantitative extrapolation is limited by three things: the relative thinness of the ice, which easily fails in bending or buckling; the warm temperature of the ice; and the limited width of the measurement devices employed. To gain a better quantitative picture of ice forces at fullscale we shall turn now to measurements made of ice interaction with wide structures in the Arctic offshore environment.

5.4 MEASUREMENTS ON WIDE STRUCTURES

5.4.1 Artificial and natural islands

5.4.1.1 Introduction

Offshore hydrocarbon exploration in the Beaufort Sea began in the early 1970s with the construction of dredged sand islands in the shallow nearshore areas of the Canadian Beaufort Sea. The first offshore well in the Beaufort Sea was drilled from the island Immerk B-48 (Fig. 5.7) during the winter of 1973–4 at a sheltered location in water depth of 3 m. The ice at Immerk was landfast. During the following 10 years some 20 further artificial sand islands were constructed, the most ambitious being Issungnak O-61, in a water depth of 20 m, which was used for drilling in 1979. A schematic cross section of the principal types of artificial island is shown in Figs 5.8(a) and (b), illustrating the use of a sacrificial beach or sandbags to minimise erosion by ice ride-up and wave action.

As it became necessary to drill in water deeper than 20 m the construction philosophy altered, partly owing to the massive quantities of sand or gravel required and also partly owing to problems encountered with wave erosion of simple sand islands. A new generation of islands, the caisson-retained islands, was born. These islands rely on a sand or gravel berm built up from the sea bed to within some 5 m of the waterline, which then has placed on it a steel or concrete caisson structure (Fig. 5.8(c)). This serves two functions: it retains a vertical-sided mound of sand, thus reducing the amount of sand required, and it allows the constructor to build rapidly through the wave zone with a structure which is resistant to wave erosion. The caisson also, of course, provides the principal defence against ice action. The first of these islands to be constucted was Tarsiut N-44, constructed in 1981 in 22 m of water (see Fig. 5.9(a)). Since then a number of alternative designs of caisson structure have been deployed in increasing depths of water, the deepest to date being the Single Steel Drilling Caisson (SSDC) at Uviluk, in 31 m of water (see Fig. 5.9(b)). Some of these structures rely on surrounding rubble piles or artificially created ice pads for protection against ice action.

In recent years (1984–7), great progress has been made in an alternative and cheaper technology: that of building a grounded island of ice. This can be achieved either by simply flooding existing ice cover with water or by spraying with water jets (Fig. 5.8(d)). The technique of spraying provides more rapid and efficient cooling and freezing, but both techniques are at their best only in air temperatures below about −20°C. Obviously, the ice island needs to be built up to have sufficient weight and freeboard that it will ground adequately in the water depth concerned. Ice islands to date have been constructed in water depths up to only 5–7.5 m. Further details of this technique are given in, for instance, Williams (1986) and Cottrill (1987). Clearly it is a technique which is limited to construction of temporary

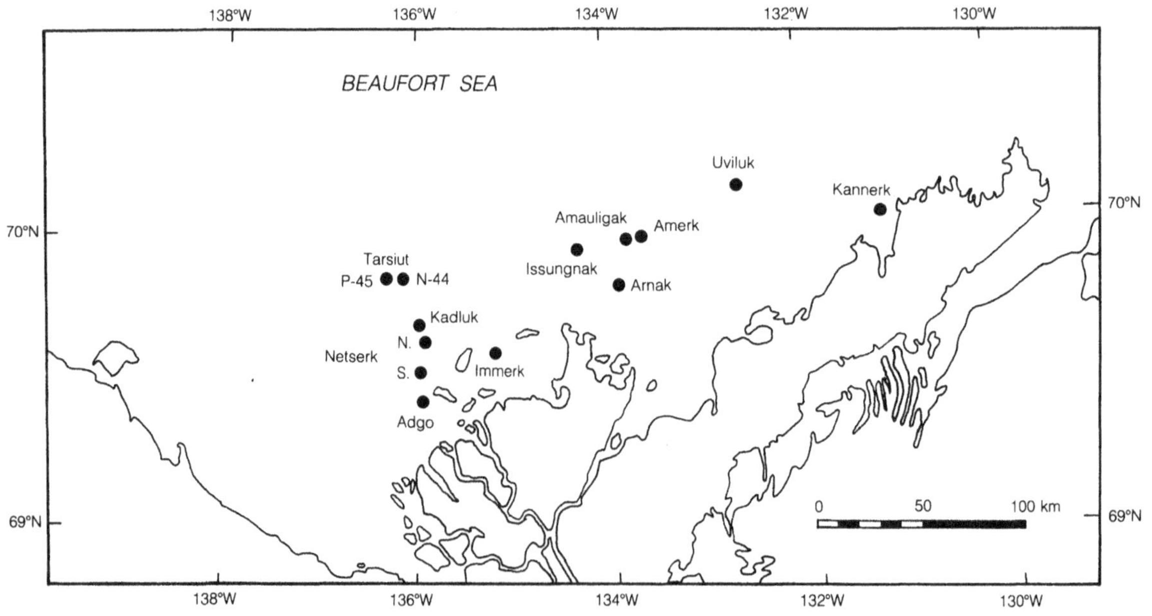

Fig. 5.7 Location of principal artificial islands and structures in the Canadian Beaufort Sea. Only those mentioned in the text are shown.

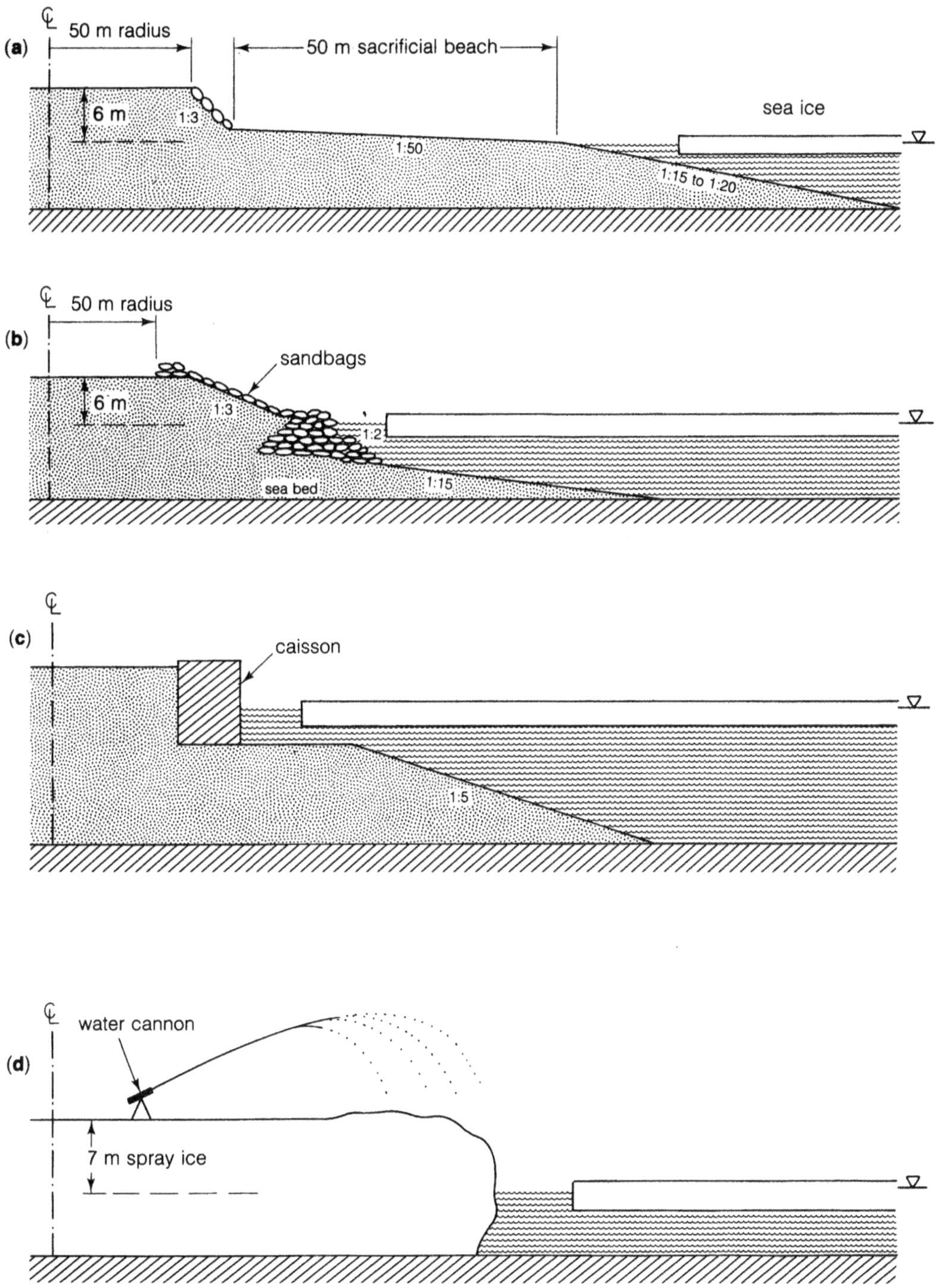

Fig. 5.8 Typical cross sections of artificial islands: (a) sacrificial beach island; (b) sandbag-retained island; (c) caisson-retained island; (d) spray-ice island. Normally such islands are surrounded by ice rubble piles. These have been omitted for clarity. Not to scale.

(a)

(b)

Fig. 5.9 *Beaufort Sea structures.* (a) Aerial photograph of Tarsiut island, N-44, spring 1982 (photograph courtesy of BP Petroleum Development Ltd). (b) Dome/Canmar's Single Steel Drilling Caisson (SSDC) at Uviluk P-66, in spring 1983. The structure is a converted oil tanker resting on a sand berm and is protected by a spray-ice pad. (Photograph by the author, courtesy of BP Petroleum Development Ltd.)

seasonal exploration islands, and it is unlikely that more permanent structures will be constructed in this way.

On many of the artificial islands and structures which have been used in the Beaufort Sea, ice observation and load measurement programmes have been carried out. These have mainly been joint industry programmes but many of them are no longer confidential. In the following pages I shall describe the most significant results.

5.4.1.2 Adgo P-25

The Adgo P-25 island was constructed in just 2 m of water and was used for drilling in the winter of 1974–5. Its dimensions were 70×140 m and ice conditions and interaction events are described by Metge (1976). The ice in the vicinity was entirely landfast first-year ice but was observed to move on several occasions during the November–January period, by distances of 0.5–1.0 m. These movements were episodic and appeared generally to be associated with winds in excess of 10–15 m s^{-1} (20–30 knots). During some of these movement episodes cracks were observed to form close to the island, but in general the island is described as being in a "frozen-in" condition for most of the winter: that is, there was no obvious tidal crack around the island. This means either that the ice was in places bottom-fast, or that any tidal movement was accommodated by flexure of ice near to the island.

The island was not extensively instrumented for direct measurements of ice pressures, but geotechnical instrumentation indicated measurable movement of the whole island during the winter, and some hindcast calculations can be made. A comparison between measurements made on 27 November 1974 and 8 January 1975 indicated a lateral movement of the island by approximately 5 cm. It is not known at precisely what time this movement occurred, but inspection of weather records indicated that it was most likely to have occurred as a result of a storm on 31 December 1974, during which winds of 20–30 m s^{-1} (45 knots gusting 65 knots) were recorded. During this period the ice thickness, h, around the island was 1.0–1.2 m, and the width of contact of ice with the island, D, was 120 m. Now, the island resistance was calculated to be in excess of 105 MN on the basis of soil overburden and yield strength. This leads to the conclusion that for the observed permanent plastic deformation of the island to have occurred, a load P in excess of 105 MN must have been applied by the ice. This is equivalent to an average indentation pressure of

$$P/Dh = 0.83 \text{ MPa}$$

over a projected contact area of approximately 130 m^2. The error on this calculation is probably in the region of 10–30%, and the calculated value is a lower bound.

Ice movement rates were also monitored, and the maximum movement during the period of interest was less than 15 cm in the course of 3 hours on 29 December. This is an average rate of less than 1.5×10^{-5} m s^{-1}, and an indentation rate U/D of the order 10^{-7} s^{-1}. Some direct *in situ* measurements of stress were also taken (Nelson and Sackinger, 1976) using sensors placed in the ice cover some 0.25–0.3 m below the surface. Measurements were taken only from 19 January 1975 onwards and measured stresses did not exceed 0.2 MPa.

5.4.1.3 Netserk South B-44

Stress sensors were also installed around the Netserk B-44 island during the winter of 1974–5 (Nelson and Sackinger, 1976). This island was constructed in 4.5 m water depth.

The stress readings obtained with these sensors cannot be regarded as reliable, owing to problems of noise, drift and calibration, and readings were taken at an insufficient number of sites to enable meaningful calculations of total load. Measured ice pressures were generally in the range 0.0–0.7 MPa, with 96 recorded instances of pressures exceeding 0.35 MPa. At one site, (site 4 to the northwest of the island) a pressure exceeding 0.7 MPa was recorded for a period of 12 hours, including one apparent peak of 1.5 MPa recorded for less than 1 hour. The high stress recordings did not appear to correlate with episodes of ice motion.

5.4.1.4 Netserk North F-40

The Netserk North island was the first island to be comprehensively instrumented in such a way that quite reliable estimates of total load could be obtained. The island was constructed during 1975 in 7 m water depth and was of diameter 125 m; however, during the early mobile ice season it became surrounded by an extensive grounded rubble pile of broken ice, which greatly increased its effective diameter. The rubble pile was of 190–250 m diameter and is shown in Fig. 5.10. The level ice effectively interacts with the edge of the

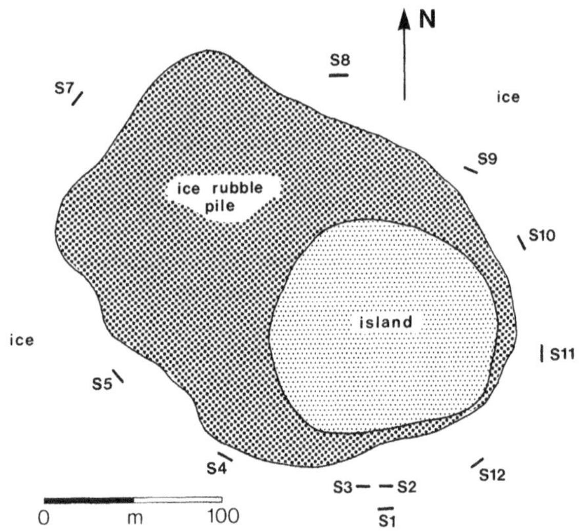

Fig. 5.10 Netserk North island (F-40), shown with its surrounding ice rubble pile and the location of stress sensors.

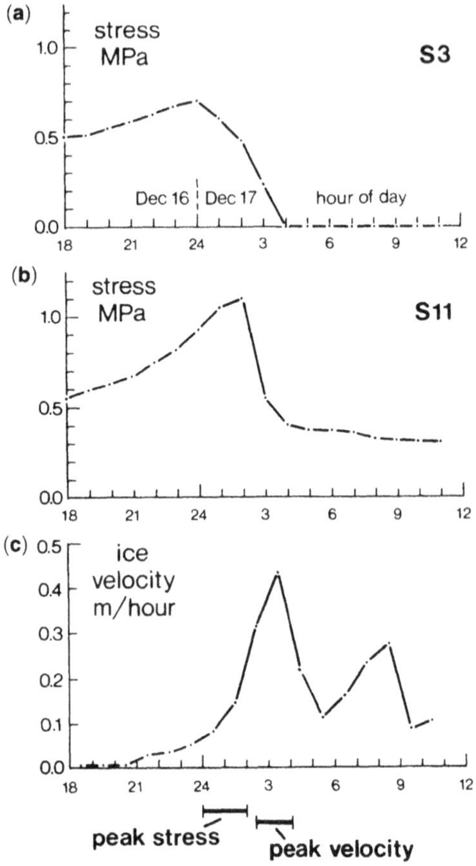

Fig. 5.11 Stresses and ice velocities on a common time base for Netserk F-40 on 16–17 December 1975. Data have been corrected for drift and calibration: (a) panel S3; (b) panel S11; (c) mean ice velocity over one hour intervals. Note that peaks of ice movement and ice stress do not coincide (data are from Strilchuk, 1977, replotted).

stable grounded rubble pile and not with the island itself.

The level ice was extensively instrumented (Strilchuk, 1977), principally with 11 flatjack pressure panels (Esso sensors) installed through the full ice thickness at nine locations, as shown in Fig. 5.10. Ice movement was monitored by wireline gauges, which are installed on the ice surface and record movement by reference to wirelines anchored to the sea bed. Pressures of around 0.7 MPa were recorded during several events, but the most significant loading event was observed in the period 16–18 December 1975, and it is worth looking at it in some detail.

During 16 December the wind was blowing at 8–10 m s^{-1} (15–20 knots) from a direction of about 190° and at night rose to 13–18 m s^{-1} (25–35 knots). Between the hours of 00:00 and 04:00 hours on 17 December, measured pressures built up slowly to maxima in the range 0.5–1.1 MPa. The ice was of thickness 1.0 m and the projected island width was about 190 m. The maximum recorded ice velocity was 0.43 m h^{-1} (1.2×10^{-4} m s^{-1}) and over the period 17:00 on 16 December to 12:00 on 18 December the total ice movement was 2.75 m from a direction of 170°.

However, there is much more to this than meets the eye, since the maximum ice movement rate did not apparently coincide with the maximum measured pressure (Fig. 5.11).

During the period 00:00 to 02:00 hours, maximum pressures of 0.7 MPa (panel S3) and 1.1 MPa (panel S11) were measured, while ice movements were in the range 0.05–0.25 m h^{-1}. By about 03:30 hours the movement rate had climbed to 0.43 m h^{-1}, but the ice pressures had dropped dramatically to 0.1 MPa (panel S3) and 0.5 MPa (panel S11). This is an important point: even though the ice was apparently deforming at indentation rates of 5×10^{-7} s^{-1}, securely in the creep regime of Fig. 4.27, pressures do not show an increase as indentation rate increases. On the contrary, stresses built up slowly to a peak and then dropped off quite rapidly as the ice moved forward.

It seems to me that we are seeing here a transition between ductile deformation and brittle behaviour, much in the same way as we do in Fig. 4.27 for small and medium-scale indentation tests, but it is occurring at a very much lower strain-rate. The picture is this: high ice pressures are recorded at low ice velocities, as the ice is loaded up in creep deformation by steadily increasing winds. At a certain strain-rate, however, here apparently about 3×10^{-7} s^{-1}, the failure mode makes a slow transition from ductile to brittle behaviour: cracks form and so weaken the ice fabric that eventually it surges forward easily and stresses are relieved. This explanation is speculation, since no direct observations of failure mode were made, but it is the only reasonable way of explaining the effect: the failure mode must have changed. We do indeed know from observation the next morning that the ice movement caused a 2 m high pile-up on the south-east side. It is likely that this occurred at 02:00–04:00 and not earlier.

The interesting thing is that the mode appears to have changed at an indentation rate (3×10^{-7} s^{-1}) and a stress (about 1 MPa) which are much lower than those expected on the basis of small-scale experiments (Fig. 4.27(a)). This observation, and the account of medium-scale experiments given in Section 4.4.3, lead us to a picture of a scale-dependent failure mode behaviour, as illustrated qualitatively in Fig. 5.12. At small scale in the laboratory, the ductile–brittle transition occurs at an indentationrate of about 10^{-2} s^{-1}; at medium scale during *in situ* full-thickness tests it occurs at about 10^{-5}–10^{-4} s^{-1}; finally, at full scale at Netserk island, it appears to occur at about 10^{-7}–10^{-6} s^{-1}.

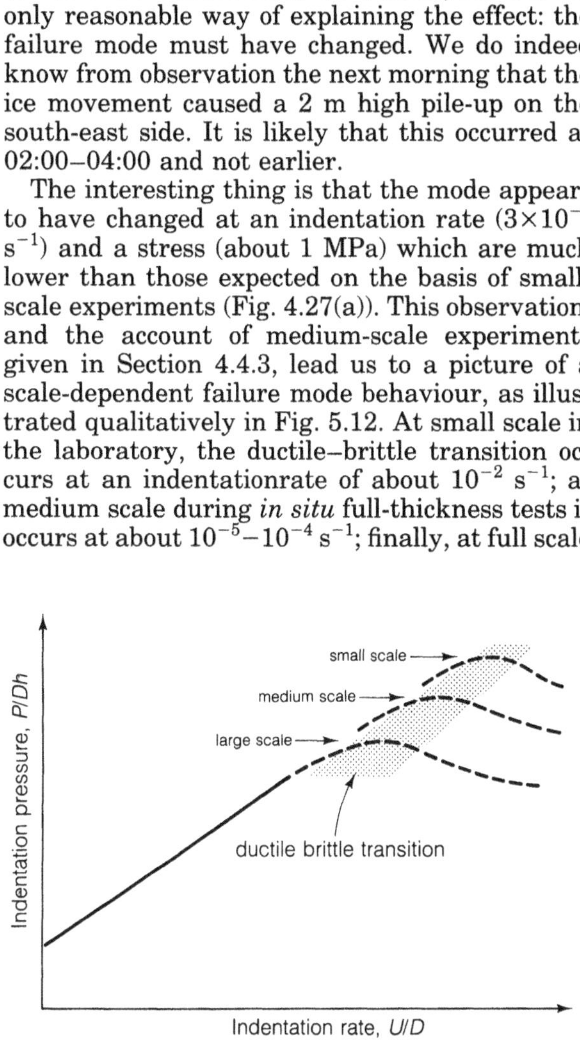

Fig. 5.12 Qualitative illustration of the scale-dependent failure mode behaviour of ice. At small scale the transition from ductile to brittle behaviour occurs at about 10^{-2} s^{-1}; at medium scale, it occurs at 10^{-5}–10^{-4} s^{-1}; at full scale it occurs at 10^{-7}–10^{-6} s^{-1}.

During the remainder of the Netserk field programme, no other loading event of similar importance was measured. However, there are some further points of interest to describe.

Movement rates. On 5 February 1976 ice moved more than 100 m towards the island, at a rate probably exceeding 10^{-2} m s^{-1}. Unfortunately, no stress sensors were operational to measure the associated stress. This is a shame, since the indentation rate was very high – about 10^{-5}–10^{-4} s^{-1}. At other times ice movement rates were generally in the range 0–18 cm h^{-1} (5×10^{-5} m s^{-1}) but more commonly less than 5 cm h^{-1} (1.4×10^{-5} m s^{-1}). Many of these more gradual movement episodes were apparently associated with overall thermal expansion of landfast ice lying between the island and the shoreline.

Stresses. Stresses in the range 0.35–0.65 MPa were measured on several other occasions, and the ice grew to a final thickness of 1.8 m. Maximum pressures were not generally observed to be coincident with peak movement rates.

Global loads. Global loads on the Netserk island were estimated (Strilchuk, 1977) by projecting theoretical stress fields back from each sensor towards the island, using standard plane strain elastic stress field equations (Timoshenko and Goodier, 1951). These were then superposed to give the effective resultant load on the island. Total loads calculated in this way were generally in the range 50–230 MN. The approach is open to criticism, since the stress sensors were actually quite close to the extensive perimeter of the island and yet the elasticity solution is strictly valid only for a line source load (that is, a very local narrow concentrated load or a very distant load). However, an alternative approach, based on an approximate line summation of average measured stresses over an entire loop surrounding the island, gives similar results, in the range 150–250 MN. During the loading event of 17 December, for which stress readings are shown in Fig. 5.10, I calculate a peak load of 160 MN. This corresponds to an average indentation pressure of 0.85 MPa over a contact area of about 190 m^2.

5.4.1.5 Kannerk G-42 and Arnak L-30

The measurement programme at Kannerk G-42, an island constructed in 8 m of water in 1976–7, yielded little useful data (Semeniuk, 1977).

A stress measurement programme was also carried out in 1976–7 at Arnak L-30, a nearby island in 8.5 m water depth (Semeniuk, 1977). The stress sensors gave poor data quality and showed drift which is difficult to correct for. Typical measured stresses were in the region of 0.35 MPa and maximum measured stresses at a sensor were in the range 1.1–1.35 MPa. The maximum quoted force on the island, again calculated by superposition of plane strain elastic stress fields, was on 25 March 1977, when a force of 227 MN was calculated. This occurred in first-year ice of thickness 1.55 m acting over an effective island width of 130 m, and represents an average indentation pressure 1.1 MPa. Ice movement rates were typically in the range 5–30 cm per day.

5.4.1.6 Issungnak O-61

The stress measurement programme carried out during 1979 at Issungnak O-61, in 20 m water depth, was plagued by instrumental and operational problems and yielded no useful data (McGonigal, 1981).

5.4.1.7 Tarsiut N-44

Tarsiut island was installed during the summer of 1981 and was the first island structure to employ prefabricated caissons (in this case constructed of reinforced concrete) to penetrate the wave zone. Four independent caissons were used, forming an octagonal plan structure of diameter 100 m, and the space between the caissons was filled with sand to form a foundation for drilling equipment and accommodation. A photograph of the island in 1981–2 is shown in Fig. 5.9(a). Water depth was 22 m, and the ice conditions were borderline between landfast ice and transition zone conditions: in fact, it is likely that the island itself formed a pinning point for the landfast ice, allowing it to remain stable during the winter for a greater distance seaward than normal. The caissons themselves were extensively instrumented with a variety of load panels and internal strain gauges, and two major ice pressure measurement programmes were carried out: in 1981–2 (Blanchet, 1983) and in 1982–3 (Anderson, 1983; Sanderson *et al.*, 1983). Both these programmes are still confidential

As at Netserk island, Tarsiut island each year quickly became surrounded by an extensive annulus of grounded ice rubble, which effectively protected the structure itself from direct ice impact. This meant, especially in 1981–2, that interpretation of measured loads on the caissons was very difficult, since it was not clear how much of the ice force applied at the perimeter was being transmitted by the rubble pile to the sand berm (Fig. 5.13).

In 1982–3 an effort was made to resolve this ambiguity. An integrated project was designed to measure forces on the island caisson structure and pressures in the surrounding rubble pile (Anderson, 1983) and correlate these with forces exerted by first-year ice loaded onto the rubble pile perimeter (Sanderson *et al.*, 1983). This gave a direct measure of the effect of the rubble pile and the extent to which it afforded protection to the structure. Forces on the caissons were measured by a combination of internal strain gauging and interfacial Medof load panels installed on the caisson walls. The measurement of applied first-year ice loads was novel in that it employed high-resolution strainmeters instead of stress gauges, as described in Section 5.2.3. Stresses were calculated by applying the assumption of secondary creep and using laws of the form given in Sections 4.2.4. and 4.2.5.

As with the Netserk programme described

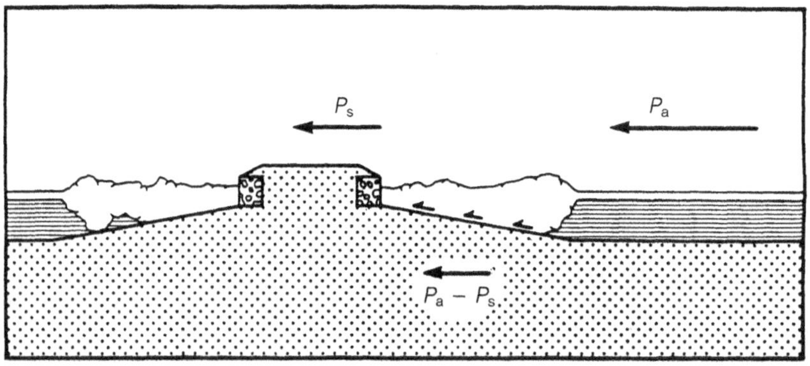

Fig. 5.13 Schematic diagram of Tarsiut island in 1982–3. Part of the applied load P_a is transmitted to the berm and sea bed by the rubble pile. The structure feels only the lesser force, P_s.

above, it was possible during a major loading event to compare detailed measurements of load (and strain-rate) with detailed ice movement rates. In addition, time-lapse photography showed the failure processes at the contact zone of the ice with the rubble pile.

Results from the programme are still proprietary at the time of writing, but a qualitative diagram of the dimensionless strain-rate field during a loading event is shown in Fig. 5.14 (after Sanderson, 1984a).

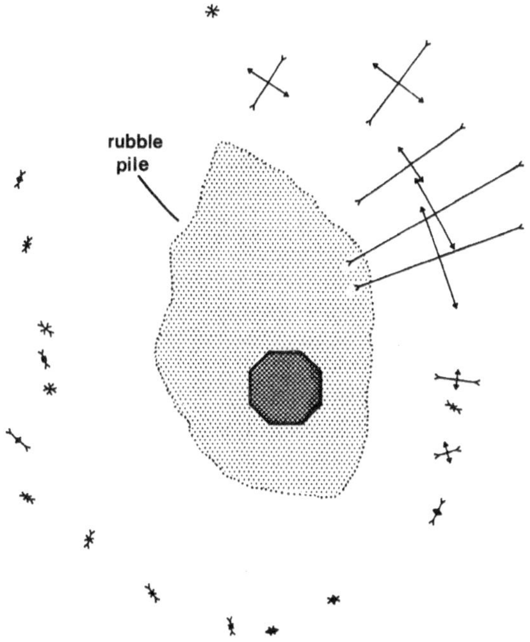

Fig. 5.14 Strain measurements around Tarsiut island in April 1983. Arrows show principal axes of strain-rate at the positions mapped. Data are from Sanderson *et al.* (1983, proprietary) with erased scales.

5.4.1.8 Adams Island, NWT

Adams Island is a small natural island some 200 m in diameter situated in Navy Board Inlet just south of Lancaster Sound. During winter the island is surrounded by landfast ice: predominantly first-year ice but containing inclusions of some 5% multi-year ice. A series of research programmes has been carried out on the island between 1982 and 1986 (Frederking *et al.*, 1983; 1986a; 1986b).

During the winter of 1985–6 a series of cylindrical biaxial stress gauges were installed around the island and enabled calculation of the total load exerted on the island. Maximum stresses measured were of the order of 0.3 MPa and the maximum total load calculated on the island was 64 MN, at the end of March 1986. Ice thickness was approximately 1.55 m acting over an effective contact width of 200 m. This amounts to an average indentation pressure of 0.21 MPa over a contact area of 310 m^2.

Ice movement was fairly systematic, showing a rate of about 8–12 cm per day throughout the winter, amounting to 15 m total motion. These movements were attributed to a tidal jacking process at the distant shoreline, and the ice at the island failed by creep buckling. No really severe loading events have been monitored at Adams Island.

5.4.1.9 Kadluk caisson-retained island

During the spring of 1984 biaxial cylindrical stress gauges were installed around the Esso caisson-retained island (CRI) at Kadluk, in 14.5 m water depth (Johnson *et al.*, 1985). The CRI consists of eight steel caissons connected together by steel cables under tension and placed on a man-made sand berm. The caissons form an octagonal ring 12.2 m high and 117 m in diameter. Further technical details are given in Mancini *et al.* (1983). At the time of the measurement programme the island was surrounded by a grounded ice rubble pile (Fig. 5.15), but level ice was present only towards the south of the island, following a major breakaway of ice cover from the north on 11 March 1984.

A total of 18 biaxial stress sensors were deployed, at six sites in level first-year ice to the south of the island (Fig. 5.15). An EPR flatjack panel was also installed for comparative purposes.

The ice during March–April 1984, at the time of installation, was well developed first-year ice of thickness 1.7 m and sensors were in general installed at depths approximately 30, 80 and 130 cm below the surface. Measured stresses were typically in the range 100–200 kPa, showing a clear diurnal thermal stress cycle of 50–100 kPa amplitude together with superimposed individual peaks of compressive stress up to a maximum of 500 kPa (0.5 MPa). The stress distribution vertically through the ice cover was found to be not at all uniform, and maximum stresses were typically measured at the uppermost stress sensor level in the ice. This is to be expected, since the upper layers of ice are generally stiffer than lower layers, owing to the lower temperatures and (typically)

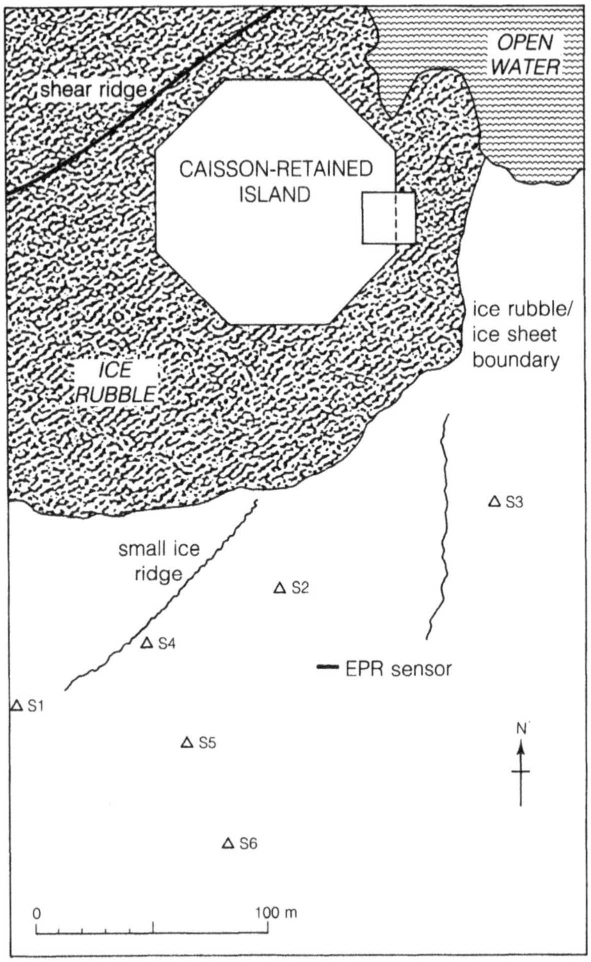

Fig. 5.15 Plan view of Kadluk caisson-retained island in April–May 1984, surrounded by ice rubble. The triangles show the position of stress gauges in May (after Johnson *et al.*, 1985).

granular rather than columnar crystallography. However, there was also some evidence of bending stress, which distorts the interpretation.

The maximum stress measured when averaged through the thickness of the ice cover was about 300 kPa and the maximum total load acting on the island/rubble pile system was calculated to be approximately 150 MN. This calculation was performed using a line integral approach on the available stress sensor network (Johnson, 1983).

Correlation of measured ice force vectors with measurements of ice temperature and wind speed and direction indicate that all the measured events were due to thermal expansion of the landfast ice between the island and shoreline (see Chapter 7, Section 7.2 for an analysis of thermal expansion). Maximum loads were correlated with a period of pronounced ice sheet warming and expansion during April. The ice was observed to fail principally by flexure, probably creep buckling (see Fig. 6.4), at the interface with the grounded rubble pile. The gross contact width with the rubble pile appears to be of the order of 300 m.

Comparison of the biaxial gauges with the EPR flatjack panel showed adequate agreement, with the EPR panel apparently showing rather higher stress readings than a biaxial sensor 130 m away.

5.4.1.10 Amerk O-09

The following season, during March to May 1985, further measurements were carried out around the CRI, in its new location at Amerk O-09 in 26 m of water (Sayed *et al.,* 1986). The caissons at Amerk rested on a sand berm built up to within 9 m of the sea surface level. Again, the island was surrounded by a grounded rubble pile which remained stable from February through to at least May. The sail-heights of the rubble were generally in the range 3.0–4.5 m above sea level and the distance from the island to the rubble pile perimeter was slightly over 100 m.

The objective of the research programme was to measure stresses within the rubble pile and determine the extent to which stresses applied at the perimeter were transmitted through to the structure itself. Stress sensors of various types, principally flat panels, were installed at several different positions in the rubble pile, and movements of the rubble pile were monitored by electronic distance meter (EDM).

During the measurement period no severe loading events appear to have occurred. Only one sensor, placed just 10 m inside from the rubble pile perimeter, gave any appreciable signals, of up to 200 kPa. These minor loading events occurred in the periods 1–3 April and 23–25 April and lasted for several days. As at Kadluk, the peak loads appear to be associated with episodes of pronounced thermal expansion, though the records also show some semi-diurnal cycling connected with tidal effects. No other sensors showed any appreciable stress records during the measurement period, and the rubble pile showed negligible movement (less than 1 cm). This indicates that stresses of up to about 200 kPa, and probably very much higher, can be sustained by the rubble pile and transmitted to the berm without any load reaching the caisson itself.

5.4.1.11 Hans Island

Hans Island is a roughly circular natural island with dimensions 1300×1700 m lying in the centre of Kennedy Channel between north-west Greenland and Ellesmere Island. Its position is indicated on Fig. 3.18. The island lies in an extremely exposed location. It is normally surrounded by stable landfast ice during the months between November and July each year, the ice consisting of about 5/10 to 8/10 thick multi-year ice. During late July and August, however, the ice conditions become highly mobile and the island is subjected to numerous impacts by large multi-year ice floes travelling southwestwards with the prevailing sea currents (Fig. 5.16).

The island has been used as the base for a series of highly successful joint industry research projects aimed at measuring forces exerted during full-scale multi-year floe impacts. These projects, operated by Dome/Canmar, were carried out in 1980, 1981 and 1983. Results from the 1980 and 1981 programmes are now publicly available (Danielewicz and Metge, 1981, 1982). The 1983 programme (Danielewicz and Cornett, 1984) remains restricted until 1989.

As already explained in Section 5.2.4, forces were calculated by means of measuring floe deceleration during impact; floe mass was estimated volumetrically and then Newton's second law was applied to derive total impact force as a function of time. Effects of added mass, Coriolis force and static driving forces of wind and current were all calculated to be negligible.

Floe impacts in general occur to the north-east perimeter of the island. On this side of the island an extensive permanent ice foot has formed, and is grounded on the sea bed (Fig. 5.17). Impacts occur not with the island itself but with the ice foot, which presents an effectively vertical face to impacting floes.

During both research programmes a series of auxiliary measurements and observations were also made. Ice failure processes and the growth of

Fig. 5.16 Aerial photograph of a multi-year floe impacting with the north-east edge of Hans Island (photograph courtesy of Dome Petroleum Ltd).

the contact zone were observed and monitored by means of time-lapse and manual photography, and measurements of temperature and salinity were made on cores taken from impacting floes.

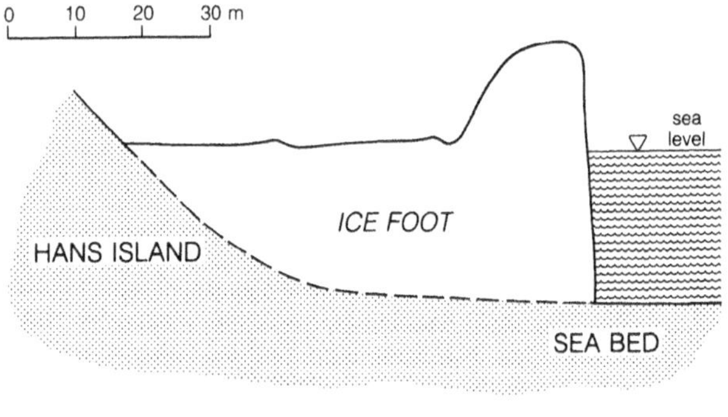

Fig. 5.17 Profile of the ice foot at the north-east perimeter of Hans Island, August 1981 (after Danielewicz and Metge, 1982).

Hans Island 1980. The first Hans Island programme, carried out in August 1980, was essentially a pilot programme, and floe decelerations were measured simply by successive manual theodolite readings from the island. Obviously, this was a technique of limited precision: the maximum practical rate of reading was of the order only one reading every 50 s, with the result that inferred velocities and accelerations are heavily filtered averages over a period of about 100 s. There is no doubt that because of this some significant peaks were missed, but the programme was nonetheless a breakthrough, in that it showed average multi-year ice forces to be radically lower than had been expected. The Hans Island measurements were the first ever made on high-velocity impact of thick ice. The earlier studies on artificial islands were limited generally to slow-moving first-year ice, and although the average stresses measured were generally, as we have seen, less than 1 MPa, there was always the fear that thicker, faster-moving ice might exert average stresses of 10–20 MPa, as might be inferred from small-scale measurements. The Hans Island measurements dispelled this fear.

Four impacts, referred to as events A–D, were measured in 1980, but of these only event A was measured with any great success.

Event A. On 15 August 1980 at about 21:00 hours a multi-year floe some 6 km in diameter impacted with Hans Island, travelling with initial velocity 0.39 m s^{-1}. The impact lasted 10 minutes, during which time the contact zone grew to a width of about 280 m. The geometry of the collision is shown in Fig. 5.18, including the positions A and B used for theodolite fixes from the island. The successive position fixes of A and B at 50 s intervals were used to calculate the trajectory followed by the centre of mass, which was then plotted as a function of time. This was then smoothed and twice differentiated with respect to time to yield velocity and deceleration.

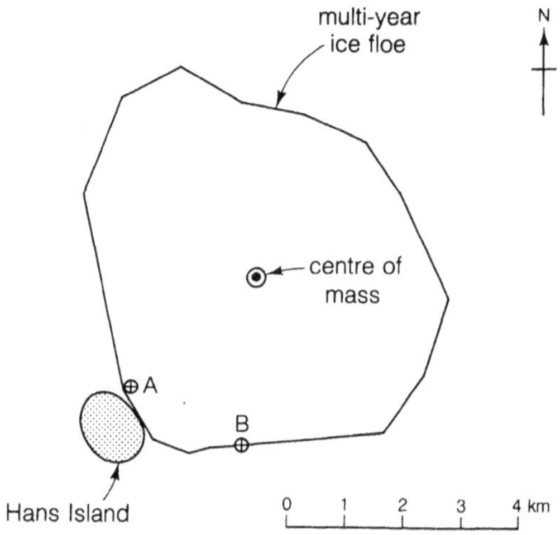

Fig. 5.18 Representation of the collision geometry of Event A, Hans Island 1980. Points A and B were used as targets for theodolite fixes (after Danielewicz and Metge, 1981).

The area of the floe was measured by theodolite fixes around its perimeter and was estimated to be 31 km^2. Thickness was determined on the basis of five auger hole measurements and surveys of freeboard, and was estimated to be an average of 5.6 m. Using an ice density of 0.9 Mg m^{-3}, the mass of the floe was calculated to be 1.6×10^{11} kg. The "effective mass", including an allowance for entrained water, was estimated to be about 1.7×10^{11} kg. The growth of the contact zone during impact was estimated by simulating the progressive penetration of the shape of the island perimeter with the floe.

Figure 5.19 shows the measured velocity, calculated total force and estimated growth of contact zone as a function of time after the first contact. The maximum force exerted by during the impact process was calculated to be 214 MN at a time 500

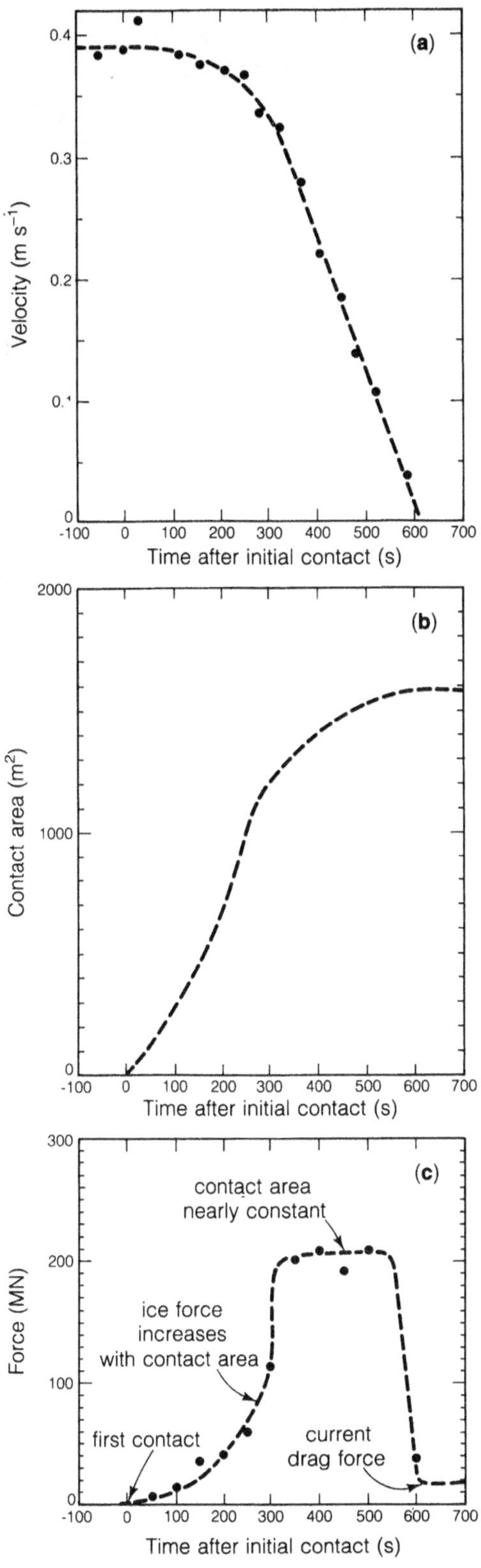

s after first contact. The contact area at this time was estimated to be 1520 m^2, yielding an average indentation pressure of 0.14 MPa. This is extraordinarily low, at least in the context of small-scale laboratory experiments, and came as a great surprise at the time. The failure mode was described as predominantly crushing, but was accompanied by spalling and flaking of massive blocks. A view of rubble at the contact zone after impact is shown in Fig. 5.20. As we have seen (Fig. 4.27) the stress associated with crushing in the laboratory is normally measured to be 100 times higher than this, in the range 10–20 MPa. A tabulation of the principal parameters during the impact is given in Table 5.1. The calculation of forces in this manner is subject to a number of errors and uncertainties, owing especially to the smoothing of peak decelerations. The expected error is probably in the region of 50–100%.

The ice of the floe was cored to 3 m below surface and characterised for crystallography, temperature and salinity. The ice was found to be predominantly granular with crystal size 0.5–3 mm. Salinity was in the range 1–2 ppt and temperature averaged about −3°C.

Events B and C. Events B and C both occurred on 16 August, at 06:10 and 09:25 respectively. These events involved the same multi-year floe as in event A: after impact on 15 August, the floe had drifted to the south-east of the island, but overnight reversed direction in response to tidal currents and returned to impact twice more with the island.

No direct measurements of these impacts were made, but velocities and decelerations were estimated by photogrammetric techniques from 8 mm time-lapse photography. This technique is very approximate and yielded estimates of mean force only. During Event B the floe impacted at initial velocity 0.26 m s^{-1} and mean force was estimated to be 48 MN, acting over a contact area of 930 m^2. Mean indentation pressure was apparently about 50 kPa. During Event C the floe impacted at initial velocity 0.11 m s^{-1} and mean force was estimated to be 18 MN over a contact area of 400 m^2. Mean indentation pressure was again about 50 kPa. These estimates are subject to errors of at least 100%.

Fig. 5.19 Impact parameters calculated during Event A at Hans Island 1980 (after Danielewicz and Metge, 1981): (a) velocity; (b) contact area; (c) total ice force.

Fig. 5.20 Failed ice at the contact zone after Event A at Hans Island 1980. Block size varied from a few centimetres to about 6 m. Thickness of the floe before failure averaged 5.5 m (photograph courtesy of Dome Petroleum Ltd).

TABLE 5.1 Summary of results from the 1980 Hans Island programme

Event	*Floe mass, m (kg)*	*Initial velocity, U_0 ($m\,s^{-1}$)*	*Average thickness, $\bar{h}$ (m)*	*Time elapsed, t (s)*	*Total force, P (MN)*	*Contact width, D (m)*	*Contact area, $D\bar{h}$ (m^2)*	*Aspect ratio, $D/\bar{h}$*	*Indentation pressure, $P/D\bar{h}$ (kPa)*
A (15 August 1980)	1.6×10^{11}	0.39	5.6	100	13	48	270	9	46
				150	24	80	450	14	53
				200	39	120	670	21	58
				250	45	177	990	32	46
				300	108	214	1200	38	90
				350	199	234	1310	42	152
				400	205	250	1400	45	146
				450	203	263	1470	47	138
				500	214	271	1520	49	141
				550	206	279	1560	50	132
B (16 August 1980)	1.6×10^{11}	0.26	5.6	mean only	48	166	930	30	52
C (16 August 1980)	1.6×10^{11}	0.11	5.6	mean only	18	71	400	13	45
D (9 August 1980)	8.7×10^{9}	0.28	1.0	mean only	8	800	800	800	10

Event D. On 9 August, early on in the programme, a large floe of area about 10 km^2 consisting entirely of highly deteriorated first-year ice about 1 m thick was monitored during impact with the island. Again, velocity was estimated photogrammetrically and the floe apparently decelerated from an initial velocity of 0.28 m s^{-1}, coming to rest in a period of 450 s. Floe mass was estimated to be 8.7×10^9 kg and average force was calculated to be 7.7 MN over a contact area of 800 m^2. This indicates an average indentation pressure of about 10 kPa.

Hans Island 1981. The second Hans Island programme, carried out in July–August 1981, used the same principles as the 1980 programme, but with major improvements in measurement techniques. Biaxial horizontal accelerometers sampling at 3 Hz were used to measure floe decelerations directly, and floe movements were monitored by electronic distance meter (EDM) instead of by simple theodolite intersections. This led to a great improvement in data quality, particularly in resolution of acceleration peaks: peaks could now be resolved at the level of about 1 Hz. As in 1980, time-lapse photography and visual observation were used to record impact geometry.

Four collisions were analysed, including a major head-on impact on 4 August, which provided the most satisfactorily monitored multi-year impact of the two programmes. The other impacts also yielded useful measurements of force, but involved more complex impact geometries and are less easy to interpret.

Event 1, 29 July. Several collisions occurred on 29 July, but only one, at 18:45, was monitored, since the camp was not at this time fully established. The collision involved two floes (Fig. 5.21): a small floe of diameter about 200 m, sandwiched against the island by a much larger floe of diameter about 2 km travelling at an initial velocity of 0.51 m s^{-1}. Both floes were estimated to be of 5 m average thickness. Only the larger floe was instrumented, with a single biaxial accelerometer, and decelerations of up to 5×10^{-3} m s^{-2} were measured during an impact process which lasted about 5 minutes. The area of the floe was estimated to be 3.5 km^2 and its mass calculated to be 1.6×10^{10} kg. The record of total force as a function of time, calculated from the mass and total acceleration records, is shown in Fig. 5.22, annotated to indicate the principal failure events.

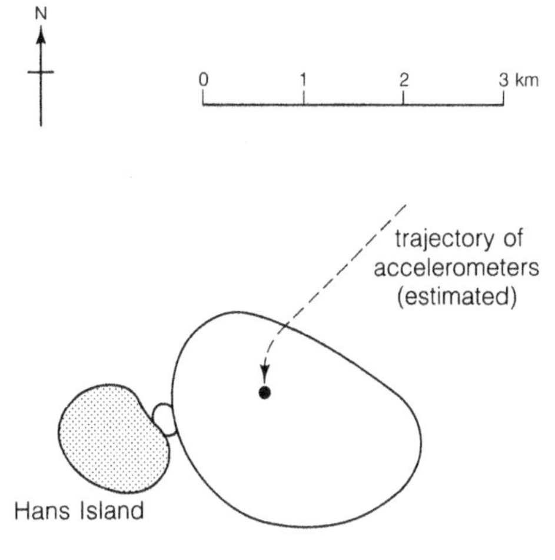

Fig. 5.21 Schematic view of Event 1 at Hans Island, 29 July 1981 (after Danielewicz and Metge, 1982).

The ice force increases gradually between elapsed times of 50 s and 70 s as the smaller floe is compressed between the large floe and the island. Between 70 s and 100 s the smaller floe fails completely and rides up on the larger floe. Following failure of the smaller floe, the contact width of the larger floe grows steadily from 100 m at 140

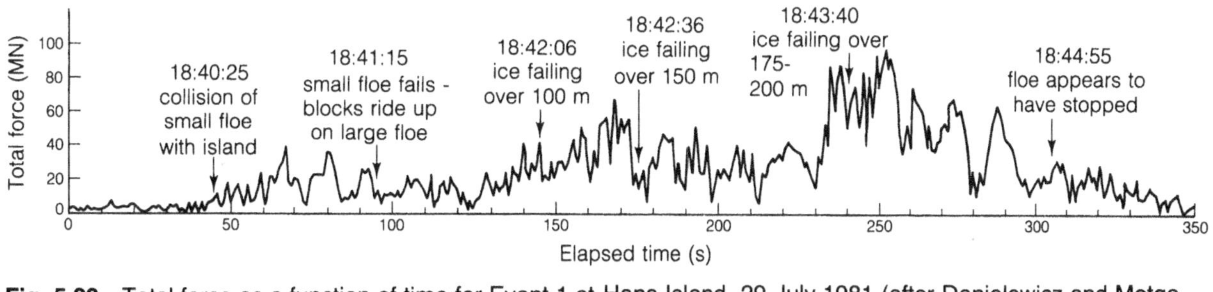

Fig. 5.22 Total force as a function of time for Event 1 at Hans Island, 29 July 1981 (after Danielewicz and Metge, 1982).

TABLE 5.2 Summary of results from the 1981 Hans Island programme

Events	Floe mass, m (kg)	Initial velocity, U_0 ($m\,s^{-1}$)	Average thickness, $\bar{h}$ (m)	Time elapsed, t (s)	Total force, P (MN)	Contact width, D (m)	Contact area, $D\bar{h}$ (m^2)	Aspect ratio, $D/\bar{h}$	Indentation pressure, $P/D\bar{h}$ (kPa)
1	1.6×10^{10}	0.51	5.0	67	38	25	125	5	300
(29 July 1981)				80	35	40	200	8	180
				140	40	104	520	21	77
				145	40	109	545	22	73
				168	65	135	675	27	96
				175	22	142	710	28	31
				235	86	187	935	37	92
				238	87	189	945	38	92
				249	92	193	965	39	95
				252	95	194	970	39	98
				261	74	196	980	39	76
				273	67	199	995	40	67
				287	63	200	1000	40	63
2	4.1×10^{10}	0.33	8.0	41	52	41	328	5	159
(4 August 1981)				47	127	57	456	7	279
				51	231	78	624	10	370
				53	269	84	672	11	400
				57	201	93	744	12	270
				66	228	98	784	12	291
				70	245	98	784	12	313
				78	264	98	784	12	337
				84	266	96	768	12	346
3A (7 August 1981)	1.2×10^{9}	0.32	6.0	54	26	30	180	5	140
3B (7 August 1981)	7.7×10^{9}	0.32	6.0	131	31	100	600	17	52

s to 150 m at 170 s and to an eventual maximum of about 200 m at elapsed time of 250 s. The maximum total force exerted on the island was 95 MN, at which time the contact area was estimated to be 970 m^2. This corresponds to an average indentation pressure of 100 kPa. Earlier on in the impact process, while the failure of the smaller floe was occurring over a more limited width of only 25–40 m, the calculated indentation pressures are higher: up to 300 kPa over an apparent contact area of about 125 m^2. The contact area is, however, subject to great uncertainties, so this calculation is not very reliable. The principal results are summarised in Table 5.2.

The force record is highly variable, and obviously not the result of continuum creep interaction processes. Rather, it is indicative of a large number of successive rapid fracture events contributing to the total force: the average force during the impact is of the order 40–50 MN but it shows repeated deviations of ±30 MN about this mean level, with an irregular period of the order about 3 s. The root mean square (r.m.s.) deviation of the force record about a smoothed 5 s average was calculated to lie in the range 10–30 MN.

Event 2, 4 August. The event of 4 August 1981 involved impact by a single floe of area 5.7 km^2 and thickness 8 m travelling at an initial velocity of 0.33 m s^{-1}. Its mass was calculated to be 4.1×10^{10} kg. The geometry of the impact is shown in Fig. 5.23. Two accelerometer units were deployed and the resulting calculation of total force is shown in Fig. 5.24.

This impact is remarkable because of the extremely rapid build-up of force. Initial contact

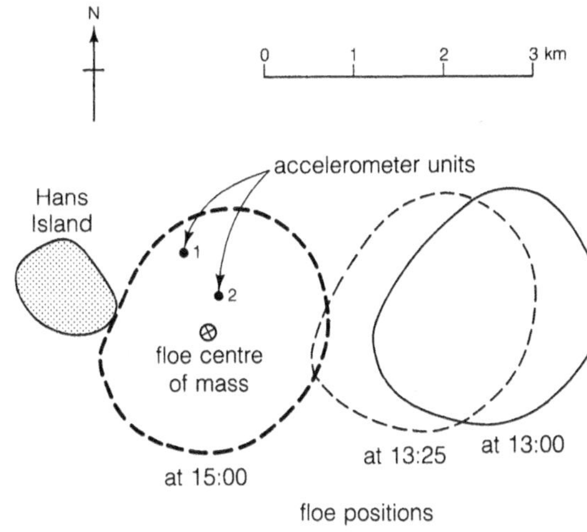

Fig. 5.23 Geometry of Event 2, 4 August 1981 at Hans Island (after Danielewicz and Metge, 1982).

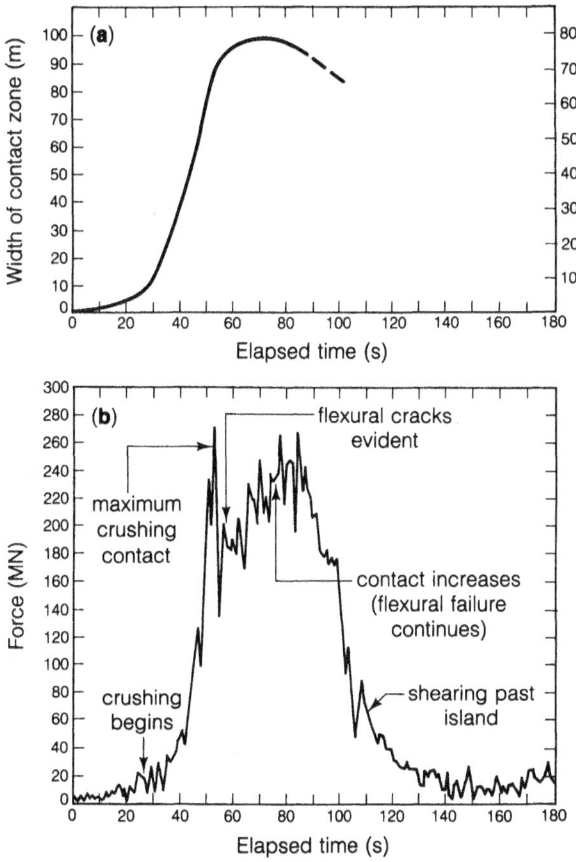

Fig. 5.24 Event 2, Hans Island 1981: (a) estimated size of contact zone (width and area): (b) one-second averages of force exerted on the island (after Danielewicz and Metge, 1982).

occurred at an elapsed time of about 40 s, as minor fragments of ice trapped between the floe and the island began to fail. The force then rapidly increased over an interval of 10–15 s as the floe itself made contact, reaching a peak of 269 MN at an elapsed time of 53 s. During this period the failure mode was described as crushing, but from elapsed time 55 s onwards flexural failure was evident and the force record is again highly variable. Initial flexural failure appeared to involve the leading edge of the floe being forced downwards, possibly under the influence of the weight of crushed rubble resting upon it. The estimated width and area of the contact zone is shown in Fig. 5.24(a). Maximum inferred indentation pressures were of the order of 300–400 kPa and are summarised in Table 5.2.

Events 3A and 3B, 7 August. On 7 August at about 01:10 hours two successive impacts occurred, involving two separate floes. Both floes were instrumented, with one accelerometer on each, and the general impact geometry is illustrated in Fig. 5.25. The first floe, which we shall call floe A, was of area just 0.22 km^2 and thickness about 6 m, with a calculated mass of 1.2×10^9 kg. The second floe, floe B, was of area about 1.4 km^2, also of thickness about 6 m, and had a calculated mass of 7.7×10^9 kg. Both floes were travelling at an initial velocity of 0.32 m s^{-1}. The calculated total force records are shown in Fig. 5.26, compiled from the separate accelerometers on the two floes. Floe

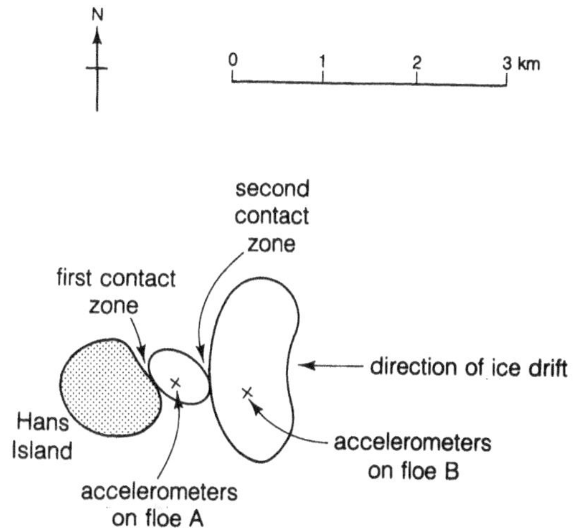

Fig. 5.25 Schematic diagram of the collision of two multi-year floes, Events 3A and 3B, at Hans Island 1981 (after Danielewicz and Metge, 1982).

A began to impact at an elapsed time of about 20 s and apparently made good clean contact with the island, its contact width growing to a maximum of just 30 m. The peak force, at an elapsed time of about 54 s, was 26 MN and it then abruptly dropped to zero as the floe came to rest. The contact area at this time was about 180 m^2 and the peak indentation pressure was therefore about 140 kPa.

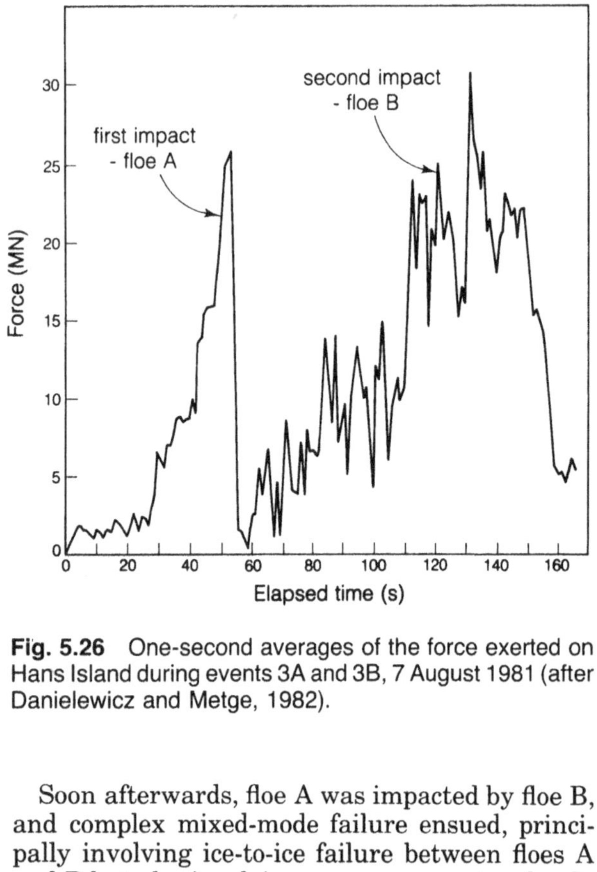

Fig. 5.26 One-second averages of the force exerted on Hans Island during events 3A and 3B, 7 August 1981 (after Danielewicz and Metge, 1982).

Soon afterwards, floe A was impacted by floe B, and complex mixed-mode failure ensued, principally involving ice-to-ice failure between floes A and B but also involving rearrangement and rafting of individual ice fragments trapped between the two floes. The maximum force between the two floes was 31 MN and the maximum contact width was estimated to be 100 m. This corresponds to a contact area of about 600 m^2 and an average indentation pressure of about 50 kPa. This low level of pressure is indicative of the average stress required to induce ridging between two multi-year floes over a contact area of 500–1000 m^2.

The results for all the impacts during the 1981 Hans Island programme are summarised in Table 5.2.

5.4.1.12 Molikpaq

The Molikpaq structure, also known as the Mobile Arctic Caisson (MAC), was constructed for Gulf Canada Resources Inc. in 1983–4 and is designed for use in Beaufort Sea water depths of 21–45 m. The structure consists of an integral steel caisson of diameter about 90 m, containing a partially drained sand-fill core and placed on a sand berm. The structure derives its resistance to horizontal loading from the weight of the sand-fill core. However, because the core is normally only partially drained and compacted it can behave as a contractive, not a dilatant sand: that is, under applied shear stress its pore pressure may increase, which can lead to partial liquefaction and instability. A plan and section of the structure are shown in Fig. 5.27: it is square in plan, with truncated corners, and presents a projected width of 89–101 m, depending on angle of impact. It has a nominal ice design load of 500 MN and the structure is designed with a safety factor of 1.4, to a total resistance of about 700 MN.

It has been deployed twice in the Canadian Beaufort Sea: at Tarsiut P-45 in 25 m water depth and at Amauligak I-65 in 32 m of water. Both these locations lie in the mobile transition zone of the Beaufort Sea and during the winter the structure has been subjected to repeated interactions with both first-year and multi-year ice of a wide variety of velocity and thickness. The structure is comprehensively instrumented with a range of different sensors designed to measure applied ice load and structural response. The most important of these are:

(a) *Medof ice panels.* 30 Medof load panels are installed at and below the waterline on the north, north-east and east external faces of the steel caissons. These measure applied ice loads directly but show limited response to loads varying on a timescale less than a few seconds. The panels are each of width 1.13 m and height 2.7 m and are deployed in seven groups of four or five sensors. Each group measures the stress over an area of about 12–15 m^2. Together, they typically cover about 10% of the ice contact area.
(b) *Strain gauges.* More than 250 strain gauges are installed in the north, north-east and east segments of the caisson annulus. They measure structural steel strain, which is calibrated by theoretical means to stress level.
(c) *Extensometers.* Each caisson is instrumented at its midpoint with a rod extensometer,

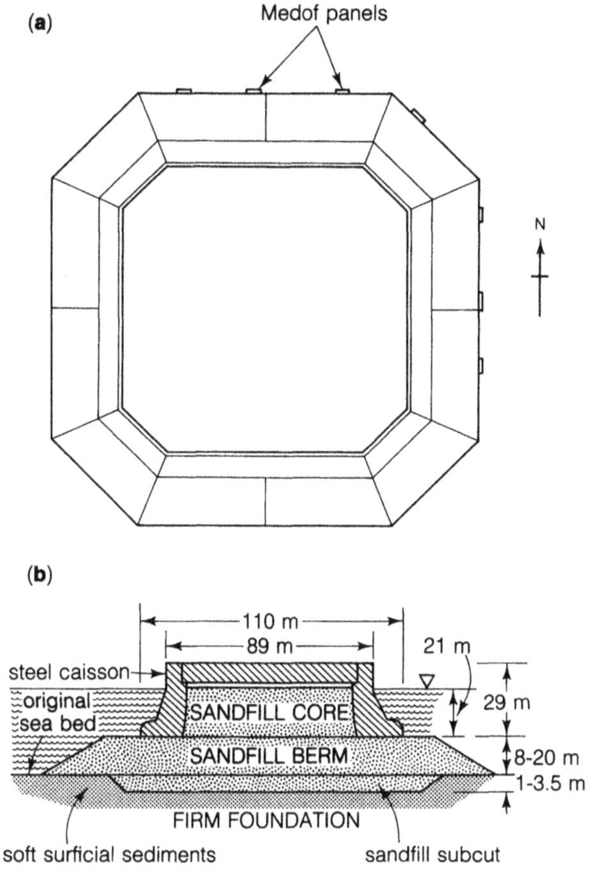

Fig. 5.27 The Molikpaq structure as deployed at Tarsiut P-45 and Amauligak I-65: (a) plan; (b) section (after Gulf, 1986, unpubl.).

measuring caisson deformation relative to the box girder deck. These provide information about the magnitude and direction of the global load applied to the caisson.

Platform response was monitored under both static and dynamic loading using inclinometers, extensometers, settlement gauges and accelerometers. Sand core fatigue was also monitored using fast-response stabilised strain-gauge piezometers. 512 channels of data were recorded on a 24-hour basis, with three sampling frequency modes, depending on the level of loading activity: 3 minute average, maximum and minimum; fast (1 Hz); and burst (50 Hz). For further details of the instrumentation, construction and deployment of this hybrid steel and sand caisson structure, refer to Bruce and Harrington (1982), Jefferies *et al.* (1985), and Rogers *et al.* (1986).

Here I shall give a summary account of the information currently available on the ice loads measured during the seasons 1984–6. Not all the information presented here has been formally published: information which has been published (principally Rogers *et al.*, 1986) is supplemented by reference to data contained in Gulf's 1986 Proposal for analysis of data from Molikpaq at Amauligak I-65 (Gulf, 1986, unpubl.), and by reference to public presentations made by R. Pilkington at the IAHR Symposium in Iowa City, August 1986. I have also relied on personal conversations and communications from Gulf personnel.

The data are remarkable and important because they are the only data available showing the stresses involved during severe impact of ice with a real "naked" structure – that is, one not protected by a rubble pile. Also, on several occasions the loads approached and even exceeded the nominal ice design load of the structure. The structure displayed, in addition, an interesting and serious dynamic response to ice loading, and on one occasion in 1986 was judged to have been within one minute of catastrophic instability.

Because the data are so fundamental to our understanding, but are not yet formally published, I have had to rely more than I would have liked on personal communications and recollections. Naturally the responsibility for any errors or misinterpretations in what follows rests entirely with me.

Tarsiut P-45. The Tarsiut P-45 location lies some 10 km to the west of Tarsiut N-44 (Fig. 5.7), in 25 m water depth. The Molikpaq structure was set down in October 1984 on a sand berm of 5 m elevation above the sea bed. The location lies in transition zone ice conditions, and during the season 1984–5 was subjected to predominantly first-year ice loading events from mobile ice of thickness in the range 0.4–1.8 m moving at velocities up to 0.4 m s^{-1} (0.75 knots). During November 1984 the structure was also impacted by several fragments of deteriorated second-year ice of thickness 2–2.4 m.

Ice loads were estimated principally from the arrays of Medof panels: the pressures measured in each group of panels were used to evaluate the local ice pressures over vertical strips of width 1.3 m (one panel width) and the results for all such strips (six for the north and east faces) were averaged to calculate an average pressure on each face. This average pressure was then multiplied by the apparent contact area of ice with each face. It seems to me that this method may potentially lead to a serious over-estimate of peak loads, since the average pressure calculated is derived only from 10% of the gross contact area. As we shall see in our overall analysis of indentation stresses as a function of contact area (Sections 5.6 and Chapter 6, Section 6.4.5), peak loads tend to scale with $1/\sqrt{A}$, where A is gross contact area, Dh. This means that measurements based on only 10% of the contact area may be too high by a factor of $\sqrt{10}$, or about 3, when extrapolated to the entire contact area. However, retrospective cross-calibration of loads calculated from load panels was performed with respect to loads calculated from extensometers and strain gauges, and apparently showed satisfactory agreement. Unfortunately, internal structural strain gauges and extensometers may also be registering unrealistically high load levels, since they are measuring structural response, not applied load. As we have seen (Sections 5.2.2 and 5.3), if the structure shows appreciable dynamic excitation under the influence of applied load then the loads inferred from structural response may well be high by a factor of about two.

With these reservations, however, let us examine the loads calculated and accept them as being, at any rate, upper bounds to the actual forces applied.

For the 1984–5 winter season a number of periods of high ice loading in unusual ice conditions were selected by Rogers *et al.* (1986) as "events" for further analysis. A total of 62 such events were analysed, and a summary of the maximum loads during these events is given in Fig. 5.28. Loads measured during events were typically of magnitude 50–150 MN, and a maximum load of 207 MN was recorded during February. Unfortunately, detailed information on the ice thicknesses associated with each loading event is not publicly available for all these events. However, information for some of the principal events is presented in Table 5.3 (B. D. Wright, pers. comm., 1986), and these provide a representative band of peak pressures measured. Ice conditions in general were reported to be normal or somewhat light, and it is likely that ice thickness for any event at a particular time of year was less than or equal to the monthly average first-year ice thickness as given in, for instance, Fig. 3.11 for the Beaufort Sea. In subsequent broad statistical analyses (Chapter 7, Section 7.6.2). I shall make the assumption that, unless otherwise stated, the ice thickness was of average seasonal thickness and that it acted over a contact width of an average of 95 m.

The maximum loading event of late February was a load of 207 MN caused by unusually thick

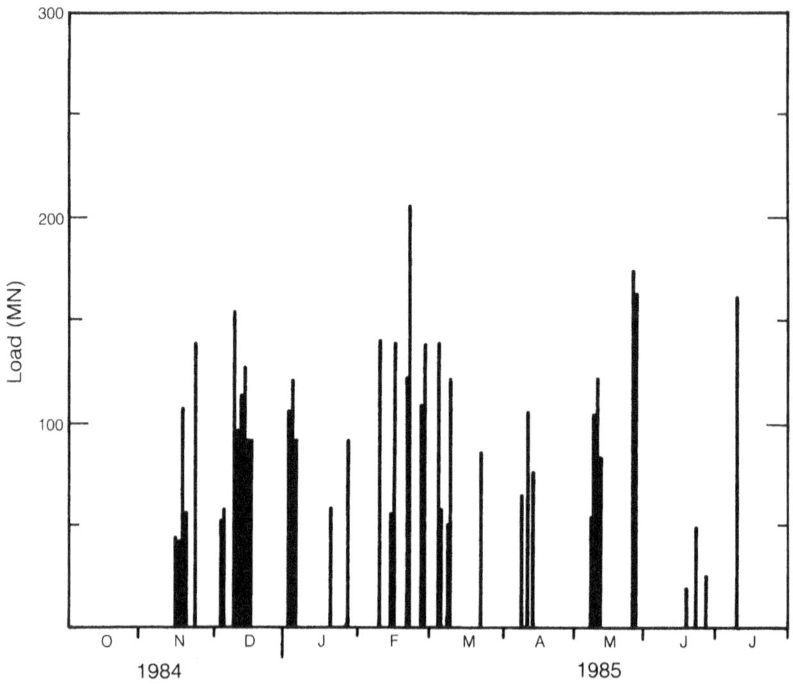

Fig. 5.28 Maximum ice loads on the Molikpaq at Tarsiut P-45 in 1984–5 (after Rogers *et al.*, 1986, redrawn). Note that a subsequent, and rather different, analysis of these data has been presented in Wright *et al.* (1986).

TABLE 5.3 Maximum loading events at Tarsiut P-45, 1984–5. Contact width is assumed to be 95 m (B. D. Wright, pers. comm., 1986)

Date	*Load (MN)*	*Ice thickness (m)*	*Gross contact area (m^2)*	*Average pressure (MPa)*
10 Dec. 1984	154	1.60	152	1.0
15 Dec. 1984	86	0.67	64	1.3
26 Jan. 1985	55	0.59	56	1.0
21 Feb. 1985	207	2.09	199	1.0
26 May 1985	165	0.90	86	1.9

ice of thickness 2.09 m. This implies an average indentation pressure of 1.0 MPa acting over a contact area of 200 m^2. Analysis of the loading events showed that the highest loads were correlated not with highest ice velocities, but with ice velocities in the range 2.5–10 cm s^{-1} (0.05–0.2 knots). At speeds in this range the ice loads showed regular pulsations at a frequency of 0.5–4 Hz, and measured ice pressures were up to twice as high as at lower speeds. It seems clear to me that this is due to resonant excitation of the structure and is not an effect caused by the ice failure itself: that is, if the structure had been perfectly rigid and unresponsive, I have little doubt that the measured loads would have been rather lower, for identical ice conditions. This resonance is an effect which we have already discussed with reference to the Cook Inlet platforms, and there is no reason to suppose that ice by itself can display regular oscillatory failure stresses during interaction with a structure. Ice failure stress may, as was seen at Hans Island, show peaks at an average, irregular period of 2–3 s (0.3–0.5 Hz), and it is likely that this stochastic application of varying stress results in the resonant excitation of the structure.

It is not, however, quite clear to me whether these high measured applied loads should be regarded as "real". If they are reliably measured by Medof panels as well as structural response they may indeed be real, though even then there are doubts about effects of averaging over the whole structure area. If they are inferred solely from structural response then the inferred loads are likely to be high by a factor of about two owing to effects of stored resonant energy. Here, I shall treat the data as real, but they need to be handled with care.

Amauligak I-65. Following the 1984–5 season at Tarsiut P-45 the Molikpaq was moved to the Amauligak I-65 site and set down in 32 m of water during the autumn of 1985. The ice conditions in this location were, as had been anticipated, more severe than at Tarsiut, and the structure was subjected to extreme loading from both multi-year ice and first-year ice. During the 1985–6 season the structure experienced approximately 18 first-year ice events with loading exceeding 100 MN, applied by ice of thickness in the range 0.7–1.8 m. In general, measured first-year ice loads were in the range 100–150 MN, and dynamic response of the structure was again sometimes observed. The predominant failure modes were crushing and flexural failure. At least two events involving second-year ice about 3 m thick were reported, in November, with loads measured in the range approximately 200–295 MN. More importantly though, a series of four severe multi-year ice loading events were observed in the period 7 March–12 April 1986, following a major invasion of multi-year ice into the area. It is worth looking at each event in detail.

(a) *7 March 1986*. On 7 March at 15:30 a multi-year floe of dimensions 200×300 m and general thickness 5–6 m approached the structure at a velocity of up to 5 cm s^{-1} (0.1 knots) from the north-west. The impact geometry is shown schematically in Fig. 5.29, and the measured force record is shown in Fig. 5.30. The floe appeared to consist of two heavily ridged multi-year floes welded together by first-year ice. During the initial stages of penetration between 15:30 and 16:00 the measured loads were an average of about 100 MN, but showing dramatic oscillations of ±80 MN at a frequency of about 1 Hz. At about 16:00 the structure interacted with a multi-year ridge of some 2 m freeboard and 40 m long. Assuming a keel-to-sail ratio of 3.3 (see Section 2.3.4.2) it is

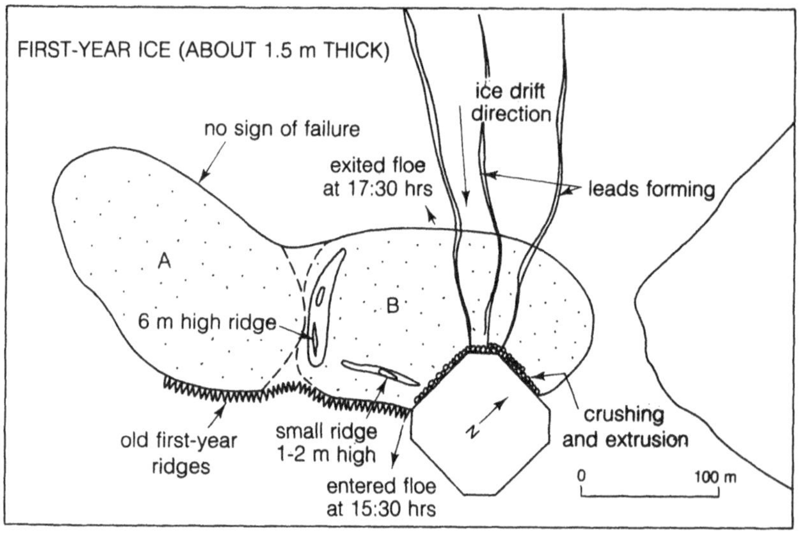

Fig. 5.29 Impact geometry of 7 March event, Amauligak I-65 (after Gulf, 1986, unpubl.). The floe appeared to be two heavily ridged multi-year floes, A and B, welded together by younger ice.

likely that this feature was of about 8–10 m thickness, probably interacting with about half the total available structure width. As this ridge interacted with the structure the load rose to an average of about 210 MN, with a dynamic component of ±60 MN. The maximum load measured was 270 MN at about 16:07. The gross projected contact width at this time was 100 m, with complete contact over the west, north-west and north faces, and it is likely that the average thickness, taking into account the presence of ridges, was in the

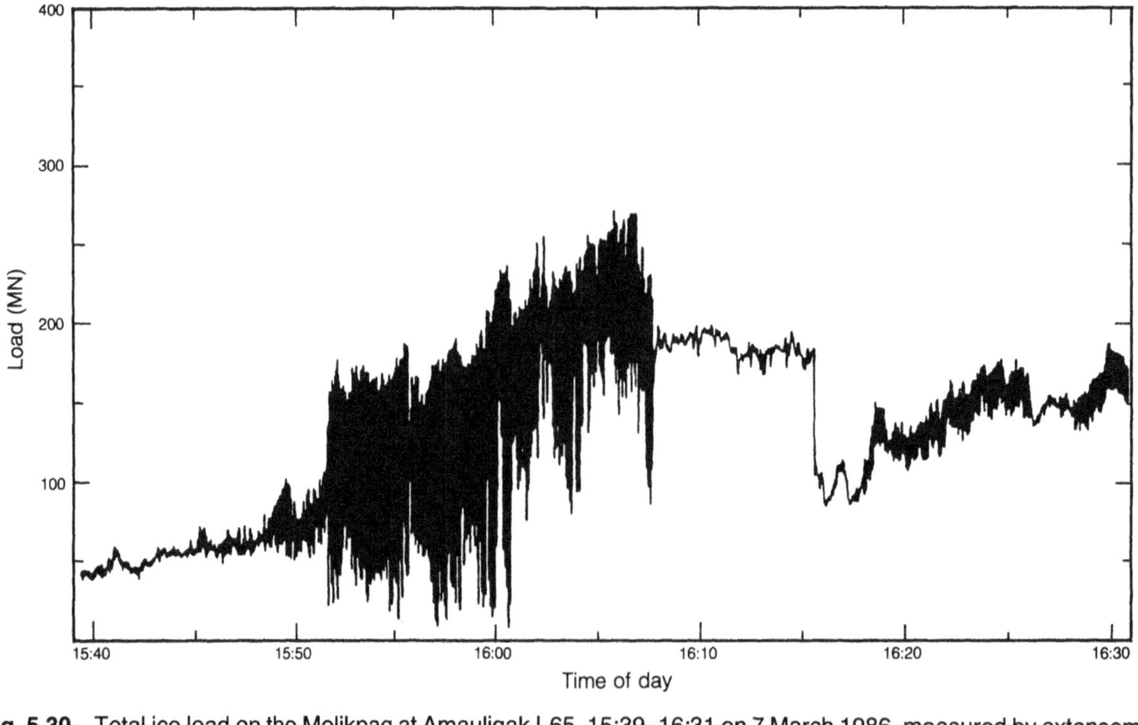

Fig. 5.30 Total ice load on the Molikpaq at Amauligak I-65, 15:39–16:31 on 7 March 1986, measured by extensometer (after Gulf, 1986, unpubl.).

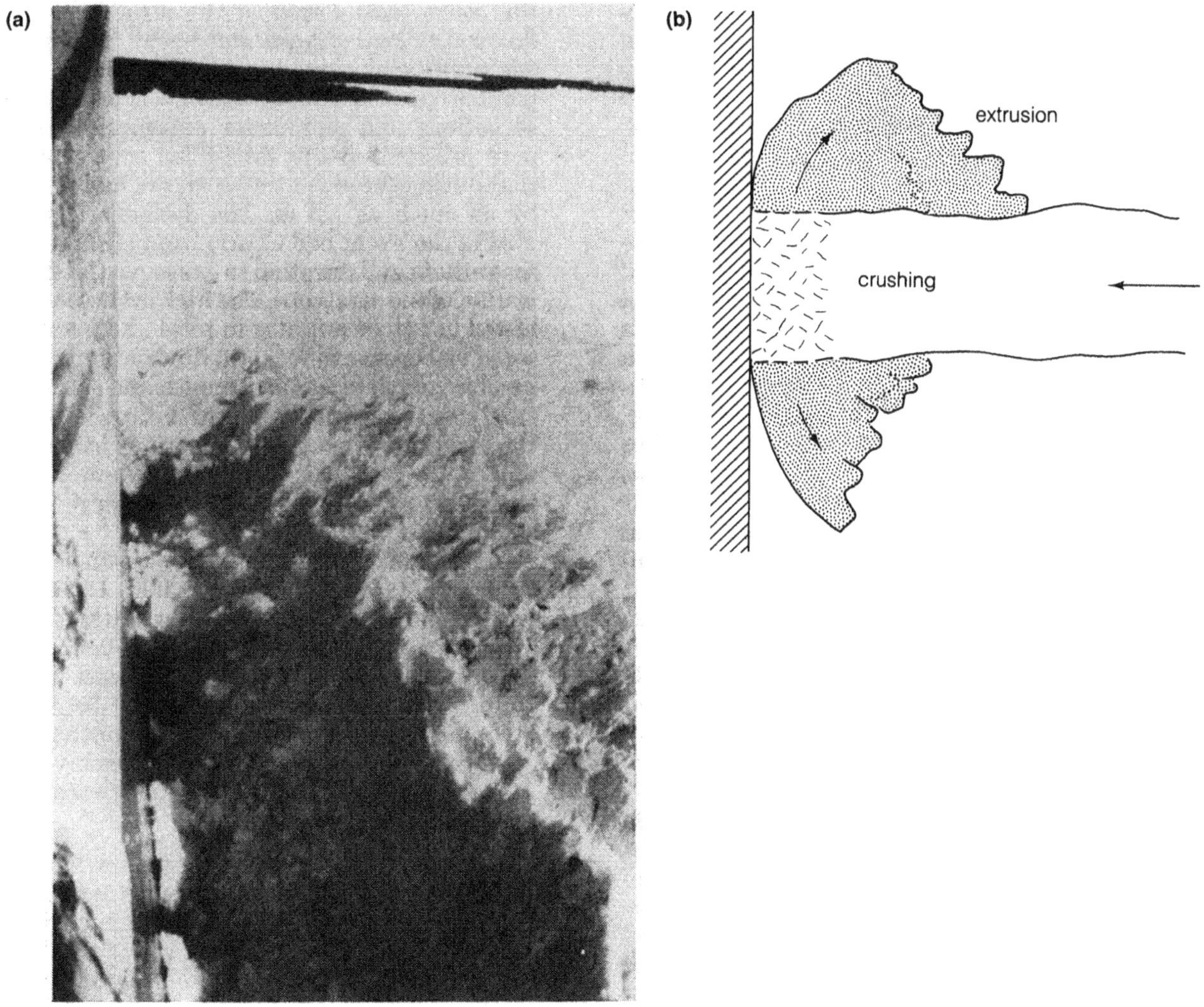

Fig. 5.31 (a) Photograph of multi-year ice crushing against the Molikpaq at Amauligak I-65 (photograph courtesy of Gulf Canada Resources Inc.). (b) Schematic diagram of multi-year ice failure. Ice is apparently crushed and pulverised in contact with the structure and then extruded upwards and downwards as a fine powder.

order of 6–7 m. This corresponds to an average indentation pressure of maximum 0.4 MPa over an area of about 650 m^2. Ice failed by crushing and is described as being totally pulverised and extruded upwards (and presumably downwards) from the point of contact. A schematic diagram of this process is shown in Fig. 5.31. Ice that has failed in this way has been described as behaving much like toothpaste.

(b) *8 March 1986*. Less detailed information is available for the event of 8 March. A multi-year floe of dimensions 1×2 km approached very slowly from the north-west and was indented slowly to a maximum penetration of 75 m. The maximum load experienced was 300 MN, with a dynamic component of 26%. (Dynamic component is here defined as peak load minus maximum load, divided by mean load.) This corresponds to an average load of the order of 265 MN with variation of ±35 MN. The thickness of the ice has not yet been published, but was probably in the range 4–8 m and presumably made full contact over a projected width of about 100 m.

(c) *12 April 1986, Event A*. Following the events of 7–8 March, the multi-year ice remained effectively stationary near to the Molikpaq structure for more than a month. The only action during this period was a slow creeping movement of ice 2.75 m thick, which failed by creep buckling. The ice close to the structure on 12 April was multi-year ice of average thickness about 4 m, but containing a major multi-year ridge of thickness 9–12 m, length in excess of 70 m, and width about 30 m. This ridge lay only about 50 m away from the east face of the caisson structure and was well aligned parallel to it. At about 08:00 the floe began moving slowly towards the structure from a direction about 280° at a speed rising to 5 cm s^{-1} (0.1 knots). The leading edge of the multi-year floe began crushing against the east, north-east and south-east faces. The ice was completely pulverised into a fine granular powder and was extruded upwards (as in Fig. 5.31) to a height of about 8-10 m against the east caisson. Just before 08:30 the ridge made contact with the structure and began crushing. It appeared eventually to contact three faces: the east, north-east and south-east, though it is not clear without submarine profiles exactly how much of the ridge was truly in contact. Loads measured by internal strain gauges at this moment showed peaks of about 500 MN with a strong dynamic component of 54%. This corresponds to an average load of about 390 MN with excursions of 110 MN oscillating at a frequency 0.8–1.4 Hz. During the period of peak loading the ice in contact with the structure was probably of average thickness as much as 10 m over a contact width of about 90 m. It is possible, however, that the thickness was no more than 4–5 m (the surrounding ice thickness), since observations are naturally limited. At about 08:32 radial cracks suddenly formed from the north-east (and presumably the south-east) corner of the structure, and flexural failure of level multi-year ice behind the multi-year ridge began. The load at this point dropped to substantially lower levels. Observations and piezometer measurements of pore pressure in the sand-filled core showed that during the event the sand core had settled by as much as 1.3 m. The dynamic loading during the event had clearly been of sufficient magnitude and duration to cause partial fluidisation of the sand core. The high cyclic loading lasted about 14 minutes in total and involved some 900 cycles. Since total fluidisation would involve complete loss of strength and since the sand core carries about 90% of the applied load, the structure was clearly in serious jeopardy, and continued application of the dynamic load would probably have resulted in large-scale sliding.

(d) *12 April 1986, Event B.* A further loading event occurred later on 12 April at 13:00–13:51, involving the same floe as on 7 March (thickness 5–6 m) moving at 15–20 cm s^{-1} (0.3–0.4 knots) towards the south-east, south and east faces. Quite steady loads were observed, of about 210 MN, and the lack of dynamic component was apparently due to the higher approach velocity, which did not excite a resonant response from the structure.

Finally, a further event is reported on 12 May 1986, when a vast floe of first-year ice 1.8 m thide containing a multi-year ice inclusion impacted the structure and generated a load of about 250 MN. However, it is not clear from the information available whether this load was associated with first-year or multi-year ice. A summary of all the loading events I know of at Amauligak I-65 is given

TABLE 5.4 Principal ice loading events on Molikpaq at Amauligak I-65 1985–6. Data are from Gulf (1986, unpubl.) and R. Pilkington (1986 IAHR Symposium lecture).
Doubtful or speculative data are enclosed in brackets. Ice is identified as first-year (fy), second year (sy) or multi-year (my)

Event	*Ice type*	*Ice velocity ($m s^{-1}$)*	*Peak load (MN)*	*Ice thickness (m)*	*Contact width (m)*	*Average pressure (MPa)*	*Dynamic component (%)*
Nov. 1985	sy	—	295	(3)	95	1.04	—
7 Mar. 1986	my	0–0.05	270	6.5	100	0.42	22
8 Mar. 1986	my	slow	300	(6)	100	0.50	26
12 Apr. 1986(A)	my	0.05	500	(10)	90	0.56	54
12 Apr. 1986(B)	my	0.15–0.2	210	5.5	100	0.38	0
12 May 1986	fy	0.2–0.0	250	(1.8)	90	1.54	25

in Table 5.4. I have left a dash where numbers are not available, but have put some reasonable guesses and inferences into brackets.

5.5 ICEBREAKERS AND FLOATING STRUCTURES

5.5.1 General

It is not the principal purpose of this book to examine details of icebreaker performance and design. A good introductory treatment of the subject is given in Ashton (1986, Chapter 9). However, it is worth looking briefly at the data which have been collected from operational icebreakers, since they are the principal public source of data on local loads during full-scale interaction processes. "Local", in this context, refers to loads sustained over areas of up to a few square metres. First, however, we will look at data on global loads on icebreakers and floating structures.

5.5.2 Global loads

5.5.2.1 Icebreakers

Global loads on icebreakers are of limited interest to the designer of fixed structures in the Arctic. There are several reasons for this. The chief reason is that icebreakers are specifically designed to break ice approaching towards the ice-breaking bow, and are practically helpless when faced with ice action from any other direction. The shape of the bows is generally designed to break ice by flexure and clear it efficiently around the hull of the vessel towards the stern, and the design is specific to forward motion of the vessel. For this reason, the data are not very relevant to static structures which are susceptible to ice action from any direction. Another reason icebreaker data are difficult to use is that assessments of global ice resistance during icebreaking do not normally correspond at all to peak instantaneous force. "Resistance" during icebreaking is in general referenced to power requirements for maintaining effectively constant forward motion, and a massive icebreaker has such large momentum that it can sustain very large peak global loads for a few seconds without coming to rest. The power requirement corresponds to average load, not peak load. The peak global loads are not normally measured and yet are of fundamental importance to the designer of a static bottom-founded Arctic structure. Thirdly, much of the data from icebreakers are collected at relatively high interaction velocities – typically several knots or metres/second – and at these velocities the forces required to submerge and clear ice around the hull normally greatly exceed the forces required to break it. Finally, the mode in which an icebreaker fails ice is rather different from the mode in which a static structure fails ice: an icebreaker relies quite heavily on its own pitching (longitudinal rocking) motion to ride up on ice cover and break it by downward pressure. In this mode the ship again relies on its massive kinetic energy. Forward motion is converted to vertical potential energy which is then used to do work breaking the ice cover. Although the peak icebreaking force may instantaneously be quite high, the ship as a whole sees only an averaged resistance to forward motion.

For all these reasons, data on global ice resistance for icebreakers are to be treated with care. An account of global ice resistance measurements on the Wind Class icebreaker and on the *USCGC Mackinaw* is given in Lewis and Edwards (1970) and Edwards *et al.* (1973). It is summarised by Ashton (1986). Figure 5.32 gives a summary of full-scale data on total ice resistance for the Wind Class icebreaker in ice of thicknesses up to 0.7 m at velocity 1.5 m s^{-1}. The beam of this vessel is 19 m. It will be seen immediately that the average resistance pressures are very low: for ice of thickness 0.7 m failing over a nominal width of 19 m

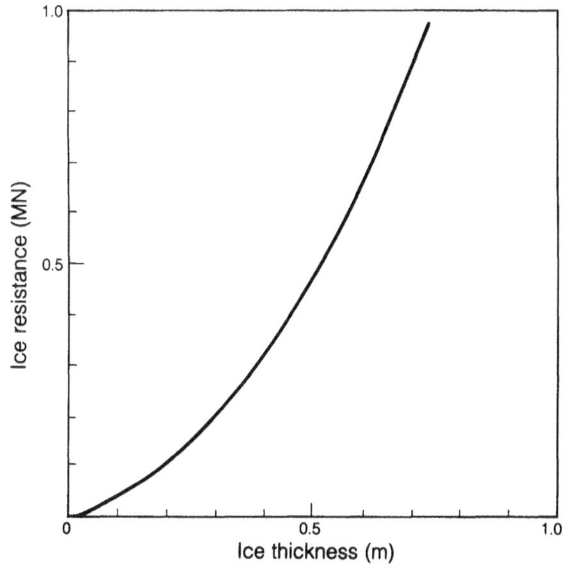

Fig. 5.32 Summary of full-scale data on ice resistance for the Wind Class icebreaker at velocity 1.5 m s^{-1} (after Ashton, 1986).

the average resistance is only about 1 MN, which corresponds to an average pressure of 75 kPa. This is a fairly typical result for ice of thickness less than 1 m.

5.5.2.2. Kulluk

The conical drillship *Kulluk* was first deployed in the Canadian Beaufort Sea in 1983. It is designed as far as possible to retain the icebreaking characteristics of an icebreaker, but is anchored on station with up to 12 wire anchor lines. The structure is essentially circular, with sloping sides, and has a diameter of about 70 m at the waterline. The structure does not itself move through the ice, but it nonetheless behaves dynamically much as an icebreaker does: it fails ice principally in bending and relies on pitch and surge motion to assist this process (Fig. 5.33). The *Kulluk* is not designed to withstand severe ice conditions: it is designed principally to withstand first-year ice of thickness up to 1.2 m, but ice of any greater thickness has to be broken up first by attendant icebreaking support vessels. A summary of *Kulluk*'s operational performance is given by Pilkington *et al.* (1986). Data on global loads are available for *Kulluk*'s 1983–4 season in the Canadian Beaufort Sea, when it experienced relatively severe ice conditions including up to 3/10 multi-year ice cover. Loads experienced when broken first-year ice was moving freely past the rig were typically in the range 0.1–1.0 MN, well within the design capacity of the system. However, higher loads than this were experienced on several occasions, either as a result of lateral confinement of the approaching ice floes, preventing them from clearing freely, or

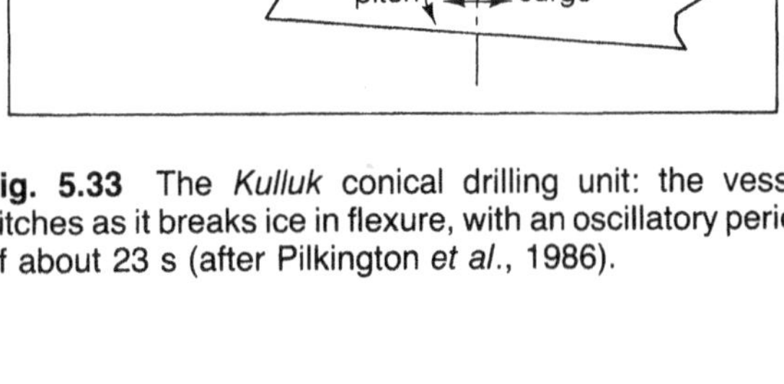

Fig. 5.33 The *Kulluk* conical drilling unit: the vessel pitches as it breaks ice in flexure, with an oscillatory period of about 23 s (after Pilkington *et al.*, 1986).

as a result of build-up of rubble around the structure. In these conditions loads in the range 2–6 MN were measured, and on several occasions the drilling unit was forced to suspend operations and move off station. The average ice indentation pressures experienced during these events are still low, as they lie in the range of about 20–60 kPa. These loads are not indicative of the maximum loads which can be developed against a static structure.

5.5.3 Local loads

Data on local loads on icebreakers are in general collected much in the same way as on a static Arctic structure, by internal strain-gauging of hull plates and interior members.

Local indentation with an icebreaker hull or a marine structure differs fundamentally from the indentation problems we have so far been looking at. The classic indentation geometry on which laboratory experiments are performed, and which represents approximately the geometry of global indentation, is shown in Fig. 4.23. The geometry is two-dimensional, and the ice is indented through its full thickness. However, during indentation with an icebreaker or a structure, local interaction processes may occur over areas very much smaller than the full thickness (Figs 5.34(a) and (b)). This means that the stress field is no longer simply biaxial, but may in the one case (Fig. 5.34(a)), approach a state of triaxial confinement, and in the other case (Fig. 5.34(b)) approach a state of simple uniaxial loading.

Measurements of peak stresses during icebreaking trials are unfortunately rarely, if ever, well correlated with observations of failure geometry: all that is typically available is a time series of stress readings on an array of pressure sensors. The most that is generally recorded about ice conditions is that it was first-year or multi-year ice of a certain general thickness. The degree of contact and the local failure area generally have to be deduced entirely from the sensor readings.

A set of pressure readings taken in this way, and plotted as a function of area of contact, is shown in Fig. 5.35 (St John and Daley, 1984) for the *USCGS Polar Sea* in the Bering and Chukchi Seas. This data set is one of the best publicly available and gives typical local pressures for both first-year and multi-year ice. It is clear here that pressure shows a very strong dependence on the area over which it is measured: the larger the area the lower the pressure. This is because individual

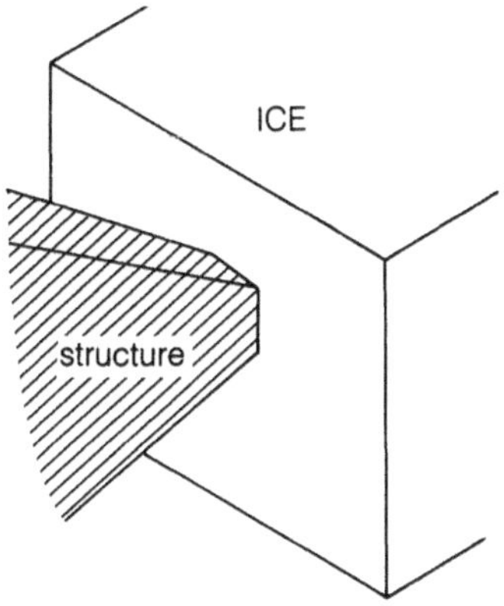

(a)

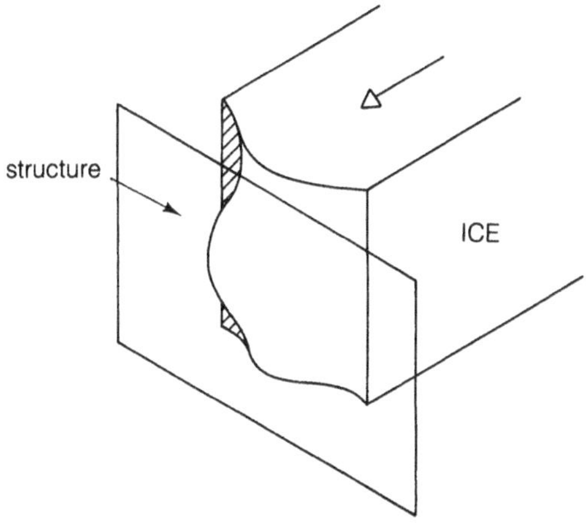

(b)

Fig. 5.34 Indentation geometry for local loads: (a) a structure or vessel with a projecting end penetrates the ice cover, and the stress-state approaches triaxial constraint; (b) a flat-walled structure is struck by an ice feature with a protruding uneven surface–in this case the stress-state is close to uniaxial.

peaks of pressure measured over small areas do not occur simultaneously with peaks of pressure elsewhere. A large area of ice is never simultaneously failing at peak local pressure, but simply shows the average effect of a large number of smaller sub-areas which may or may not be at peak pressure. We shall investigate the formal theory behind this much more thoroughly in Chapter 6 (Section 6.4.5).

During the series of ramming trials carried out in 1983 with the *USCGS Polar Sea*, the highest pressure measured on a single pressure cell (area 0.16 m^2) was 11.3 MPa. The highest pressure recorded over a relatively large area was 1.2 MPa measured over 4.1 m^2. This occurred while ramming multi-year ice in the North Chukchi Sea.

5.6 A PRESSURE–AREA CURVE FOR ICE

It will have become obvious, from the account I have given of a wide range of experiments on indentation of ice, that ice does not have a simple "failure pressure". Ice in the laboratory fails at 10–20 MPa and yet at large scale in the field it appears to fail at about 1 MPa; direct measurements of failure pressures on icebreakers (Fig. 5.35) show a similar effect. It is clear that, for some reason, failure pressure depends on contact area. This is a fact of central importance, and will form the basis of theoretical developments in Chapter 7. For the moment, however, we shall restrict ourselves simply to looking at a general raw compilation of all the data available.

5.6.1 Full-thickness indentation

We begin with data from tests in which full-thickness edge indentation was carried out. This is the classic indentation geometry of Fig. 4.23. A wide range of data for this type of indentation geometry are available, and are shown plotted in Fig. 5.36(a). This plot shows peak pressure $\hat{\sigma}$ (peak total load $\hat{P}$ divided by gross contact area Dh) as a function of gross contact area Dh. The data set covers a wide variety of ice types (freshwater S-2 ice, first-year ice and multi-year ice) and a wide variety of test conditions (laboratory experiments, *in situ* jacking tests, and interaction with bridge piers, lighthouses, offshore drilling structures and islands). It includes all the indentation data discussed so far in Chapters 4 and 5 and some additional data from lighthouses and bridge piers. All data are quite raw and have not been in any way corrected for temperature or salinity. Despite such indiscriminate plotting of disparate data types there is a useful general trend illustrating the pressure–area effect, as highlighted by the shaded areas in Fig. 5.36(b). The data points fall naturally into four principal clusters as identified by the shaded areas, and come from the following data sources:

(A) *Laboratory tests.* These are small-scale laboratory tests, generally performed on uniform bubble-free samples of S-2 ice with the

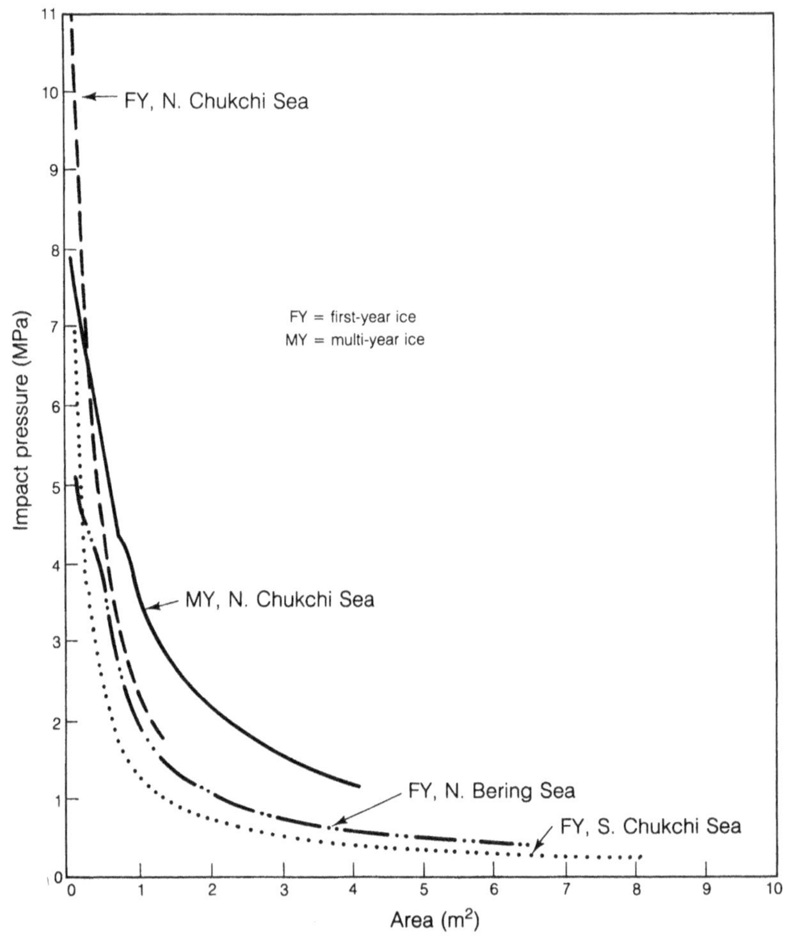

Fig. 5.35 Impact pressure as a function of contact area for four extreme impact events of the *USCGS Polar Sea* in the Bering Sea and Chukchi Sea (after St John and Daley, 1984).

c-axis oriented in the plane of the applied indentation force. The data come from Hirayama *et al.* (1974), Frederking and Gold (1975), Zabilansky *et al.* (1975), Michel and Toussaint (1977), Kry (1979b), Michel and Blanchet (1983), and Timco (1986a). They cover a wide range of aspect ratio, D/h, from 0.21 to 83. Note that data from another relevant study by Kry (1979a) are not included in Fig. 5.36, since they were performed with ice adfrozen to the indentor. They are, however plotted separately later, in Fig. 5.37, together with icebreaker data.

(B) *Medium-scale tests.* The next set of data comes from medium-scale tests, generally through the full thickness of natural floating ice cover. These include various *in situ* jacking tests on first-year ice such as the "Nutcracker" tests (Croasdale, 1970, 1971), the Exxon large-scale strength tests (Lee *et al.*, 1986) and Inoue and Koma (1985); tests on floating freshwater lake ice (Croasdale, 1971; Taylor, 1973; Miller *et al.*, 1974; Nakajima *et al.*, 1981); measurements on river bridge-piers (Lipsett and Gerard, 1980); and sea-ice data from lighthouses (Määttänen, 1981; Engelbrektson, 1983; Hoikkanen, 1984) and Cook Inlet structures (Blenkarn, 1970).

(C) *Full-scale islands.* This data group comes from Beaufort Sea artificial islands (Metge, 1976; Strilchuk, 1977; Johnson *et al.*, 1985; Rogers *et al.*, 1986; Gulf, 1986, unpubl.) and from the first and second Hans Island programmes (Danielewicz and Metge, 1981, 1982). Results from the third (1983) Hans Island programme (Danielewicz and Cornett, 1984) are still proprietary, and should be added to the plot when they become available.

(D) *Meso-scale models.* In the course of modelling the large-scale overall dynamics of the Arctic

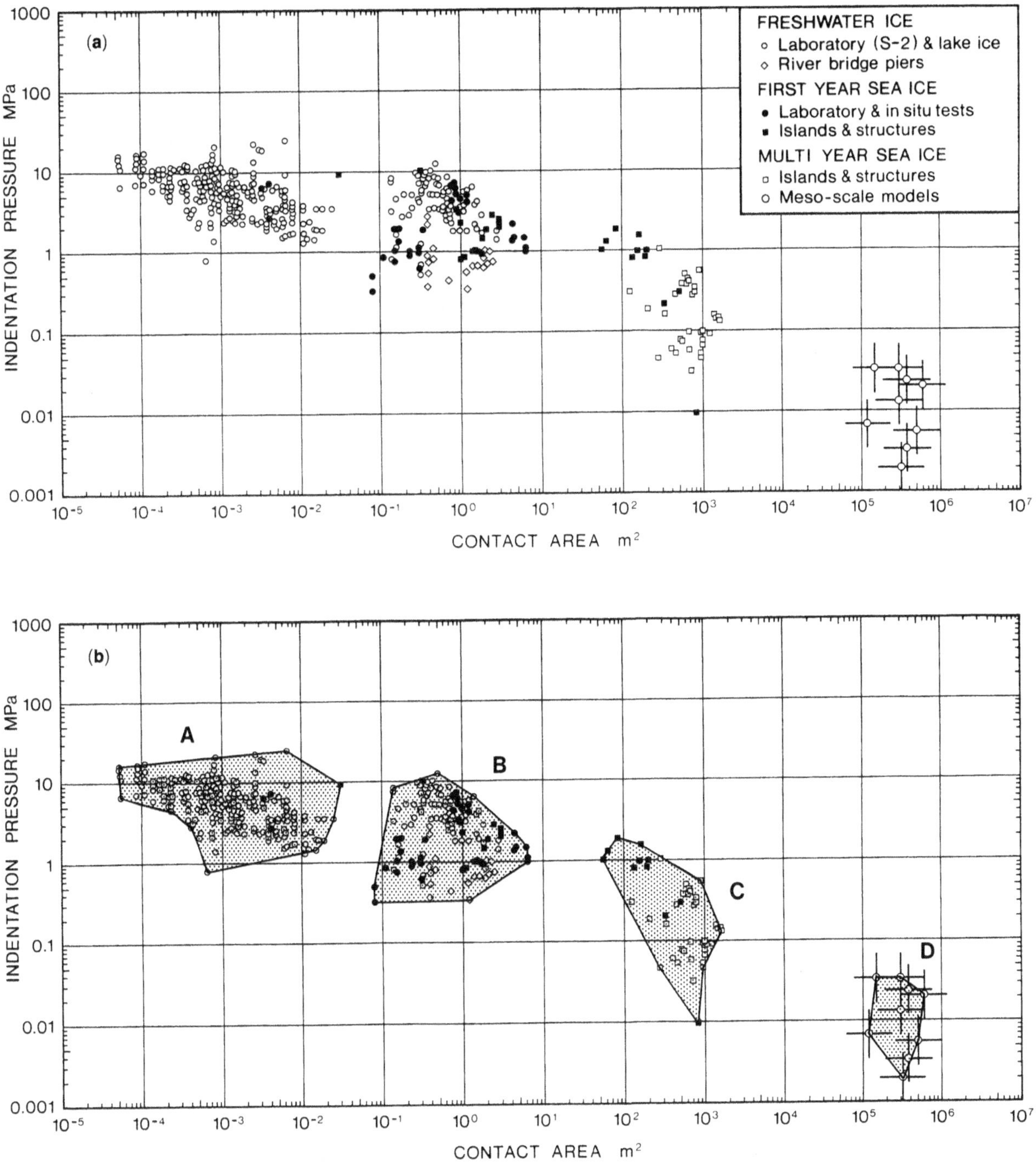

Fig. 5.36 Pressure–area curve for edge indentation of ice. Peak indentation pressure *P/Dh* is plotted against gross contact area *Dh*. (a) Raw data (key as in diagram). (b) Data groups A, B, C and D (see text for explanation and references). No corrections have been applied.

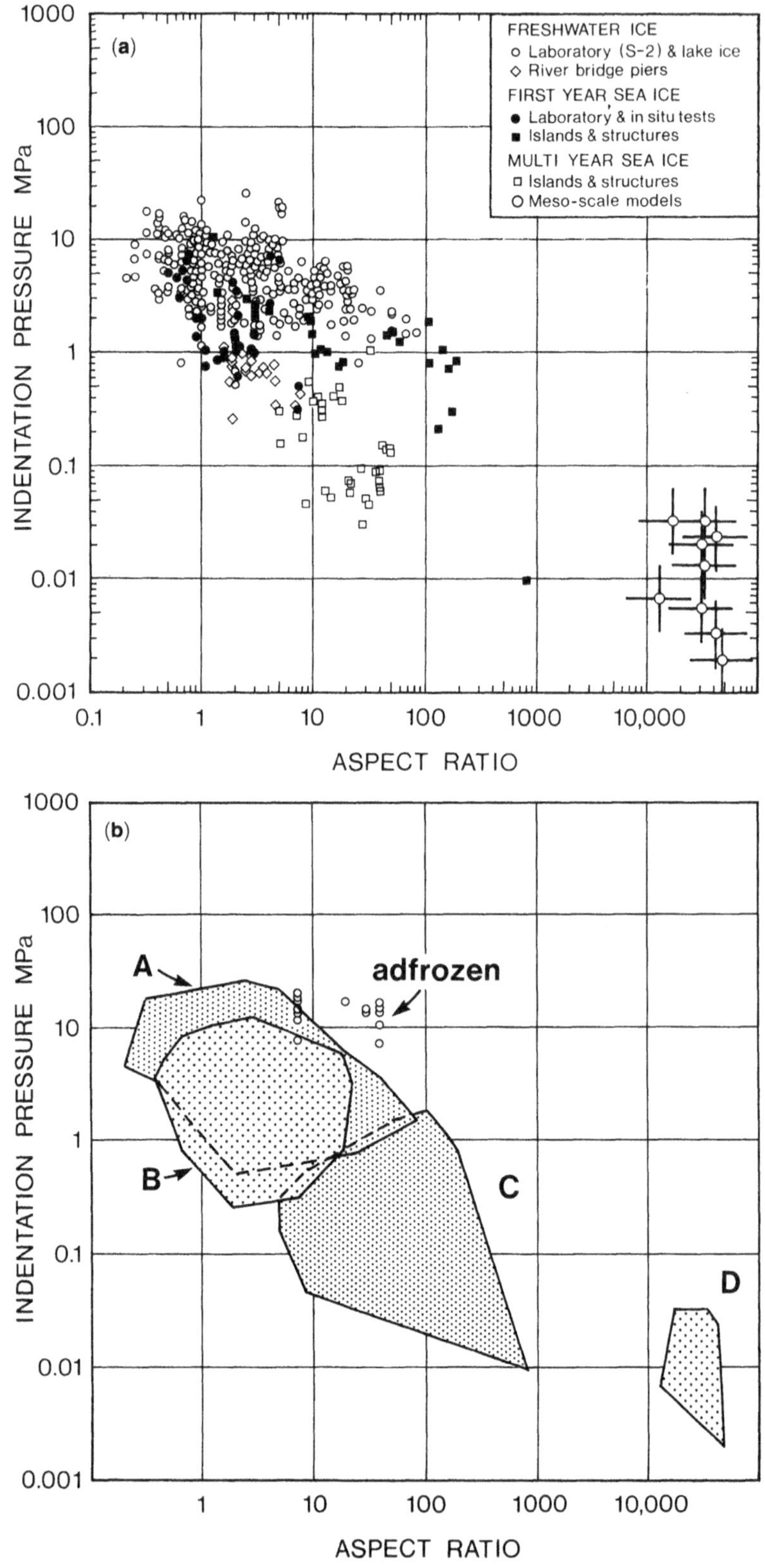

Fig. 5.37 Plot of peak indentation pressure against aspect ratio *D/h* for edge indentation of ice. (a) Raw data. (b) Data groups A, B, C and D, with addition of data from Kry (1979a), with adfrozen indentor. Data sources are as in Fig. 5.36.

Ocean and its ice cover it is necessary to make model assumptions about the gross "compressive strength" of polar pack ice (Pritchard, 1977, 1980; Hibler, 1980a,b). The models used are computer-based and generally use a grid-size of 40–125 km in which ice movement is modelled as a function of winds and sea current. "Compressive strength" (expressed as load per unit length of ice cover) is then a parameter which has to be inserted into the governing equations and adjusted in order to obtain a plausible match with observed behaviour. Values obtained in this way by matching models to reality generally lie in the range 5×10^3 to 10^5 Nm^{-1}. These results refer to pack ice of highly variable thickness. The mean modelled ice thickness typically lies in the range 2.6–4.0 m, though it is likely that gross ice strength is actually controlled by areas of much thinner ice of thickness as little as 30 cm (Rothrock, 1975). It is also not clear over what distance the large-scale ice strength should be effectively averaged, and the data are thus rather difficult to plot on a pressure–area curve. For present purposes, however, we shall assume that the data refer to a length equal to the model grid spacing, and that the thickness is simply equal to the average pack-ice thickness. This means that the strength values used refer to areas Dh of the order 10^5 m^2. On Fig. 5.36 I have plotted ice strength values from Pritchard (1977, 1980), Hibler (1980a,b), and Tucker and Hibler (1981). Because there is considerable doubt over the ice thickness and distance scale to which these data refer, the data points have been plotted with error bars representing uncertainty by a factor of two.

Some direct measurements of large-scale pack ice forces have now been performed (Croasdale *et al.*, in press). During April/May 1986 a pilot field programme was carried out in the southern Canadian Beaufort Sea, in which a multi-year floe of size about 5 km by 2 km was instrumented with an array of stress sensors at its centre while it drifted freely and underwent pack ice action at its perimeter. The thickness of the floe was generally in the range 1.7–3.0 m, and the thickness at the stress sensor site was about 1.8 m. On 22 April the floe was subjected to convergent pack ice action, accompanied by active ridging of surrounding pack ice along the floe's perimeter of about 5 km. The principal action lasted about 2 hours, and during this period a maximum stress increase of about 19–29 kPa was monitored by the stress sensors. A simple interpretation of this ridging event leads to the conclusion that a force of about 170–260 MN was being applied to the floe, which is equivalent to a load intensity per unit width of about $3\text{–}5\times10^4$ N m^{-1} applied over the full floe width of 5 km. This lies nicely in the middle of the range implied by the theoretical models considered above, but it is important to recognize that it refers to a scale of only about 5 km rather than 40–125 km. If we assume that the pack ice undergoing ridging was of average thickness about 1.5 m then this would represent an average pressure of 0.02–0.03 MPa applied over a contact area of about 7500 m^2. This new data point has not yet been added to Figs 5.36–5.38; it would lie between data groups C and D, and is low relative to the general trend.

5.6.2 Aspect ratio or area?

The plots of Fig. 5.36 provide an apparently clear picture of dependence on size: pressure seems to depend on area of contact. However, the question arises, could this instead be an effect of aspect ratio rather than size? The answer is unfortunately not quite clear-cut. Figure 5.37(a) shows all the same data as on the pressure area curve (Fig. 5.36) plotted instead as a function of aspect ratio D/h. The picture is more confused than that of Fig. 5.36, but again shows a perceptible trend, with peak pressures generally falling off with increasing aspect ratio.

It is not, however, an unambiguous dependence on aspect ratio, since the only points available for very high aspect ratio (D/h greater than 100) happen also to be points from very large size tests (areas exceeding 100 m^2). This can be seen in Fig. 5.37(b), where the points of Fig. 5.37(a) are subjected to the same grouping (A, B, C and D) as in Fig. 5.36(b). The results from Kry (1979a) are also separately included, where ice was adfrozen to the indentor. The following points emerge from Figs 5.36 and 5.37:

(1) Data groups C and D show generally lower pressures than groups A and B, but it is ambiguous whether this is due to higher aspect ratio or to larger contact area.
(2) Aspect ratio is certainly not the only factor responsible for the stress reduction observed. For instance, at aspect ratio approximately 10, peak stresses anywhere between 0.05 MPa and 6.5 MPa are observed – a factor of more

than 100 difference. Furthermore, the lower stresses are generally associated with increasing contact area.

(3) The data from Kry (1979a), in which ice was adfrozen to the indentor, appear not to show any dependence on aspect ratio. This is significant, and lends weight to the argument that scale effects, whether as a function of size or of aspect ratio, may be due to effects of imperfect contact and non-simultaneous failure.

5.6.3 Triaxial indentation: icebreakers and impact hammers

As we have seen in Section 5.5 and Fig. 5.34, tests involving icebreakers do not involve the same stress geometry as simple edge indentation. If the geometry resembles the form shown in Fig. 5.34(a) then we might expect higher stresses than those measured during edge indentation. Figure 5.38 shows a variety of tests from icebreakers and from impact hammer tests, superimposed on the general data groups of Fig. 5.36. Data come from icebreakers impacting first-year ice (St John and Daley, 1984; St John *et al.*, 1984; Kujala and Vuorio, 1985) and icebreakers impacting multi-year ice (Ghoneim and Keinonen, 1983; Glen and Blount, 1984; St John and Daley, 1984; St John *et al.*, 1984; Edwards *et al.*, unpubl.). Results for impact hammer tests are taken from Glen and Comfort (1983). In addition, I have superimposed the data from Kry (1979a), in which ice was adfrozen to the indentor.

A series of medium-scale controlled ramming tests have also been carried out by the industry, during which multi-year ice and iceberg ice was indented by hydraulic rams of area up to about 1 m^2 (Geotechnical Resources Ltd, 1986). Unfortunately these data are still proprietary and cannot be shown here.

Interestingly, peak pressures measured on icebreakers and impact hammers as a function of contact area do not appear to depart significantly from the trend already established for simple edge indentation. The upper bound of pressures in Fig. 5.38 is broadly the same as that of the data groups A, B, C and D already treated: at small areas (less than 0.1 m^2) very local pressures are found to be very high, but at larger areas (greater than 1 m^2) the data conform to the existing trend. This is perhaps surprising, since triaxial constraint conditions might be expected to raise peak failure pressures. However, it seems to me that the reason for the degree of accord between the diverse data sets may be that true triaxial constraint is rarely achieved during real full-scale interactions. The geometry of Fig. 5.34(a) is, after all, rather artificial except for head-on ramming by the bows of an icebreaker. In practice, the local impact geometry is much more likely to resemble Fig. 5.34(b). This geometry, which is characterised by a protruding lump of ice interacting with a flat structural surface, rather than *vice versa*, is not very significantly different from a simple biaxial, or even uniaxial stress state.

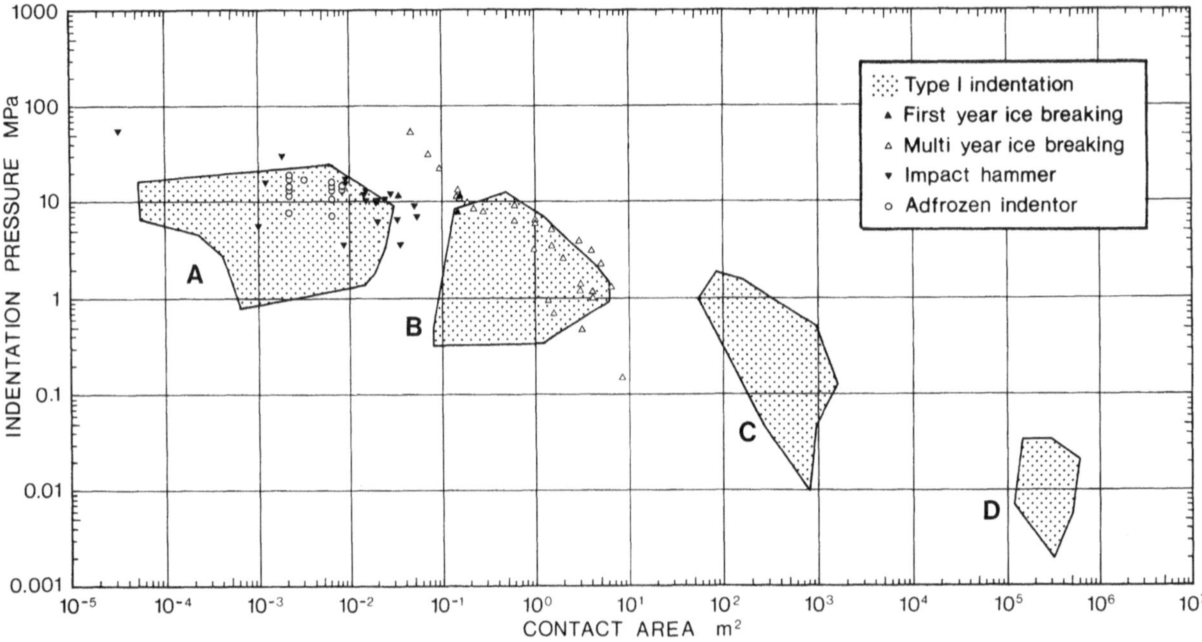

Fig. 5.38 Icebreaker and impact hammer data superimposed on the trend of Fig. 5.36(b).

Chapter 6

Theoretical Analysis of Ice Failure

6.1 INTRODUCTION

In this chapter we develop a theoretical basis for understanding the most important aspects of the experimental data described in Chapters 4 and 5. This theoretical understanding will provide the tools of the trade for tackling Chapter 7, where we shall begin to calculate design ice forces for offshore structures.

It should be clear by now that the distinction between continuum behaviour and fracture behaviour is crucial. Continuum behaviour is quite easy to understand, while fracture behaviour is quite difficult. Most difficult of all, however, is the transition between them. Unfortunately, this transition turns out to be most important. Here, we will begin by looking in more detail at analysis of continuum behaviour (elastic buckling, creep buckling and ductile indentation) and then move on to tackle diverse modes of failure by fracture.

6.2 CONTINUUM BEHAVIOUR

6.2.1 Elastic buckling

Simple in-plane elastic deformation of ice under load is not of great importance generally, since creep deformation rapidly overtakes it in magnitude: if a sample of ice is subjected to a stress of 10 MPa its elastic strain deformation is about 10^{-4} while its secondary creep strain-rate is about $4\times10^{-4}\ \mathrm{s}^{-1}$ (equation (4.7)). This means that creep strain exceeds elastic strain after only 0.25 s. Even at a lower stress of 1 MPa, creep strain exceeds elastic strain after only 25 s.

However, it is worth looking at the problem of out-of-plane elastic buckling, since this is a failure mode frequently observed in laboratory tests and model experiments. The geometry of the problem is shown in Fig. 6.1, which is designed to represent interaction of a floating sheet of ice with a structure. A truncated wedge-shaped plate of elastic material (ice) floats on an elastic foundation (sea water) and is loaded at its edge by a load P acting over a width D. The seawater "foundation" can be modelled as an elastic foundation because it provides buoyancy forces: sea water is not, of course, an elastic material, but if a block of ice is either depressed or lifted from floating equilibrium it experiences a restoring force back towards equilibrium of magnitude proportional to its displacement. This is a simple analogue of Hooke's Law and allows us to treat the sea as an elastic continuum half-space.

The plate of ice is modelled as a wedge for two reasons: generality and realism. In practice, buckling of ice sheets during indentation experiments is almost invariably preceded by the formation of radial cracks at angles of around 45° (Fig. 6.2).

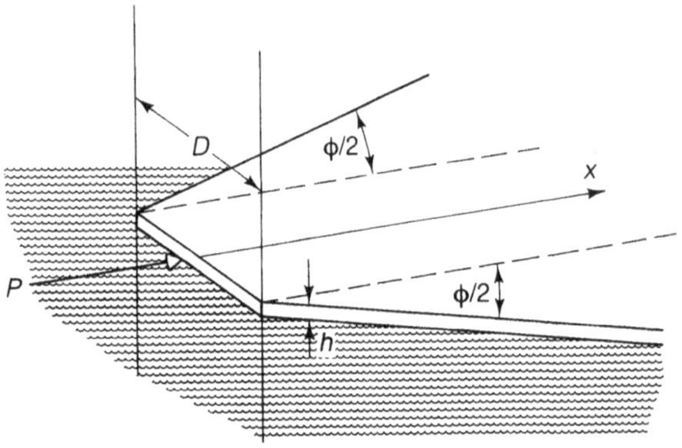

Fig. 6.1 Geometry for analysis of elastic buckling. A floating truncated wedge of ice, with half-angle $\phi/2$ interacts with a structure of diameter D (after Kerr, 1978).

The same phenomenon is frequently observed around full-scale Arctic structures, and it serves to divide the floating ice plate into a series of wedge-shaped plates. The analysis we look at here is for a variable half-angle of wedge, $\phi/2$. This reduces to the case of a parallel-sided floating plate (uniaxial loading) if $\phi = 0$. As ϕ tends to 180° we approach the geometry of an infinite half-plane and, as we might expect, plate buckling is inhibited and requires prohibitively high loads.

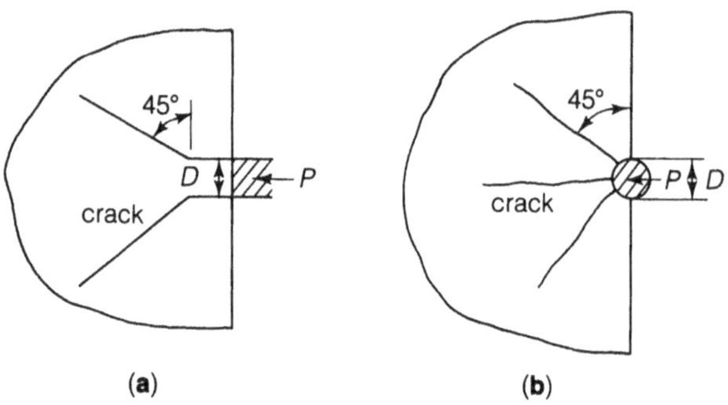

Fig. 6.2 Observed radial crack patterns during model experiments with different shapes of indentor (after Kerr, 1978): (a) rectangular indentor; (b) cylindrical indentor. Similar patterns are observed around full-scale Arctic structures.

The buckling problem can be treated fairly straightforwardly by classical analysis of the behaviour of elastic plates on elastic foundations, and some approximate analytic solutions are given by Kerr (1978). These provide the critical buckling load P_b as a function of ice thickness h, loading width D and wedge angle ϕ, for ice with Young's modulus E and Poisson's ratio ν. The sea water has density ρ_w. More complex treatments are also available, which treat non-linear post-buckling behaviour (Kerr, 1980), but these are not important to the argument here.

The solutions given by Kerr (1978) depend quite strongly on the boundary condition at the loaded edge (that is, whether it is clamped or simply-supported), but for the case of a simply-supported end, which is the most realistic case, the buckling load P_b is given by:

$$P_b = 5.3\,B_f\kappa\left(\kappa D + 2\tan\frac{\phi}{2}\right) \tag{6.1}$$

where B_f represents the flexural rigidity of the ice cover,

$$B_f = \frac{Eh^3}{12(1-\nu^2)} \tag{6.2}$$

and we define

$$\kappa = \left(\frac{\rho_w g}{4B_f}\right)^{1/4} \tag{6.3}$$

For the case of a clamped end the buckling load is higher, and the solution is

$$P_b = 8B_f\kappa\left(2\kappa D + 2\tan\frac{\phi}{2}\right)$$

It is, however, difficult to conceive of situations in which a clamped-end boundary condition is appropriate: the case of an extremely effective adfreeze bond between ice and structure is the only one which springs to mind, and this is seldom observed.

Only a small modification to equations (6.1)–(6.3) is necessary in order to account for variable elastic properties with depth through the ice cover (Kerr and Palmer, 1970). Properties do vary appreciably with depth: owing to the variation of ice temperature and brine volume with depth, Young's modulus E may reduce substantially in the warmer lower layers of a floating ice cover. This effectively lifts the neutral axis of the ice and slightly reduces the overall flexural rigidity. Poisson's ratio ν does not show significant variation with depth. Now, if E is a function $E(z)$ of depth z through the ice cover, then equations (6.1) and (6.3) still hold, but with the following revised expression for flexural rigidity:

$$B_f' = \frac{1}{(1-\nu^2)}\int_{-h}^{0} z^2 E(z)\,\mathrm{d}z$$

This integral can easily be performed for a number of representative variations of $E(z)$ with depth. For illustration, consider a simple linear variation as shown in Fig. 6.3. This corresponds approximately to realistic floating sea ice of salinity 6–8 ppt under cold winter conditions: $E(z)$ varies from $E_0 = 9.5$ GPa at the upper surface, where brine volume is negligible, to about $0.2\,E_0$, or 1.9 GPa at the bottom surface. In this case the revised expression for flexural rigidity is

$$B_f' = \frac{0.51Eh^3}{12(1-\nu^2)}$$

or roughly half that for uniform pure ice (equation (6.3)). This is equivalent to using an effective Young's modulus of 5 GPa instead of 9.5 GPa. Equations (6.1)–(6.3) then apply as before.

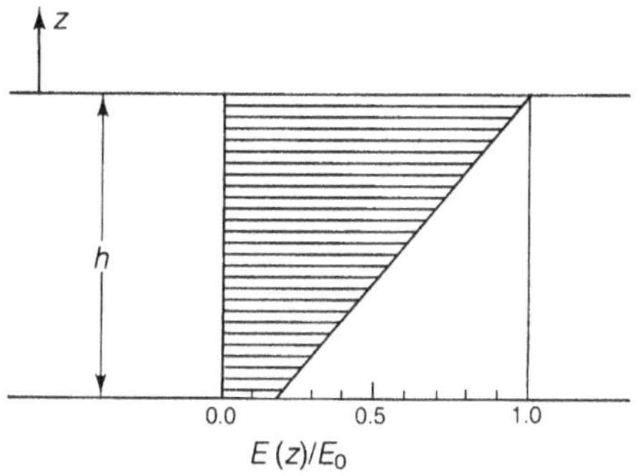

Fig. 6.3 Schematic variation of Young's modulus E with depth. The variation is due to increasing temperature and brine volume towards the bottom.

The interesting thing about equation (6.1) is its functional dependence on ice thickness, h. For simplicity, consider the case of a parallel-sided plate (that is, $\phi = 0$). The expression for buckling load is then simply:

$$P_b = 5.3B_f\kappa^2D$$

Substituting equations (6.2) and (6.3), we find

$$P_b = \left[\frac{0.59\rho_w gE}{(1-\nu^2)}\right]^{1/2} Dh^{3/2} \tag{6.4}$$

showing that the load for buckling depends on the 3/2 power of thickness h. This means that the average contact pressure required for buckling, P_b/Dh, depends on $\sqrt{h}$. This dependence of buckling load upon thickness has important consequences, since it means that buckling becomes progressively more and more difficult as thickness increases. It means, in fact, that only quite thin ice will ever fail in the elastic buckling mode: for thick ice, as we shall see, the crushing strength of ice is reached well before the ice can buckle elastically. In this case, of course, the ice will fail by crushing instead of buckling.

Quantitatively, consider a comparison of the force required to crush a plate of pure ice with the force required to buckle it. The force P_c required to crush a parallel-sided plate of ice is at most

$$P_c = Dh\sigma_c \tag{6.5}$$

where σ_c is a value for the uniaxial crushing strength of pure ice. Now, for buckling to occur before crushing occurs we require

$$P_b < P_c$$

or, substituting equations (6.4) and (6.5):

$$h < \frac{\sigma_c^2\,(1-\nu^2)}{0.59\rho_w gE} \tag{6.6}$$

As we have seen in Chapter 4, the uniaxial crushing strength of freshwater ice is typically of the order 5 MPa. Taking, then:

$$\sigma_c = 5 \text{ MPa}$$
$$E = 9.5 \text{ GPa}$$
$$\nu = 0.3$$
$$\rho_w = 1028 \text{ kg m}^{-3}$$

and

$$g = 9.81 \text{ m s}^{-2}$$

we find that

$$h < 0.4 \text{ m}$$

is an approximate condition for elastic buckling to occur. For more realistic sea-ice conditions we might use instead a value of 2–4 MPa for the uniaxial crushing strength σ_c and a value of $E = 5$ GPa. This leads to much the same conclusion, that ice thickness must be less than about 0.1–0.5 m for it to be susceptible to buckling. This accords well with observations in the field: arching and buckling of ice is commonly observed for ice sheets less than 0.3 m thick, but crushing is the limiting factor when thickness exceeds about 0.6 m (see, for instance, Rose, 1947).

The value of 0.4 m is a liberal maximum thickness for elastic buckling to be theoretically permissible, for two reasons: firstly, if we change the geometry of interaction to that of a wedge with $\phi/2 \sim 45°$ then equation (6.1) supplies an even higher critical buckling load; secondly, as we have seen in Chapter 5, the actual failure strength of ice at largescale, which is our principal concern here, appears to be more in the region of 1 MPa, not 5 MPa.

For these reasons, I shall not give any further consideration to the question of elastic buckling: it occurs in the laboratory and in the model ice tank, where ice is suitably thin, but is rarely important to the design of real offshore structures, where ice of thickness in the range 1–10 m is more likely to be the controlling factor.

TABLE 6.1 Theoretical elastic buckling pressure for floating ice of thickness *h* interacting with a structure 100 m wide. Wedge geometries of $\phi = 0$ (parallel-sided and $\phi/2 = 45°$ (more realistic) are both considered for edge conditions (a) simply-supported, and (b) clamped. A dashed line shows the limit of 5 MPa stress. At thicknesses exceeding this, elastic buckling is most unlikely

Ice thickness h (m)	*Buckling pressure, P_b/Dh (MPa)*			
	Simply-supported		*Clamped*	
	$\phi/2 = 0°$	*$\phi/2 = 45°$*	*$\phi/2 = 0°$*	*$\phi/2 = 45°$*
	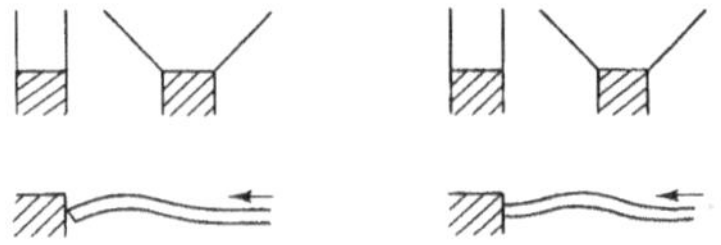			
	(a)		*(b)*	
0.01	*0.6*	*0.6*	*1.7*	*1.7*
0.05	1.3	1.3	3.8	3.9
0.1	1.8	1.9	5.4	5.6
0.2	2.5	2.9	7.7	8.2
0.3	3.1	3.6	9.4	10.2
0.4	3.6	4.3	10.9	12.0
0.5	4.0	5.0	12.2	13.6
1.0	5.7	8.0	17.2	20.7
1.5	7.0	10.9	21.1	26.9
2.0	8.1	13.6	24.3	32.8
3.0	9.9	19.1	29.8	43.8

For interest, Table 6.1 shows critical buckling pressures, defined as P_b/Dh, for the cases $\phi = 0$ (parallel-sided plate) and $\phi = 45°$ (wedge-shaped plate) for a variety of ice thicknesses interacting with a structure 100 m in diameter. The cases of a simply-supported contact and a clamped edge are both considered. For Young's modulus a value of $E = 5$ GPa is used, to reflect effects of increasing brine volume through a realistic ice cover.

The table illustrates that for realistic ice conditions, such as 1.5 m thick first-year ice in a wedge formation interacting with a structure 100 m wide, the average stress needs to reach 10.9 MPa before elastic buckling should occur. This is 5–10 times higher than any stress ever measured under such conditions, and illustrates why elastic buckling is not an important failure mode.

6.2.2 Creep buckling

Having dismissed elastic buckling as unlikely to occur in ice of thickness greater than about 0.5 m, it is nonetheless worthwhile spending some time considering creep buckling.

Creep buckling certainly does occur, and indeed is one of the most striking failure modes against Arctic structures: two examples of it are shown in Fig. 6.4, at successive stages of development. Creep buckling typically occurs only during very slow loading processes, especially, for instance, during late spring (April–May) when landfast ice undergoes thermal expansion. During such periods, steady ice movement rates of about 0.5 m per day are typical, and buckle features develop slowly over a period of several days. A schematic diagram of the formation of a creep buckle is shown in Fig. 6.5. The axis of the buckle may extend about 50–100 m.

The formation of a creep buckle is fundamentally different from an elastic buckle, in the following respects:

(1) It requires an initial imperfection (which then grows under load). If no imperfections existed then the creep buckling load would be equal to the elastic buckling load, since that is the load required to form an initial imperfection.
(2) Like pure creep itself, creep buckling is a rate-dependent, load-dependent process. The fact that a creep buckle begins to form at a certain rate under a certain load does not mean that a higher load could not be sustained if applied. If a higher load is applied, the rate of creep buckling simply increases (and, incidentally, different wavelength buckles may become dominant). In this sense, the creep buckling load is not a "failure" load at all, any more than a stress causing a uniaxial strain-rate is a failure load. Elastic buckling leads immediately to unstable failure, but creep buckling does not. So, even though a creep buckle may be incipient and growing the ice can still sustain much higher loads if they are applied: as high, in fact, as the elastic buckling load for the plate (taking account of any imperfections which exist), or, of course, the crushing load if that is lower.

Fig. 6.4 Two examples of creep buckling. (a) Formation of a tensile crack at the surface, stabilised by the overall compressive stress field (Uviluk, P-66, April 1983). (b) Later stages of a creep buckle (Tarsiut island, N-44, April 1982). (Photographs by the author, courtesy of BP Petroleum Development Ltd.)

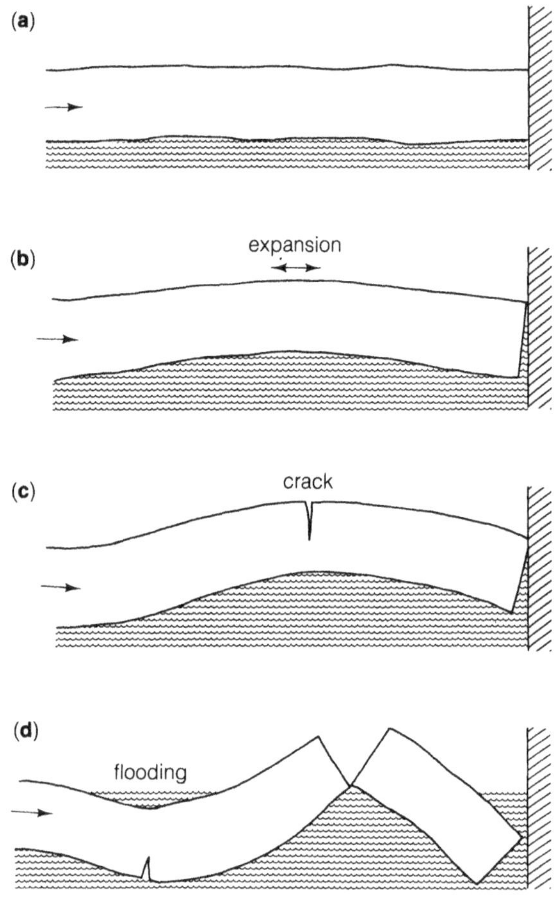

Fig. 6.5 Four stages of growth of a creep buckle: (a) initially stable ice with slight imperfections; (b) growth of imperfection, often assisted by differential thermal expansion at the surface; (c) formation of a stable tensile crack at the upper surface; (d) final stages, accompanied by flooding of submerged areas.

The process of creep buckling has been satisfactorily analysed theoretically, through an approximate analytic treatment (Calladine, 1982, unpubl.) and through the use of numerical methods (Sjölind, 1984, 1985). The problem of creep buckling of a non-linear creep beam floating on an elastic foundation turns out to be not difficult to solve if certain simplifications are made. The basic simplification is this: even though the constitutive law for ice is non-linear, of the form (see Section 4.2):

$$\dot{\varepsilon} = B_i \sigma^n \qquad (6.7)$$

nonetheless the bending-moment/curvature-rate relationship under axial load is linear for small loads. This simplification is quite a usual one in the analysis of creep buckling problems, and is described by Calladine (1962). It arises because there is a linear relationship between small perturbations in stress and strain-rate about any mean value. The following analysis follows Calladine (1982, unpubl.).

Consider a beam of rectangular cross section, of width D and of thickness h, under an axial compressive force P. The mean compressive stress $\bar{\sigma}$ is given by P/Dh, and the associated mean strain-rate $\bar{\dot{\varepsilon}}$ is given by equation (6.7). Differentiating equation (6.7) we find

$$\delta\dot{\varepsilon} = nB_i \bar{\sigma}^{(n-1)} \delta\sigma$$

but substituting equation (6.7) back into this, we find the simple equation:

$$\delta\dot{\varepsilon} = n\frac{\bar{\dot{\varepsilon}}}{\bar{\sigma}}\delta\sigma = C_i \delta\sigma$$

where C_i represents an effective creep-rate compliance. Note that this equation is linear.

Applying standard beam theory, if the beam is subjected to a bending moment M then it undergoes a curvature change at rate $\dot{\xi}$, given by

$$\dot{\xi} = \frac{C_i}{\mathcal{I}} M$$

where $\mathcal{I}$ is the second moment of area of the cross section, here given by

$$\mathcal{I} = Dh^3/12$$

Now consider such a beam resting on an elastic foundation of stiffness k_f and subjected to axial load P along the horizontal x-axis. The vertical deflection w is then governed by the equation:

$$\left(\frac{\mathcal{I}}{C_i}\right)\frac{d^4\dot{w}}{dx^4} + P\frac{d^2w}{dx^2} + k_f w = 0 \qquad (6.8)$$

where $\dot{w}$ is the differential of w with respect to time. Note that this is an analogue of the corresponding equation for time-independent elastic behaviour:

$$E\mathcal{I}\frac{d^4w}{dx^4} + P\frac{d^2w}{dx^2} + k_f w = 0$$

Now suppose that at time $t = 0$ we have an initial small displacement $w(t)$ of the form

$$w(0) = w_0 \sin(\pi x/\lambda_0) \qquad (6.9)$$

where λ_0 is an arbitrary half-wavelength. The creeping behaviour of the material will tend to make any such perturbation grow with time, and we need to find time-dependent solutions which satisfy equation (6.8). This can be done by educated trial and error, since the growth of such a process is normally exponential. We try a general solution of the form

$$w(t) = w_0 e^{t/T_0} \sin(\pi x/\lambda_0) \tag{6.10}$$

where T_0 is a time constant to be determined. Substituting into equation (6.8) we find

$$\frac{1}{T_0} = \frac{C_i}{\mathcal{J}}\left[P\left(\frac{\lambda_0}{\pi}\right)^2 - k_f\left(\frac{\lambda_0}{\pi}\right)^4\right] \tag{6.11}$$

This effectively provides the growth-rate time constant as a function of wavelength. Now, all wavelengths will grow with time, but there will generally be a wavelength λ_0 which grows faster than all the others. This wavelength of deformation will soon dominate the situation and form the principal creep buckle. It is the wavelength for which $1/T_0$ is maximised (that is, the minimum value of T_0), and it is found by differentiating equation (6.11) so that

$$\frac{d(1/T_0)}{d\lambda_0} = 0$$

This occurs when

$$\lambda_0 = \pi\sqrt{P/2k_f}$$

and this is the dominant creep buckle half-wavelength.

The time constant for this dominant wavelength is then found by equation (6.11) to be

$$T_0 = 4k_f\mathcal{J}/C_iP^2$$

Now for the case in question we have

$$P = Dh\bar{\sigma}$$

$$C_i = n\bar{\varepsilon}/\bar{\sigma}$$

$$\mathcal{J} = Dh^3/12$$

and

$$k_f = \rho_w g D$$

where ρ_w is the density of sea water. This gives us, finally, the wavelength and time constant for the dominant creep buckle under the conditions described. They are:

$$\lambda_0 = \pi\sqrt{\bar{\sigma}h/2\rho_w g} \tag{6.12}$$

$$T_0 = \rho_w gh/3n\bar{\varepsilon}\bar{\sigma} \tag{6.13}$$

We can now insert reasonable parameters into these equations and calculate what sort of creep buckling behaviour is theoretically expected. Consider, for example, a sheet of sea ice of thickness $h = 1.5$ m subjected to an average stress $\bar{\sigma} = 0.5$ MPa. By equation (4.9) this corresponds to an average strain-rate $\bar{\varepsilon}$ of about $3\times10^{-7}\,s^{-1}$, assuming the sea ice to be of salinity 6 ppt and temperature −10°C. Inserting these figures into equations (6.12) and (6.13) we find that the creep buckle half-wavelength is

$$\lambda_0 = 19.2 \text{ m}$$

and the time constant is:

$$T_0 = 3 \text{ h}$$

The same calculation performed for a stress of $\bar{\sigma} = 0.25$ MPa gives:

$$\lambda_0 = 13.6 \text{ m}$$
$$T_0 = 45 \text{ h}$$

These figures accord very well with observations. As might be expected, the buckle half-wavelength is not very strongly dependent on stress, but is typically about 10 times the ice thickness; the buckle rate, however, is very sensitively dependent on stress, and the calculation agrees with the observation that a creep buckle normally forms in a matter of hours or days.

The subject has also been examined numerically by Sjölind (1984, 1985). He models the ice as a viscoelastic plate on elastic foundation and describes the properties of the ice by a non-linear power creep law similar to equation (6.7). Deformation patterns are then found by numerical integration using the finite-element method. These models display the most important features of creep buckling, but require finite initial perturbations at a variety of wavelengths in order to allow a creep buckle to form. The models as presented deal with ice of thickness only 0.2 m, with rather large initial perturbations of 0.05 m. I am not aware of any numerical studies which attempt to simulate the creep buckling of ice of thickness 1–2 m with realistic initial perturbations.

In any event, although creep buckling is an interesting and real phenomenon it is generally associated with rather slow loading processes and is unlikely ever to constitute the design condition for offshore structures.

6.2.3 Indentation analysis

Consider now the problem of a plate of ice of thickness h moving at velocity U past a vertical indentor of width D. This is the classical indentation geometry we have already looked at in Chapters 4 and 5, and is shown in Fig. 4.23. The geometry as shown is somewhat idealised, since the ice is represented as a perfect semi-infinite sheet presenting its free edge exactly parallel to, and in perfect contact with, a flat-punch indentor. However, it is worth looking at the formalism for this simplified problem; we shall then look later at the principal refinement needed to make it cope better with reality. The principal refinement required turns out to be that of imperfect contact, and beside this problem refinements to such details as indentor shape (for instance, a circular indentor) are of minor importance.

In the absence of fracture, (that is at very slow movement rates) the problem we are trying to solve is shown schematically in Fig. 6.6. Ice, a creeping material, flows around the indentor placed in its way. What we are interested in is the load P exerted on the structure as the ice moves past at velocity U. Of course this is in general unrealistic, since fracture is almost invariably observed in practice, but for the present analysis we shall assume the deformation to be strictly a continuum process, and then look at the onset of fracture.

Because ice is a non-linear creeping material there is unfortunately no exact solution to the creep indentation problem. There are, however, two major formal theoretical treatments of the

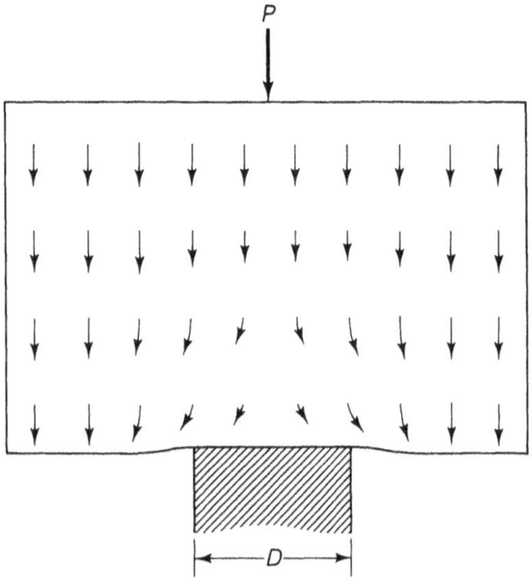

Fig. 6.6 Schematic diagram of creep flow field around an indentor.

indentation problem which provide good approximate solutions. They are:

(a) *Plastic limit analysis* (Ralston, 1977b, 1978, 1979). This approach idealises ice as a perfectly plastic material by calculating stress at a given strain-rate and adopting this as a plastic "yield stress". This is an imperfect model of ice behaviour, but has the advantage that all the considerable literature on plastic limit analysis can be applied to the problem. It takes no account of brittle behaviour, but treats all failure as a plastic yield.

(b) *Reference stress method* (Ponter *et al.*, 1983). The reference stress approach has the advantage that a proper formulation of realistic continuum deformation properties of ice can be incorporated into the analysis. The disadvantage is that the method itself is an approximation. Again, it takes no account of brittle behaviour but treats all failure as creep deformation.

I shall treat each of these approaches in turn: they each have limitations but have gained general currency and are worth explaining in some detail. A further alternative treatment is given by Bruen and Vivatrat (1984), who present a method of force calculation based on velocity strain-rate fields. Again, this is a continuum creep analysis. It adopts plane strain, linear viscous flow fields and calculates stresses using a realistic creeping flow law for ice. Solutions have not yet been presented for plane stress conditions nor for the geometry we have defined here, and I shall not consider it further.

6.2.3.1 Plastic limit analysis

The first problem in applying plastic limit analysis is that a plastic yield stress for ice must be chosen. This is a major difficulty, since ice does not have a plastic yield stress: it is a non-linear creeping material and it has behaviour which is intermediate between viscous Newtonian flow and purely plastic yield (Fig. 6.7). In terms of Norton's general flow law, uniaxial strain-rate $\dot{\varepsilon}$ is related to stress σ by a law of the form

$$\dot{\varepsilon} = (\sigma/\sigma_Y)^n$$

where σ_Y is a normalising stress. Newtonian flow (purely viscous) corresponds to $n = 1$, while plastic yield behaviour corresponds to $n = \infty$, with σ_Y representing the yield stress. A purely plastic material undergoes no deformation at all at stresses less than σ_Y, but yields catastrophically when the yield stress is reached. For ice, as we have seen in Chapter 4, $n \approx 3$.

Since ice "yields" to some extent at all stresses, however low, it cannot be said to have a yield stress. Instead, it has a certain finite rate of deformation associated with any applied stress. As ice deforms in contact with a structure it develops a complicated stress field: stress in the ice varies with distance from the structure and at a given point the ice deforms in response to the stress at that point. So not all the ice is deforming at the same stress or at the same rate.

However, the problem can be simplified and expressed in terms of a single stress corresponding to a single strain-rate. These are representative values and are an abstraction. (Formally they correspond to the reference stress and reference strain-rate of Ponter *et al.* (1983), which we shall look at in Section 6.2.3.2.) If the single representative stress and strain-rate are properly chosen, then the indentation problem may be successfully analysed using plastic yield formalism.

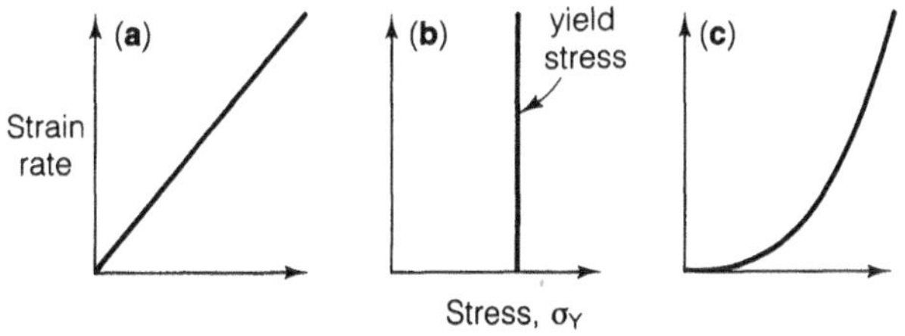

Fig. 6.7 Principal types of deforming material: (a) Newtonian (viscous, $n = 1$); (b) plastic ($n = \infty$); (c) power-law creep ($n = 3$).

First, it is worth giving a brief introduction to principles of plasticity analysis; for a full introduction, however, refer to Calladine (1969). As we have seen, the fundamental tenet of plasticity analysis is that a material will yield if anywhere it is subjected to a stress exceeding the yield stress. This is shown for a uniaxial stress in Fig. 6.7(b). For a more general stress-state we have to define a yield criterion involving all three principal axes of stress, σ_{11}, σ_{22} and σ_{33}. There are two common ways of doing this, both of which involve assessing the deviation of stress from pure hydrostatic: loosely speaking, if the stress field deviates from hydrostatic by more than the yield stress then yield occurs. The two criteria are as follows:

(a) The *Tresca* criterion: yield occurs if

$$(\sigma_{max}-\sigma_{min}) > \sigma_Y$$

where σ_{max} and σ_{min} are the maximum and minimum principal stresses.

(b) The *von Mises* criterion: yield occurs if

$$\tfrac{1}{2}[(\sigma_{11}-\sigma_{22})^2+(\sigma_{22}-\sigma_{33})^2+(\sigma_{33}-\sigma_{11})^2] > \sigma_Y$$

where σ_{11}, σ_{22} and σ_{33} are the principal axes of stress.

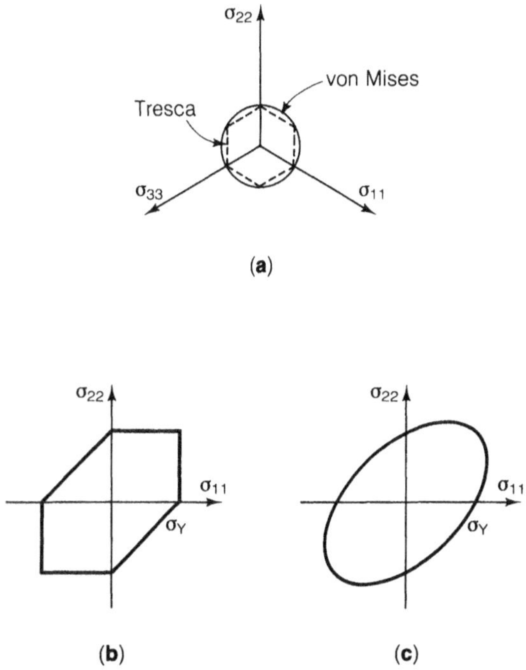

Fig. 6.8 Plasticity analysis: the Tresca and von Mises yield criteria: (a) looking down the axis of pure hydrostatic stress, $\sigma_{11} = \sigma_{22} = \sigma_{33}$; (b)Tresca criterion under biaxial conditions, $\sigma_{33} = 0$; (c) von Mises criterion under biaxial conditions, $\sigma_{33}=0$.

Both these criteria may be visualised by imagining we are looking down the axis $\sigma_{11} = \sigma_{22} = \sigma_{33}$ in principal stress space (Fig. 6.8). This line, which passes through the origin, represents the line of pure hydrostatic stress. The von Mises criterion then looks like a cylinder centred on the hydrostatic axis, and if a stressstate falls outside the cylinder then yield occurs. The Tresca criterion, on the other hand, is an inscribed hexagonal-shaped "cylinder", again centred on the hydrostatic axis. In pure biaxial conditions, when $\sigma_{33} = 0$, the two criteria appear as in Figs 6.8(b) and (c). They are saying very much the same thing:

the von Mises criterion is more physically realistic, while the Tresca criterion is easier to handle analytically.

Plasticity analysis then proceeds by the use of two central theorems, the Lower Bound Theorem and the Upper Bound Theorem.

The Lower Bound Theorem states the following: if any stress distribution throughout a loaded body can be found which is everywhere in internal equilibrium and balances the external applied loads, and at the same time does not violate the yield criterion, then these loads will be safely carried by the body. This is a lower bound to the collapse load.

The Upper Bound Theorem states the following: if an estimate of the plastic collapse load of a body is made by equating internal rate of energy dissipation to the rate at which external forces do work according to any postulated deformation mechanism, then the estimate will be either high or correct. It is an upper bound to the collapse load.

Note that the postulated stress fields and deformation fields do not need to be the actual ones. They merely provide upper and lower bounds to loads, and can be chosen by trial and error. The art is to find upper and lower bounds which are close to each other and thus closely define the actual load which lies between them. In certain classic cases (for instance the Prandtl solution, see below) the two bounds coincide and the solution is exact.

Now, in applying plastic limit analysis to the ice indentation problem we proceed as follows.

Indentation pressure. When we apply a load P to ice of thickness h over a width D the indentation pressure is, by definition, P/Dh. If we are dealing with a purely plastic isotropic material then plastic limit analysis gives classical solutions to this indentation problem in terms of the plastic yield stress σ_Y and shows that for plane strain conditions (Fig. 4.23(b), $D \ll h$):

$$P/Dh = \frac{\pi+2}{\sqrt{3}}\sigma_Y = 2.97\sigma_Y \qquad (6.14)$$

That is, the indentation pressure under plane strain conditions may reach almost three times the yield stress. This is the Prandtl solution using an isotropic von Mises yield criterion (see, for instance, Calladine, 1969). It also assumes a condition of free slip at the indentor surface. If the Tresca criterion is used instead (Croasdale *et al.*, 1977), the solution is:

$$P/Dh = \frac{\pi+2}{2}\sigma_Y = 2.57\sigma_Y$$

Here we shall use the von Mises criterion throughout; results obtained using the Tresca criterion are in general up to 15% lower.

For plane stress conditions (Fig. 4.23(a), $D \gg h$), which is the case of interest for wide structures, another exact solution exists (Croasdale *et al.*, 1977). It is, again for the von Mises criterion:

$$P/Dh = 2\sigma_Y/\sqrt{3} = 1.15\sigma_Y \qquad (6.15)$$

Now, equations (6.14) and (6.15) are both of the form

$$P/Dh = I\sigma_Y \qquad (6.16)$$

and the numbers $I = 2.97$ and $I = 1.15$ correspond to the two limiting cases of aspect ratio D/h: plane strain ($D \ll h$) and plane stress ($D \gg h$). The factor I is known as the *indentation factor*, and can also be calculated for intermediate values of D/h (Morgenstern and Nuttall, 1971). For $D = h$ it is 1.55, and a graph of indentation factor I as a function of aspect ratio D/h is given in Fig. 6.9(a). The dashed line provides a similar analysis for the case of the rough indentor for which a no-slip condition exists at the contact surface. Both curves refer to an isotropic perfectly plastic material.

The indentation factors which apply for an anisotropic material such as columnar-grained ice are different, though not straightforward to calculate. For an orthotropic material such as columnar ice, as described in Chapter 4 (Section 4.2), the indentation factors exceed those of an isotropic material at the same aspect ratio. Numerical analyses by Ralston (1978) and Prodanovic (1979) have been carried out to determine the indentation factors which would be expected to apply to anisotropic columnar ice. Their analyses use a generalised anisotropic von Mises yield criterion, but the anisotropic function they define is rather different from that defined in my Section 4.2.5 on constitutive equations. In particular, they specify unequal strength in tension and compression, and sensitivity to hydrostatic pressure. While this may be true for loading at high strain-rates, when fracture influences the behaviour, it is not true for proper continuum creep. If fracture occurs, then, as we have seen in Chapter 4 (Section 4.3), tensile fracture occurs more easily than compressive fracture, and fracture can be suppressed by applying an overall hydrostatic pressure. Pure creep, however,

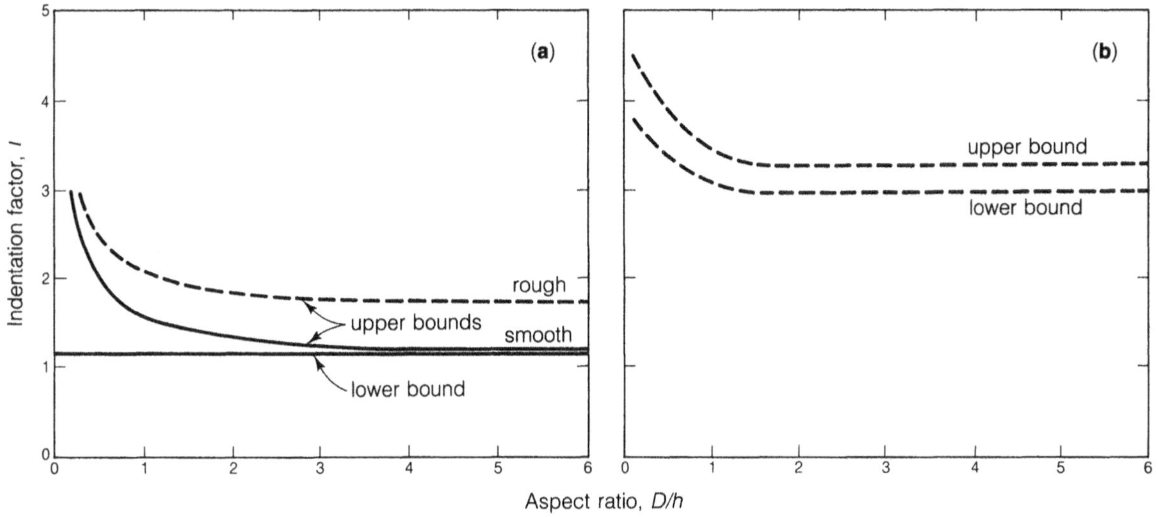

Fig. 6.9 Theoretical indentation factors I as a function of aspect ratio D/h. (a) Isotropic material, for rough and smooth indentor. (b) Anisotropic material, modelled to represent horizontal loading of columnar sea ice against a smooth indentor (after Ralston, 1978).

is effectively independent of hydrostatic pressure and is equal in tension and in compression. For these reasons, the analyses by Ralston and Prodanovic should be treated with some caution, since they are designed to apply in some way to transitional creep-fracture behaviour. The results are that $I \sim 4$ for plane strain and $I \sim 3$ for plane stress, and a graph of indentation factor as a function of aspect ratio is shown in Fig. 6.9(b).

Experimental data for S-2 ice indentation can be used to support either these anisotropic factors (Ralston, 1978) or the simple isotropic factors (Taylor, 1973). The data are not conclusive, mainly because it is impossible to define a universal uniaxial compressive strength to which indentation pressures can be normalised. The calculated factors might also be slightly different if an analysis were done using the anisotropic parameters defined in Section 4.2.5. I have not done this.

A summary of indentation factors as derived by Ralston (1978) is given in Table 6.2. In many applications we might expect to be dealing with unknown degrees of anisotropy, unknown relative proportions of granular and columnar ice and with geometries which are intermediate between plane stress and plane strain. All things considered, it seems sensible to adopt a value of $I = 3$ for most circumstances.

Yield stress. We still have to select a yield stress σ_Y. As we have seen, ice has no such thing as a yield stress in the proper sense of the term. Stress is simply a function of the rate at which the ice is being deformed, and ice deforms at all stresses, however low. For plasticity analysis the customary approach to selecting a yield stress is to adopt the stress corresponding to a "representative" ice strain-rate during the indentation process (Michel and Toussaint, 1977; Ralston, 1979). This can be obtained through a semi-empirical approach by comparing experimental indentation data with the large body of uniaxial compression data available.

TABLE 6.2 The parameters I and Ψ to be used in continuum indentation analysis. Values of I are from Ralston (1978). Values of Ψ (isotropic case) are from Ponter *et al.* (1983). For an anisotropic material, Ψ is assumed to approach the value for an isotropic material under plane strain (see Section 6.2.3.2)

	Granular ice (isotropic)		*Columnar ice (anisotropic)*	
	I	Ψ	I	Ψ
Plane strain, $D \ll h$	2.97	0.45	4.12	0.45
Plane stress, $D \gg h$	1.15	0.39	3.13	0.45

During indentation the ice is not being subjected to one uniform strain-rate, as it is during uniaxial loading, but from dimensional considerations on the indentation geometry it is clear that the ice strain-rate must be proportional to either U/D or U/h. It is generally postulated that D, not h, is the appropriate length parameter, since this is certainly true for the same geometry in the continuum elastic or viscous case. Note, however that this may cease to be true when the material fails by fracture and continuum conditions no longer apply. Spalling and flexural fracture, for instance, are probably dependent on thickness h rather than diameter D, but are anyway outside the scope of this present continuum indentation analysis. Fracture processes are treated later in this chapter (Section 6.4).

So, let us assume that indentation strain-rate is

$$U/\alpha D$$

where α is some constant to be determined empirically. (Notice that for this analysis we have not shown that α truly is a constant: in the next section, on reference stress analysis, we shall see that it actually depends on aspect ratio.) The empirical fit to determine α is done simply by plotting indentation data for S-2 ice on the same graph as comparable uniaxial results, and varying the stress and strain-rate axes until a good fit is obtained. For the uniaxial data we plot uniaxial stress against uniaxial strain-rate. For the indentation data, however, we plot indentation pressure divided by indentation coefficient, that is,

$$P/IDh$$

against strain-rate

$$U/\alpha D$$

and we vary α to obtain the best fit. For I we use 3 as a universal value (following Michel and Toussaint (1977) and Ralston (1979)), and we find that a good fit can be obtained with $\alpha = 4$ (Fig. 6.10(a)). The fit is quite good, but not highly sensitive, and Ralston (1979) adopts $\alpha = 2$ as being a more conservative assumption (Fig. 6.10(b)). These graphs are effectively empirical universal curves for S-2 ice and allow uniaxial data and indentation data to be plotted onto the same curve. The normalisation has, however, little theoretical founding, and is not unique.

Application of method. In summary, the plastic limit methodology for calculating forces during indentation is as follows:

(1) From the velocity U of the ice and the diameter D of the structure, calculate

$$U/2D$$

Adopt this as indentation strain-rate $\dot{\varepsilon}$.

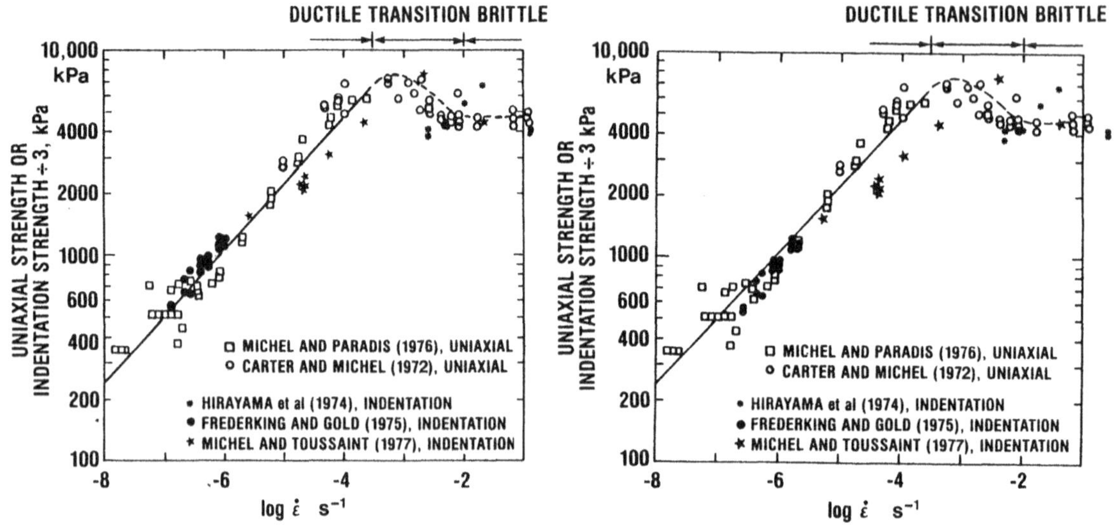

(**a**) $\dot{\varepsilon}_{eff} = U/4D$ (unconservative) (**b**) $\dot{\varepsilon}_{eff} = U/2D$

Fig. 6.10 Proposed universal curves for indentation of S-2 ice (solid symbols) plotted to coincide with uniaxial compression data (open symbols). Indentation pressure is divided by a factor $I = 3$ and indentation strain-rate is defined alternatively as (a) $U/4D$ (Michel and Toussaint, 1977), and (b) $U/2D$ (Ralston, 1979).

(2) Calculate the *uniaxial* stress which would correspond to this strain-rate if it were a uniaxial strain-rate. This can be done using the equations of Chapter 4 (equations (4.9) and (4.10) as appropriate). We denote this stress by

$$\sigma_Y = F_\sigma\{\dot{\varepsilon}\} = F_\sigma\{U/2D\}$$

The expression $F_\sigma\{\dot{\varepsilon}\}$ means "the stress corresponding to strain-rate $\dot{\varepsilon}$"; similarly, the symbol $F_{\dot{\varepsilon}}\{\sigma\}$ may be used to denote "the strain-rate corresponding to stress σ".

(3) Adopt σ_Y as a plastic yield stress in the formalism of plasticity analysis.

(4) Calculate load P as

$$P = IDhF_\sigma\{U/2D\} \qquad (6.17)$$

where $I \approx 3$ but, strictly, takes values between 1.15 and 4 as defined by Table 6.2.

Criticism. There are two drawbacks to this methodology.

(a) Ice is not a perfectly plastic material, and there is no clear justification for assigning it a strain-rate dependent yield stress.

(b) The strain-rate adopted, $U/2D$ *or* $U/4D$, is entirely empirical and is derived from indentation data (Michel and Toussaint, 1977; Frederking and Gold, 1975; Hirayama *et al.*, 1974) in which the aspect ratio varies between 0.2 and 4. There is no proof that α is independent of aspect ratio. It may vary significantly over this limited D/h range and even more so when $D/h \sim 100$, as it may be for full-scale interaction.

Fortunately, the paper by Ponter *et al.* (1983) has helped to resolve these difficulties by use of an entirely different approach. It enables the problem to be put into secure theoretical framework.

6.2.3.2 Reference stress method

General. The engineering problems associated with creeping materials are not peculiar to ice. They arise also, for instance, in the design of high-temperature steel pressure vessels, which may involve the creep of complex-shaped solids under complex stress fields. Since the mid-1960s, a powerful tool has been available to tackle these problems. This is the reference stress method (Anderson *et al.*, 1963; Sim, 1968; MacKenzie, 1968; Pomeroy, 1978).

The method is approximate and generally works by expressing the behaviour of a creeping material of power law n (equation (6.7)) in terms of two limiting cases: normally $n = 1$ (the linear viscous case, for which elastic solutions can be used) and $n = \infty$ (the perfectly plastic case). For many stress geometries, exact solutions exist for these two cases and allow an approximate analytical solution to be obtained also for intermediate values of n. Typical values of the creep index n for structural steels lie in the range $1 < n < 6$, and for ice we have $n = 3$.

The reference stress method does *not* assume ice to be viscous or plastic. It incorporates correct creeping material properties into the formalism and determines other parameters in the equation (analogues of I and α in Section 6.2.3.1 above) by matching the creep case to the limiting cases of plastic and viscous behaviour. The advantage of this approach over the plastic limit analysis above is that the various factors in the equation are derived by the reference stress formalism and are not empirical. The results can be shown, as above, to agree with indentation tests in laboratory conditions (e.g. Michel and Toussaint, 1977), but the formalism justifies the extension of the theory to other aspect ratios outside the limited range covered by experiment.

The basis of the reference stress method is explained by Ponter *et al.* (1983). In outline it is as follows. Consider a body of creeping material subjected to a stress (in our case, being indented). The deformation of this body, plotted against time, is broadly similar in form to that of the same material undergoing uniaxial creep, and it can be shown that deformation may be approximated by multiplying the uniaxial creep strain-rate at a certain stress level by an appropriate scaling factor (Goodman and Goodall, 1978). The important point is this: it is possible to choose a stress level – the so-called reference stress – and a scaling factor, in such a way that any error due to variation of flow parameter n is minimised (Sim, 1968). This is achieved by choosing these two parameters to be exact for two estimates of n either side of the real value. If we choose $n = 1$ and $n = \infty$, then classic solutions often already exist.

Method. We begin by writing a fundamental work equation for the stresses and strain-rates in a body as it is subjected to a load P moving at velocity U. The rate of work is:

$$PU = \int_V \sigma_{ij}\dot{\varepsilon}_{ij}\, dV$$

integrated over the volume V of the whole body. This is a nightmare to evaluate precisely for a statically indeterminate body, so we simplify to:

$$PU = \sigma_R F_{\dot{\varepsilon}}\{\sigma_R\} V_R \quad (6.18)$$

where σ_R is a reference stress, $F_{\dot{\varepsilon}}\{\sigma_R\}$ is the strain-rate corresponding to this reference stress (using the material's real stress/strain-rate relationship), and V_R is a reference volume which can be determined analytically for certain easy values of n.

Now the reference stress σ_R is given according to plastic yield solutions as (Sim, 1968):

$$\sigma_R = P(\sigma_Y/P_Y)$$

where P_Y is the rigid-plastic collapse load for a material of yield stress σ_Y. Note that the ratio

$$\sigma_Y/P_Y$$

is dependent only on geometry for a given loading pattern. It is independent of σ_Y, so the choice of σ_Y is arbitrary: we are not assigning ice a yield stress.

The reference volume V_R is then calculated by applying equation (6.18) to another easy value of n (generally $n = 1$, the viscous case), calculating $F_{\dot{\varepsilon}}\{\sigma_R\}$ for this case and solving for V_R. Loosely speaking, V_R is the volume within which most of the work is done. We have then determined the two unknowns σ_R and V_R in equation (6.18) and can apply it directly to ice (or any other material) simply by inserting $F_{\dot{\varepsilon}}\{\sigma_R\}$ calculated for the uniaxial reference stress, using the laws of Chapter 4.

Indentation. For the indentor problem we can calculate the reference stress σ_R by using the plastic indentation solutions already discussed in Section 6.2.3.1 and we find, by equation (6.16), that:

$$\sigma_R = P/IDh \quad (6.19)$$

where I is the same plastic indentation factor as before. (In the notation of Ponter *et al.*, it is denoted Φ, but it is the same as I in Ralston's notation.)

Solving for V_R is less simple, since the $n = 1$ (elastic/viscous) solution to the flat indentation punch problem has an unfortunate instability (Sadowsky, 1928). This means that an alternative solution for another value of n has to be found. Ponter *et al.* (1983) give a finite-element solution at $n = 3$ using the initial strain method (Ponter and Brown, 1978). This supplies V_R and we proceed to the final formula. It is helpful to introduce a new variable, Ψ, known as the compatibility factor, defined by:

$$\Psi = V_R/D^2hI \quad (6.20)$$

The design load equation is then found by rearranging equations (6.18), (6.19) and (6.20) to give:

$$U/I\Psi D = F_{\dot{\varepsilon}}\{P/IDh\}$$

or, more usefully,

$$P = IDhF_{\sigma}\{U/I\Psi D\} \quad (6.21)$$

where $F_{\sigma}\{\dot{\varepsilon}\}$ is the stress corresponding to strain-rate $\dot{\varepsilon}$.

Notice the remarkable similarity between equation (6.21) and that obtained by Ralston by quite different means (equation (6.17)). It differs only in the quantity $U/I\Psi D$ instead of $U/2D$ as the representative strain-rate.

The parameter I is as defined before and tabulated in Table 6.2: Ponter *et al.* (1983) do not provide calculations for the anisotropic case, but it seems reasonable to adopt the values calculated by Ralston (1978). The value of Ψ is provided by Ponter *et al.* for the isotropic case, and ranges between 0.39 and 0.45. For the anisotropic case it is reasonable to suppose that anisotropy will make the stress conditions approach more closely to plane strain (restricted vertical movement), and hence I suggest that a constant value of $\Psi = 0.45$ should be adopted, irrespective of aspect ratio. Bear in mind that in the case of a material which is highly anisotropic (very stiff in the vertical direction, and the parameter a in equation (4.21) tending to infinity), the distinction between plane stress and plane strain disappears entirely. This is, no doubt, why the value of I for anisotropic plane stress conditions ($I = 3.13$) is very close to the value for isotropic plane strain conditions ($I = 2.97$).

It is readily calculated from Table 6.2 that the quantity $I\Psi$ for an isotropic material takes values between 0.44 (plane stress) and 1.33 (plane strain), according to aspect ratio. For an anisotropic material, $I\Psi$ ranges between 1.41 and 1.85.

We can check the validity of equation (6.21) by plotting P/IDh against $U/I\Psi D$ for available indentation data, just as Michel and Toussaint (1977) and Ralston (1979) plot P/IDh against $U/\alpha D$ for $\alpha = 2$ and $\alpha = 4$ in Fig. 6.10. Ponter *et al.* do this, and the result is shown in Fig. 6.11. They use the isotropic indentation factors of Table 6.2, not the anisotropic factors derived by Ralston (1978) and Prodanovic (1979). The agreement is good. The reason that two different plots (Figs 6.10 and 6.11) both apparently fit well is that many of the tests are fairly close to plane strain conditions. For plane strain $I\Psi = 1.33$, and since stress is propor-

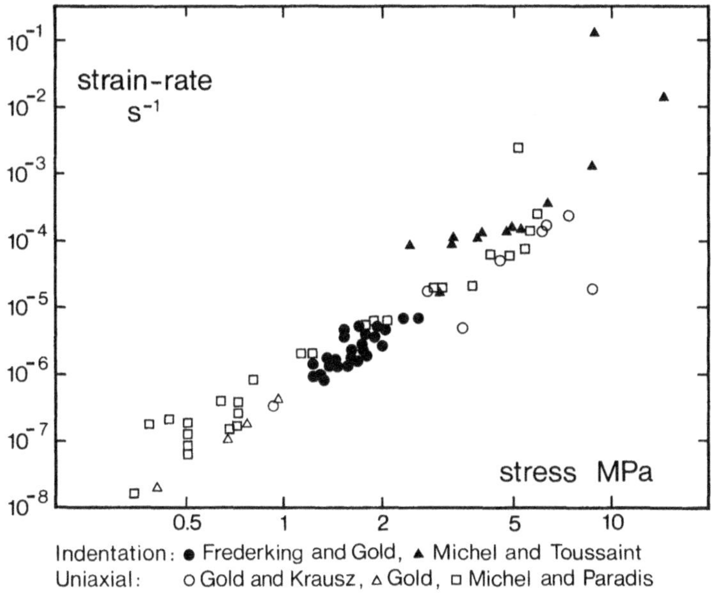

Fig. 6.11 Universal curve for indentation based on the reference stress analysis (Ponter *et al.*, 1983). Indentation pressure is divided by the parameter *I* and indentation strain-rate is defined as $U/I\psi D$. Both *I* and ψ are dependent on aspect ratio *D/h*. Data are from Frederking and Gold (1975), Michel and Toussaint (1977), Gold and Krausz (1971), Gold (1977) and Michel and Paradis (1976).

tional to the cube root of strain-rate, the use of equation (6.21) rather than equation (6.17) leads to stresses differing only by a factor of

$$(2/1.33)^{1/3} = 1.15 \quad \text{or } 15\%$$

This is negligible on the scale of the graphs. However, the equations for plane stress in an isotropic material differ substantially, by a factor of

$$(2/0.44)^{1/3} = 1.65 \quad \text{or } 65\%$$

Criticisms. The method is approximate. However, it has been successfully applied to many other diverse creep problems and it is well backed up by experiment (Pomeroy, 1978). A shortcoming in its application to the ice problem is that it has not yet been shown to cope properly with anisotropic creep. Both parameters *I* and Ψ in equation (6.21) and Table 6.2 should be properly recalculated for creep of anisotropic columnar ice. The anisotropic yield function adopted by Ralston and Prodanovic for the calculations in Table 6.2 is not necessarily appropriate to pure creep, but is the best answer available to date.

The most important thing to grasp about creep indentation is that we are not actually dealing with ice "failure". All we have calculated is the force exerted in order for deformation at a certain rate to take place. If we increase the deformation rate, the force increases and continues to be borne by the ice (even though it is deforming at a higher rate). In this sense the ice has not failed. For failure we have to look further, to fracture processes which cause the ice to lose its load-bearing capacity. This will be the subject of Sections 6.3 and 6.4.

6.2.3.3 Imperfect contact

Consider once more the indentation geometry we have been analysing. It is shown again, in plan view, in Fig. 6.12(a). A perfectly straight-edged plate of ice is perfectly aligned with a perfectly straight-edged indentor, and moves against it at a constant velocity, achieving perfect uniform contact. Is this realistic? Of course it is not quite realistic, and in fact it departs from reality in some highly significant ways. The most important of these is, I believe, the contact condition, which we shall look at briefly here. Another significant departure from reality is the assumption of suddenly applied and perfectly constant velocity: in practice, ice movement tends to develop gradually from a standstill, and we shall examine the implications of this in Section 6.3.2.

Here, however, let us look at some more realistic contact geometries. Two such geometries are shown in Figs 6.12(b) and (c). The first, Fig. 6.12(b), represents ice with a slightly irregular leading edge profile interacting with a flat-sided structure; the second, Fig. 6.12(c), represents ice which has developed around an irregular rubble pile or spray-ice island. In both cases, contact is only imperfect and is achieved typically over a length d_r which is only a small fraction of the total diameter D. Even in the case shown in Fig. 6.12(c), where the advancing ice has superficially the same shape as the indentor, contact is far from total: although the ice may have grown *in situ* around the indentor, in practice the presence of a tidal crack and the influence of any small lateral movements mean that the mating of shapes is most unlikely to be perfect. This can be seen conceptually by considering two interlocking pieces of a jigsaw puzzle (Fig. 6.12(d)): although they are nominally of exactly the same shape, in practice it will be found that if they are pressed together they make contact over only a very limited zone.

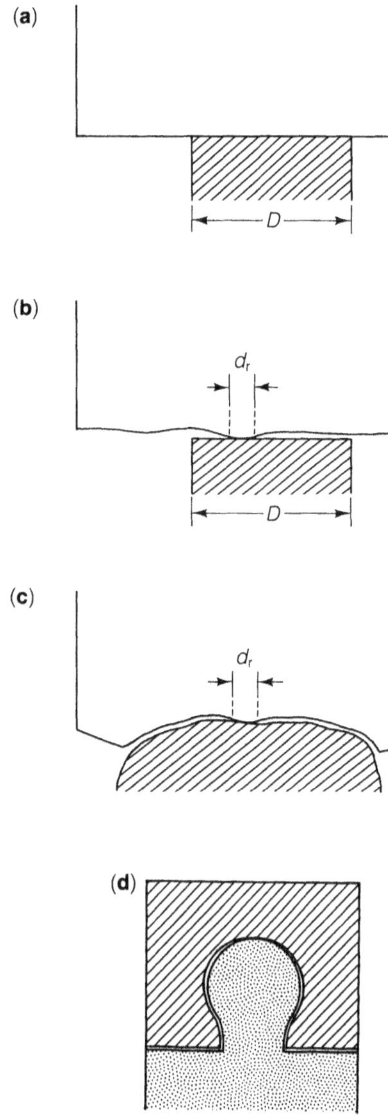

Fig. 6.12 Contact conditions: (a) ideal perfect contact; (b) irregular ice edge meeting a flat indentor; (c) irregular ice edge mating poorly with irregular indentor; (d) the jigsaw puzzle analogy.

Perfect contact of ice with an indentor is probably only ever achieved under those rare conditions when ice is adfrozen to the indentor. If, more realistically, the contact is imperfect, we have a geometry in which the relevant indentation width in equations (6.17) or (6.21) is d_r not D. In this case the reference stress equation for indentation load becomes

$$P = Id_r hF_\sigma\{U/I\Psi d_r\} \qquad (6.22)$$

For a given ice velocity U (and assuming $n = 3$) this force is lower than that predicted by equation (6.21) by a factor of

$$(d_r / D)^{2/3}$$

So if, for instance, ice is actually contacting over only 10% of the gross contact diameter D, we would expect measured loads to be only about 22% of those which would be calculated for full contact.

The problem of contact has been to some extent recognised in the literature on ice forces, and is enshrined in Korzhavin's equation (Korzhavin, 1962) in the form of a "contact factor" k (we shall look at Korzhavin's equation in some detail in Section 6.5). However, there seems to be confusion over its application: there is a tendency to assert that if contact is occurring over only a limited fraction k of the total diameter then we can simply modify equation (6.21) by writing

$$P = IkDhF_\sigma\{U/I\Psi D\}$$

This is wrong. If contact is really occurring over only a limited diameter kD then of course the expression

$$U/I\Psi D$$

must be altered to

$$U/I\Psi kD$$

This is the case for imperfect contact over only one single zone. It can easily be generalised to imperfect contact in which multiple zones are active (Fig. 6.13). Let there be n_d zones of contact diameter d_r $(r = 1, \ldots, n_d)$. The total force P is then

$$P = \sum_{r=1}^{n_d} Id_r hF_\sigma\{U/I\Psi d_r\} \qquad (6.23)$$

There are some interesting special cases of this, which are worth looking at:

(a) *n_d zones of width $d_r = \lambda_1 D$*. Consider the case where contact is achieved over n_d zones of constant width $d_r = \lambda_1 D$, where λ_1 is some proportion of the total diameter. Then

$$P = n_d I d_r h F_\sigma\{U/I\Psi d_r\}$$
$$= n_d I \lambda_1 D h F_\sigma\{U/I\Psi\lambda_1 D\}$$

In terms of the total contact factor k (the proportion of the diameter D which is in contact), we have

$$k = n_d d_r/D = n_d\lambda_1$$

and here:

$$P = IkDhF_\sigma\{U/I\Psi\lambda_1 D\} \qquad (6.24)$$

(b) *n_d zones of width $d_r = \mu_1 h$*. Alternatively, consider the case where the width of a contact zone d_r is a function not of diameter D but of ice thickness h. If $d_r = \mu_1 h$, where μ_1 is a constant, then we can write:

$$P = n_d I \mu_1 h^2 F_\sigma\{U/I\Psi\mu_1 h\}$$

In terms of contact factor k we would have

$$k = n_d\mu_1 h/D$$

and hence we could write:

$$P = IkDhF_\sigma\{U/I\Psi\mu_1 h\} \qquad (6.25)$$

Note that both these equations, (6.24) and (6.25), are merely theoretical illustrations, and physical evidence is required to show whether they are good models of reality. They illustrate, however, that if a contact factor k is to be used in the creep indentation equation then some physical assumption has to be made about how the imperfect contact is geometrically arranged: it affects the argument of F_σ, as shown in the two example cases of equations (6.24) and (6.25).

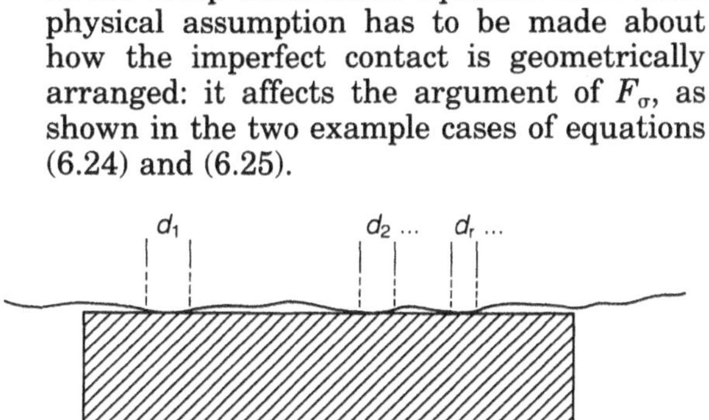

Fig. 6.13 Imperfect contact: n_d zones of width d_r ($r = 1,\ldots,n_d$) interact with a structure of width D.

6.3 TRANSITION TO FRACTURE

We move now to the question of fracture of ice, and at what point and for what reason the continuum treatment of ice becomes inadequate.

The onset of fracture is of prime importance to ice loading. If it did not occur, then the continuum indentation analysis we have given would imply that forces mononotically increase without limit as indentation rate increases. Further, fracture is obviously the dominant failure mode in nature: Arctic sea ice around an offshore structure presents a chaotic picture of multifarious and fantastically fractured blocks (see for instance, the cover photograph and Fig. 2.8(b)). There is very little apparent evidence of plastic or creep deformation modes; they can be found, it is true (Fig. 6.4), but you have to look and they are not the norm. What is more striking is the preponderance of fracture, in a wide array of manifestations: fine crushing, spalled blocks, delaminated slabs, blocks failed in flexure, extensive radial cracks. Not all of these are thoroughly understood, and in Section 6.4 we shall examine the formalism available to treat them. In this present section we shall look at two aspects of the onset of fracture: imperfect contact and loading-rate dependence.

6.3.1 Imperfect contact: local fracture

In Chapter 4 (Section 4.4) we saw that in the laboratory ice tends to make a transition to fracture behaviour when the indentation rate U/D exceeds about 10^{-2} s^{-1}. We also saw that there is some evidence that this "critical strain-rate" for onset of fracture may be scale-dependent: data from medium-scale and large-scale experiments (for instance, Netserk island, Section 5.4.1.4), suggest that the critical rate may be as low as 10^{-7} s^{-1} at full scale. This has been illustrated schematically in Fig. 5.11.

It seems to me that this must be connected with the phenomenon of imperfect contact. In the laboratory, perfect contact normally is achieved with a fair degree of precision, and hence fracture occurs when U/D exceeds about 10^{-2} s^{-1}. Consider, however, what happens during full-scale indentation with imperfect contact (Fig. 6.13 and equation (6.23)). It is clear that the onset of fracture will occur if *any one* zone of contact d_r reaches an indentation rate U/d_r which exceeds about 10^{-2} s^{-1}. If this occurs in any one zone then it is likely that the zone will fail in fracture, cease to be load-bearing, and will shed load onto the remaining zones of contact. In this way the process is likely to progress rapidly until the whole front of contact, D, is failing by fracture in a series of discrete zones.

It is difficult to be quantitative about this process. As a mechanism it is no more than speculation, albeit reasonable speculation, since the geometry of imperfect contact has not yet been properly analysed. However, it is a mechanism which at least provides one reasonable explanation of the apparent scale-dependence of the onset of fracture. If, for instance, contact actually occurs in a number of discrete zones of width d_r of the order of the ice thickness h, then we would expect fracture behaviour to begin when

$$U/h > 10^{-2}\ \mathrm{s}^{-1}$$

For an island of width 100 m interacting with ice of thickness 1 m, this represents a critical velocity of $U = 10^{-2}\ \mathrm{m\ s^{-1}}$ (or $36\ \mathrm{m\ h^{-1}}$), which is 100 times lower than the critical velocity of $U = 1\ \mathrm{m\ s^{-1}}$ ($3600\ \mathrm{m\ h^{-1}}$) which would be anticipated if the true physical criterion were

$$U/D > 10^{-2}\ \mathrm{s}^{-1}$$

Although this is a plausible explanation, note that it is by no means the only one. It is not sufficient, for instance, to account for the critical velocity of only $2.25\ \mathrm{m\ h^{-1}}$ apparently observed at Netserk F-40 (Section 5.4.1.4).

6.3.2 Loading-rate conditions

There is another facet to the onset of fracture which may also help to explain the low thresholds at which fracture apparently occurs in nature. This concerns loading-rate.

As we discussed earlier, in Section 4.2, when ice deforms under load its total strain ε may be written as:

$$\varepsilon = \varepsilon_e + \varepsilon_d + \varepsilon_v$$

where ε_e is elastic strain, ε_d is delayed elastic (primary creep) strain and ε_v is permanent viscous (secondary creep) strain. Laboratory observations of crack formation (Gold, 1972; Sinha, 1982) suggest that the formation of first cracks is simply a function of the amount of delayed elastic strain ε_d. That is, it is independent of elastic and secondary creep strain. Cracks begin to form in S-2 ice of grain size 4.5 mm if the delayed elastic strain exceeds 1.0×10^{-4}. This critical delayed elastic strain, denoted ε_d^c, is inversely proportional to grain size, and it may be used to predict the stress at which first cracks will form for a variety of loading conditions.

Now, under conditions of constant applied stress σ, the delayed elastic strain at time t after application of the load is given by equation (4.1):

$$\varepsilon_d = \frac{C\sigma}{Ed}\{1 - \exp[-(a_T t)^q]\}$$

with parameters as defined in Section 4.2. (The parameter d here denotes grain size and is nothing to do with d_r, the width of a contact zone used above.)

Equation (4.1) treats the case of sudden application of a constant stress, and gives the type of curve which has been shown in Fig. 4.2. This is the behaviour to be expected in a laboratory dead-load test. A more interesting and more realistic case, especially for offshore Arctic applications, is that of a gradually increasing stress; this corresponds, for instance, to a large expanse of floating ice surrounding a structure and acted on by a slowly increasing wind. As the wind increases from 0 to $15\ \mathrm{m\ s^{-1}}$ (0–30 knots) the stress in the ice against the structure gradually and monotonically increases. Under such conditions we do not see a sudden application of constant stress.

This case of monotonically increasing stress has been treated by Sinha (1983a), and can be solved by stepwise numerical integration. Here we shall treat the special case of constant stress-rate (linearly increasing stress) and examine its implications for real loading conditions. Equation (4.1) above enables us, through the delayed elastic strain criterion, to predict exactly when first cracks should appear in a laboratory sample of ice under any applied stress conditions. Once ε_d exceeds ε_d^c, cracks begin to form.

Let us now make a further assumption: at full scale, in real field conditions, when inhomogeneous natural sea ice 1.5 m thick is interacting with a structure which may be 100 m wide, I propose that formation of first cracks is effectively synonymous with failure of the ice. That is, once first cracks appear, the ice makes a rapid transition from ductile to brittle behaviour. This is likely to occur since significant stress concentrators will in general be present and, if contact is imperfect, load will be transferred to other zones, encouraging progressive failure. Under well controlled laboratory conditions (using precisely milled flaw-free specimens) this is not quite true, and specimens continue to support load after cracks initiate; however, it seems reasonable to suppose that in real field conditions, brittle failure will follow soon after the formation of first cracks. We make this assumption.

Consider now the application of a gradually increasing stress to ice in contact with a structure. Assume that the stress increases from zero at a constant rate $\dot{\sigma}$. The delayed elastic strain ε_d at time t can then be evaluated by stepwise integration, and after N interval steps of Δt it is (Sinha, 1983):

$$\varepsilon_d = \frac{C}{Ed} \sum_{i=1}^{N+1} \dot{\sigma}\Delta t\{1 - \exp[a_T(N+1-i)\Delta t]^q\} \qquad (6.26)$$

When this exceeds ε_d^c we assume that brittle failure occurs and the stress is relieved. For a given rate of stress increase we can compute the time taken to failure (first cracks), and the maximum stress reached before this occurs.

Sanderson and Child (1986) have done this for a range of stress-rates. The results are shown in Fig. 6.14, where stress is plotted against time for tests involving a variety of constant stress-rates. They are done for horizontal loading of S-2 ice with 4.5 mm grain diameter at −10°C. In the shaded region of the stress–time space, we see formation of cracks and a departure from continuum behaviour. (This plot may be inappropriate for high stress-rates, where other brittle mechanisms may become important.)

We can also plot the maximum stress attainable before cracks occur, for a range of stress-rates. This is shown in Fig. 6.15, for S-2 ice as before, but also for representative sea ice of 7 ppt salinity, 20 mm grain diameter and temperature −10°C. The correction for sea ice is made simply by altering the net section stress due to brine volume, as described in Chapter 4, and inserting a larger grain size into equation (6.26). The results are similar. Grain size does not in fact alter the stress for formation of first cracks, since the critical delayed elastic strain ε_d^c is also grain-size dependent (Sinha, 1982). It may be calculated approximately, for S-2 ice, from the empirical equation

$$\varepsilon_d^c = (4.7\times10^{-7})/d$$

where d is grain size (m). For a grain size of 20 mm, $\varepsilon_d^c = 2.34\times10^{-5}$. The difference of the two curves in Fig. 6.15 is due entirely to brine volume effects.

The important point to notice at once is that high stresses of 5–6 MPa can only be obtained at very high loading rates, of the order 1 MPa s^{-1}, such as are frequently used in the laboratory. If the loading rate is lower than this then first cracks occur at a much lower stress. Let us consider now what stress-rates are likely to exist in nature.

Some data are available on the rate of stress increase experienced by real Arctic structures. Observations at Netserk F-40, in the Canadian Beaufort Sea during winter 1975–6, show stresses increasing from 0.5 to 1.0 MPa over the course of 8 h (see Section 5.4.1.4). Stresses measured by two

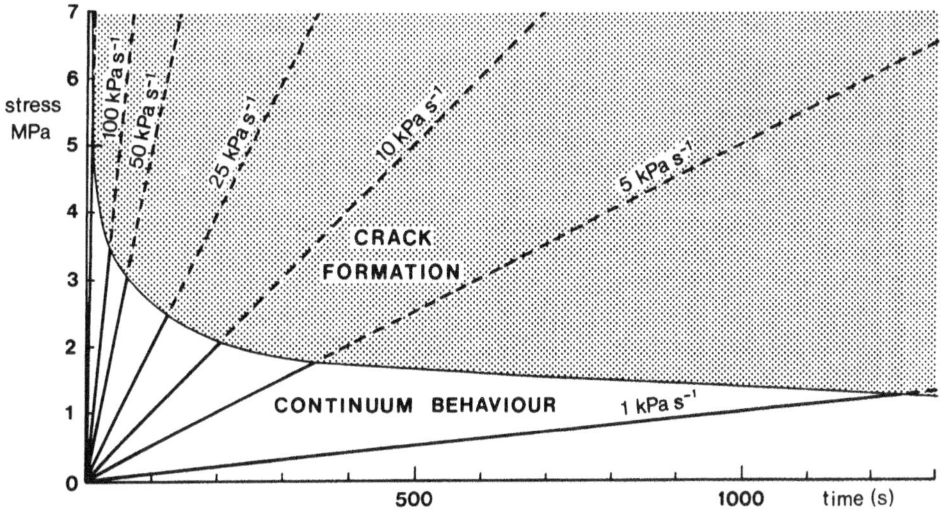

Fig. 6.14 Formation of first cracks during tests at constant strain-rate (after Sanderson and Child, 1986). The graph shows stress against time for a variety of stress-rates, and the shaded section represents crack formation. Notice that for low stress-rates the stress reaches only 1–2 MPa before cracks appear.

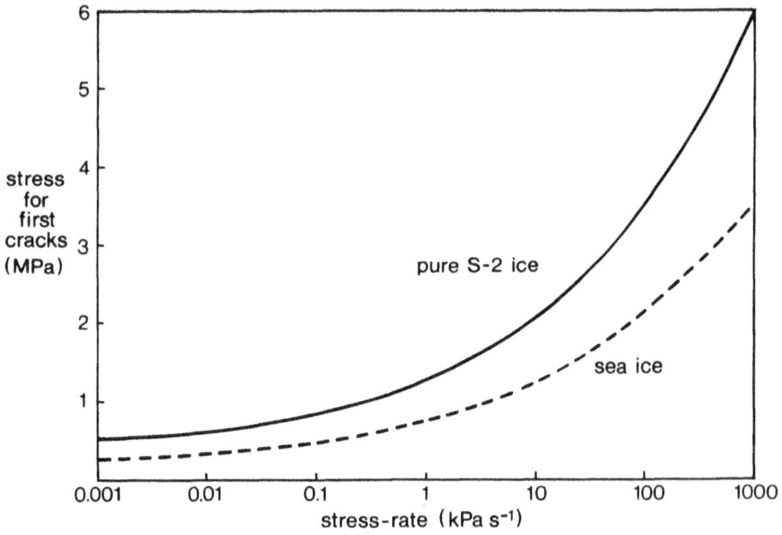

Fig. 6.15 Stress at which first cracks appear, as a function of stress-rate, for pure S-2 columnar ice of 4 mm grain diameter at −10°C (solid line), and saline columnar ice of 7 ppt salinity, 20 mm grain size at −10°C (dashed line) (after Sanderson and Child, 1986).

pressure panels have been shown in Fig. 5.10 together with ice velocity. The rates of stress increases are in the range only 0.01–0.035 kPa s^{-1} (5–18 psi h^{-1}). Similarly, measurements at Reindeer Island in 1979 (Ralston, 1979) show stress-rates of approximately 0.02 kPa s^{-1} (10 psi h^{-1}).

Referring to Fig. 6.15 it appears that, for these stress-rates, the stress theoretically attainable before first cracks appear is of the order of only 0.6–0.7 MPa for pure S-2 ice and 0.4 MPa for sea ice. This compares quite well with the data from Netserk (Fig. 5.10), in which peak stresses of 0.70–1.1 MPa are immediately followed by high movement rates and falling stresses. It may reasonably be suggested that this activity was associated with the onset of crack formation, enabling the ice to move rapidly forward in a brittle failure mode. Stresses experienced at Reindeer Island were similar, showing a peak of 0.72 MPa during the event of 7 March 1979 (Ralston, 1979). These stresses and stress-rates appear to be typical for winter loading conditions, though there is little justification for excluding the possibility of higher stress-rates and hence, higher stresses.

I should point out that this, again, is only one possible interpretation of the mechanism whereby the transition to fracture takes place at large scale in the field. It is not yet clear whether imperfect contact or stress-rate limitations are responsible or, indeed, whether some statistical effect of flaw-size distribution is responsible.

6.4 FRACTURE ANALYSIS

We turn now to the formal analysis of fracture failure modes: crushing (and the effects of flaw-size), radial cracking, spalling and flexural failure. Following this we will look at the phenomenon of non-simultaneous failure and we shall attempt to simulate a realistic dynamic impact event.

6.4.1 Crushing: the effects of flaw-size

We have already, in Section 4.3.3.2, introduced the formalism available for treating crack propagation and crushing failure under compression. It is interesting now to consider how we might cope with populations of flaws of different sizes, and to consider what light this throws on the apparent scale effect for compressive strength.

It is easy to call up general arguments to demonstrate why a brittle material such as ice should display a scale effect (Iyer, 1983; Sanderson, 1984a; Bazant and Kim, 1985). I have already briefly done so for the case of tensile fracture, in Section 4.3.2.2. The arguments generally concern, either explicitly or implicitly, the existence of flaws in the material, and tacitly make the assumption that flaws increase in size as sample size increases. The sample thus becomes weaker. It is, however, worth making it perfectly clear what these assumptions are, since they are often hidden.

Consider the compressive failure of a laboratory sample of ice containing a population of flaws. The

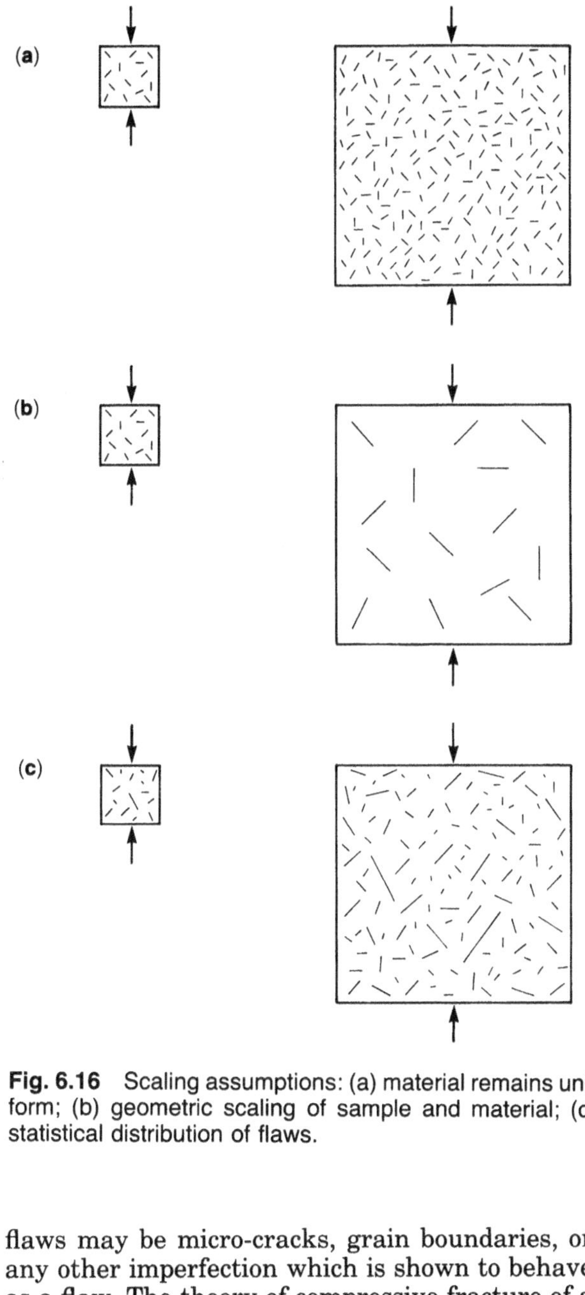

Fig. 6.16 Scaling assumptions: (a) material remains uniform; (b) geometric scaling of sample and material; (c) statistical distribution of flaws.

flaws may be micro-cracks, grain boundaries, or any other imperfection which is shown to behave as a flaw. The theory of compressive fracture of a flawed sample is as given by Ashby and Hallam (1986) and Hallam (1986), and explained in Section 4.3.3.2.

For illustration let us suppose the sample to be cubic of side 10 cm, to have regular flaws of size 5 mm and a compressive strength of 11 MPa. The sample is then tested under uniaxial compression. As we have seen, a strength of this magnitude can be explained well by linear elastic fracture mechanics (Section 4.3.3.2, equations (4.31) and (4.35)). Now, consider that an identical test is performed on a sample 100 times the size. There are three simple assumptions which can be made about scaling of the sample. They are shown in Fig. 6.16.

6.4.1.1 Constant flaw size

Firstly, assume that the material of which the sample is made really remains precisely the same (Fig. 6.16(a)) and continues to contain a population of regular flaws of size 5 mm. In this case, a simple fracture mechanics analysis would show that the compressive strength remains unchanged, at 11 MPa: flaws remain the same size and so the stress required to propagate them remains the same.

6.4.1.2 Geometric scaling

If, instead, the whole sample is strictly scaled up geometrically, its imperfections included (Fig. 6.16(b)), then the larger sample will contain flaws 100 times larger, that is, 0.5 m long. In this case a fracture mechanics analysis (or indeed a simple dimensional analysis) will show that if the linear dimensions of the sample increase by a factor ζ then the fracture strength will decrease by a factor $\zeta^{-1/2}$, (see, for instance, Bazant and Kim, 1985). The strength of the sample should therefore decrease by a factor of 10 (This can be verified simply by re-evaluating equations (4.31) or (4.35). The predicted strength is 1.1 MPa, instead of the previously predicted strength of 11 MPa.)

Although this is a convenient demonstration of a strength reduction of the order required, it is facile to claim that it is a scientific explanation. It contains the bold physical assumption that a larger sample contains larger flaws. This may in fact be true, but if it is, it requires experimental proof that doubling sample size automatically precisely doubles flaw size. This would in fact be most odd, as it actually implies that the material has changed. Nonetheless, it is interesting to note that this explanation is still not quite sufficient to explain the observed reduction of stress with area: simple scaling would imply

$$\sigma \propto \zeta^{-1/2} \quad \text{or} \quad A^{-1/4}$$

while in fact we observe approximately (as in Figs 5.36 and 5.38):

$$\sigma \propto A^{-1/2}$$

6.4.1.3 Statistical effects

Another class of scaling is shown schematically in Fig. 6.16(c). This contains the more realistic starting assumption that any material contains a statistical population of flaws of various sizes. On selecting a sample of larger size there is then a higher chance of it containing flaws from the tail-ends of the statistical distribution. There is thus a higher chance of it containing larger flaws. This argument, developed originally by Weibull (1951) is apparently powerful, but was intended principally for tensile fracture where the "weakest link" concept applies and a single controlling flaw is enough to lead to complete failure. This is not obviously true of compressive fracture. However, if compressive strength were indeed to behave in the same way we should expect strength to be a function of volume V of the form:

$$\sigma \propto V^{-1/b_f} \quad \text{or} \quad \zeta^{-3/b_f}$$

where b_f is a statistical parameter characterising the flaw-size distribution (see e.g. Dieter, 1981). The parameter b_f is not well known for ice, but for glass it lies in the range 2–5 and for rocks and ceramics it is about 15. Lavrov (1973) gives tests on freshwater ice which indicate a factor of $b_f \sim 15$, while direct statistical measurements by Gold (1972) for S-2 ice place b_f as low as 1 or 3. Jellinek (1958) presents a wide range of data on small-scale tensile strength of ice and derives values for b_f of between 1 and 2.

Now, in terms of contact area A, we expect A to be proportional to the square of the linear dimension ζ, so if $A \propto \zeta^2$ we should expect failure stress to have dependence defined by:

$$\sigma \propto A^{-3/2b_f}$$

In order to make this agree with the observed dependence on $A^{-1/2}$ we should conclude that b_f is of the order of 3, if the Weibull treatment is appropriate. This is consistent with most of the directly measured data already quoted. Figure 6.17 shows the pressure–area curve of Fig. 5.38, with superimposition of "theoretical curves" generated by both the geometric scaling hypothesis and the Weibull statistical hypothesis with $b_f = 3$.

The Weibull statistical argument is beguiling: it provides a pleasing general argument and apparently predicts ice strengths quite well. However, a lot is hidden behind the parameter b_f, and there has been no rigorous justification yet given to the use of such statistics in compression rather than tension. Let us therefore examine quantitatively the size and separation of flaws required.

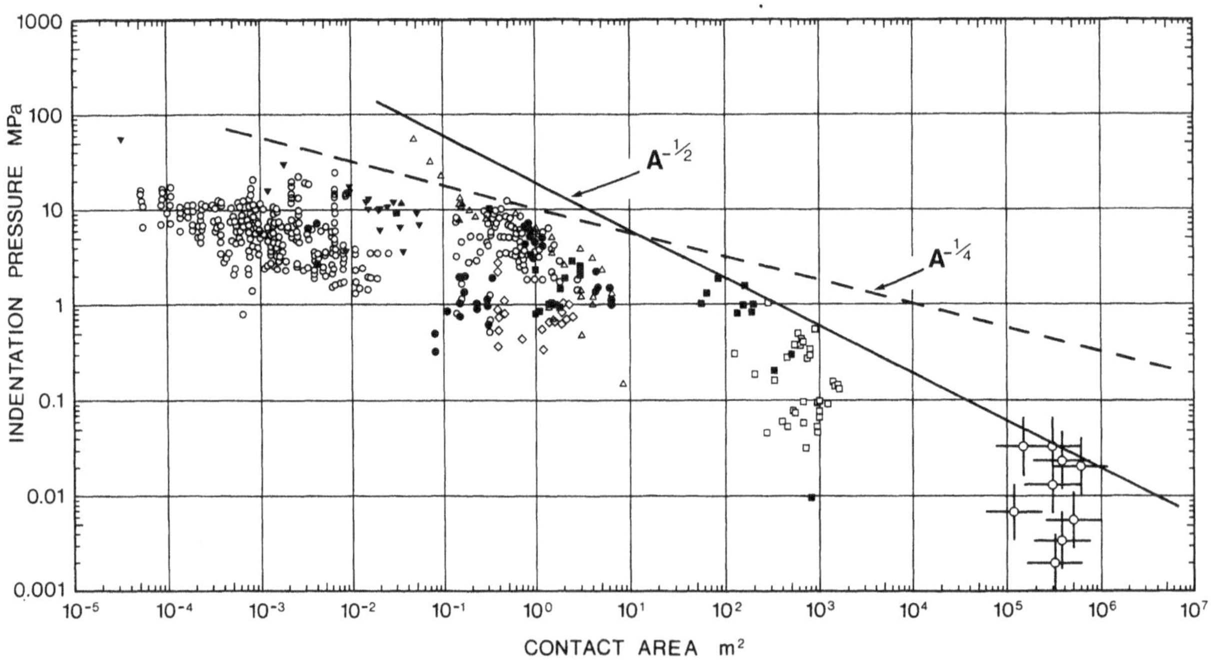

Fig. 6.17 The pressure–area curve of Fig. 5.38, with superimposition of theoretical curves generated by the geometric scaling hypothesis (dashed line) and the Weibull analysis with $b_f = 3$ (solid line).

Equation (4.34) is of general application, and allows us to calculate failure stresses for any combination of flaw-size $2a$ and flaw separation Δ_f. It is (in approximate terms):

$$\sigma = 3.5\frac{K_{IC}}{a}\sqrt{\frac{\Delta_f}{2}} \qquad \text{[equation (4.34)]}$$

For example, flaws of size $2a = 20$ mm separated by $\Delta_f = 40$ mm give a failure pressure of 4.9 MPa; flaws of size $2a = 1$ m separated by $\Delta_f = 10$ m lead to $\sigma = 1.6$ MPa. In Table 6.3, I have tabulated various interesting combinations of values, using equation (4.34) with $K_{IC} = 0.1$ MPa m$^{1/2}$. The table shows that compressive strength is highly flaw-size dependent. Note that very large flaws, such as 10 m cracks, drastically reduce ice strength even if they are very widely separated: at 1 km separation, a population of 10 m cracks would lower the theoretical strength to 1.6 MPa.

TABLE 6.3 Theoretical failure stress for ice under compression containing flaws of length $2a$ separated by distance Δ_f. Equation (4.34) is used

Length of flaws, 2a (m)	*Separation of flaws, Δ_f (m)*	*Theoretical failure stress (MPa)*
0.001	0.001	15.7
0.001	0.002	22.1
0.01	0.01	4.9
0.01	0.05	11.1
0.1	0.1	1.6
0.1	0.5	3.5
1	1	0.5
1	5	1.1
1	10	1.6
1	100	4.9
10	10	0.2
10	100	0.5
10	1 000	1.6
100	1 000	0.2
1 000	10 000	0.05
10 000	100 000	0.02

It is worth looking at the approximate flaw size required, on this model, to explain the pressure–area curve.

(a) Small- to medium-scale tests (groups A and B of Fig. 5.36(b)) show failure pressures of around 10 MPa. This is consistent, as we have seen, with flaws nucleated at boundaries of grains of diameter $d = 5$ mm.

(b) Large-scale tests, group C, show failure pressures of about 0.5 MPa – a 20-fold reduction. From Table 6.3, this is consistent with, for instance, 1 m length flaws at 1 m intervals or 10 m flaws at 100 m intervals. There is however, no positive evidence that a sufficiently large number of flaws of this size do indeed exist in naturally occurring first-year and multi-year ice covers. If the argument from flaws is to be believed, the existence of these flaws needs to be established.

(c) Meso-scale models, group D, show strength of the order 0.01 MPa, or 1000 times less than groups A and B. This would imply flaws with a size of several kilometres. Large as this may seem, it is actually quite realistic since, on the scale of the entire Arctic Ocean, it is likely that open leads and individual floes of size several kilometres do indeed behave as flaws. Any flight across the Arctic Ocean will confirm that a sufficient number of such flaws apparently exist.

6.4.2 Radial cracking

The indentation process, both at laboratory scale and at large scale, is almost invariably accompanied by radial cracking. This is particularly so at large aspect ratio. The fracture mechanical approach can be used to analyse the growth of such cracks (Palmer *et al.*, 1983). The analysis is approximate, and is carried out for a circular indentor, not a flat punch.

Consider the indentation geometry shown in Fig. 6.18(a). Under loading conditions in which only an axial load P is applied along the x-axis, the net force applied in the lateral y direction is, of course, zero. However, the biaxial constraint conditions as the ice interacts with the indentor generate lateral compressive stresses close to the indentor which are balanced by tensile stresses in the far field (Fig. 6.18(b)). (The lateral compressive stresses in the vicinity of the indentor are, indeed, precisely those which lead to an indentation factor of $I > 1$ in the creep indentation analysis (Section 6.2.3): the local presence of a compressive stress σ_{22} raises the stress σ_{11} required to achieve a certain level of strain-rate $\dot{\varepsilon}_{11}$.)

For present purposes, we do not need to know the precise form of the stress field close to the indentor, nor the precise shape of the indentor. In all cases we expect a zone to develop near the indentor which is under compression (and which may be undergoing crushing, spalling or any other

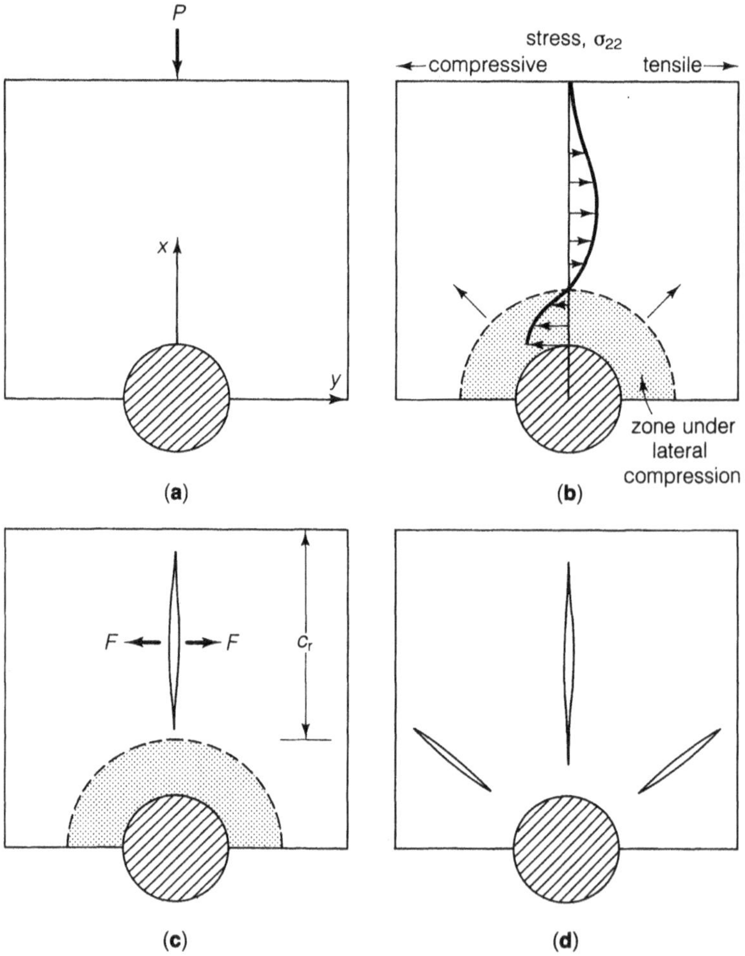

Fig. 6.18 Radial crack analysis: (a) geometry; (b) tensile and compressive stress fields (schematic); (c) formation of crack, propagating to length c_r; (d) subsequent formation of further cracks at 45°–60° separation.

local failure process). Outside this zone the lateral stresses are tensile, and if a crack exists or if the stress is sufficient to nucleate one then we can calculate how far the crack should propagate. In practice, it is probably unnecessary to reach a critical nucleation stress: in real field conditions involving a large expanse of natural ice cover it is practically certain that a pre-existing crack or flaw of some form will already be available within the tensile stress field.

The total far-field tensile force F (Fig. 6.18(b)) must be sufficient to balance the total lateral compressive force F existing in the ice close to the indentor. This force is generally of the same order as the axial applied load P, and investigations by Lawn *et al.* (1980) indicate that $F \sim P/2$ is an appropriate approximation. For generality, we write

$$F = \alpha_1 P$$

where α_1 is a constant to be specified.

Now consider the stability of a crack of length c_r in the tensile zone of the ice cover (Fig. 6.18(c)). The crack can be idealised as an edge crack in a semi-infinite plate wedged open by opposing crack-opening forces F. The stress intensity factor K_I for this geometry is given by Sih (1973) as

$$K_I = 2.59F/h\sqrt{\pi c_r}$$

where h is the ice thickness. This means that, once initiated, the crack in Fig. 6.18(c) should extend until the stress intensity at the crack tip falls below the fracture toughness K_{IC}. That is,

$$c_r = 2.13(\alpha_1 P/hK_{IC})^2 \qquad (6.27)$$

If we now take $K_{IC} = 0.1$ MPa $m^{1/2}$ and $\alpha_1 = 0.5$, we can calculate the load P required to propagate radial cracks of a certain length. A tabulation of these is given in Table 6.4, for ice of thickness 1.5 m. The average indentation pressure is also tabulated, on the basis of a 100 m wide structure sustaining the load.

TABLE 6.4 Radial crack length propagated for various loads P against a structure 100 m wide. Ice thickness is 1.5 m

Load, P (MN)	*Crack length, c_r*	*Indentation pressure (MPa)*
1	24 m	0.007
5	600 m	0.03
10	2.4 km	0.07
50	60 km	0.3
100	240 km	0.7

The numbers are remarkable: an average indentation pressure of only 0.3 MPa is apparently enough (in the absence of any externally applied lateral confining stresses) to propagate a tensile crack for 60 km. Although this may seem surprising, it is not unrealistic. Radial cracks of very great length (certainly exceeding 10 km) are very commonly observed in Arctic sea ice, especially around structures. Indeed, it is generally difficult to locate a crack tip; they are so extensive. Note that the existence of even quite small remote lateral stresses in the ice cover has a dramatic effect on the growth of a crack. If a tensile stress exists the crack growth may become unstable, while a compressive transverse stress exerts a stabilising influence. This has been analysed formally by Palmer *et al.* (1983).

The process of crack formation described is not limited to just one radial crack along the line of symmetry, but can be repeated in each of the new separated sectors formed by the initial crack. It is common for cracks to form around an indentor at spacings on the order of 45°–60° (Fig. 6.18(d)). The existence of such cracks does not of itself imply that the ice cover around the structure has failed and ceases to apply load. The cracks do, however, grossly alter the interaction geometry, and their presence should be taken into account, if possible, when analysing creep indentation or buckling processes.

6.4.3 Spalling

Spalling consists of the formation and propagation of an in-plane cleavage crack at some depth within the ice cover. This is shown schematically in Fig. 6.19 and an example of the resulting fracture surfaces is shown in Fig. 6.20. Spalling cracks tend to initiate at some distance from the point of contact with the indentor and then propagate unstably, generally deviating upwards or downwards to meet the surface. This results in spalling-off of chunks of ice of roughly semi-circular plan shape (Fig. 6.19(b)).

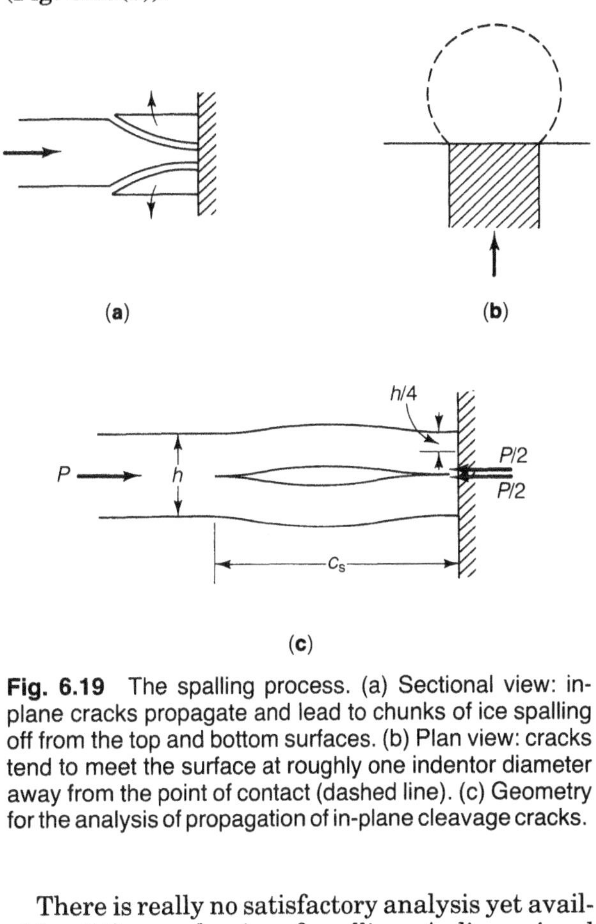

Fig. 6.19 The spalling process. (a) Sectional view: in-plane cracks propagate and lead to chunks of ice spalling off from the top and bottom surfaces. (b) Plan view: cracks tend to meet the surface at roughly one indentor diameter away from the point of contact (dashed line). (c) Geometry for the analysis of propagation of in-plane cleavage cracks.

There is really no satisfactory analysis yet available of the mechanics of spalling. A dimensional analysis of the initiation of spalling cracks has been presented by Evans *et al.* (1984), but this has not yet been properly calibrated by experimental data (Hallam, 1986). The second stage of spalling – the propagation of an existing in-plane crack – has been tackled by Wierzbicki (1985) and by

Fig. 6.20 An example of spalling and in-plane cleavage, from the Canadian Beaufort Sea, spring 1983. The cleavage crack at the top surface has formed along an existing plane of weakness and has then broken off in flexure. The spalling crack at the lower surface has curved to meet the surface (photograph by the author, courtesy of BP Petroleum Development Ltd).

Thouless *et al.* (1985), but in a sense this is not very much help, since it is almost certain that initiation, not propagation, constitutes the limiting step.

The geometry for analysis of propagation is shown in Fig. 6.19(c). The analysis originates with Kendall (1978), and the compressive force P is conceptually divided into two forces $P/2$ acting at points a distance $h/2$ in from the top and bottom surfaces. These two forces lead to tensile forces at the crack tip, and if the two halves of the ice cover are treated as eccentrically loaded columns it can be shown that the average stress σ required to propagate a crack is independent of initial crack length, c_s, and is given by Wierzbicki (1985) as:

$$\sigma = \frac{K_{IC}}{\sqrt{3(1-\nu^2)h}} \tag{6.28}$$

where K_{IC} is the fracture toughness and ν is Poisson's ratio. If $K_{IC} = 0.1$ MPa m$^{1/2}$ and $\nu = 0.3$ we have

$$\sigma = 0.06/\sqrt{h} \quad \text{(MPa)}$$

This would imply a failure stress of only 60 kPa for ice of thickness 1 m and of 30 kPa for ice of thickness 4 m.

These are appealingly low stresses, but the argument is deceptively simplistic and should not be applied uncritically: although the propagation force is apparently independent of initial crack length c_s, in fact the very foundation of the analysis is that the two halves of the ice cover can be analysed using simple beam theory. In effect this means that

$$c_s \gg h/2$$

so that, in practice, the analysis should not be applied unless an initial crack of several metres length exists in the ice. Such cracks do sometimes exist, especially at boundaries between ice layers of different morphology (Fig. 6.20), but of course they cannot be relied upon. In addition, the conceptual division of the force into two components at the centre of the ice is hardly realistic. It applies only if the ice is subjected to a line load at its centre; if the load is evenly distributed throughout the thickness then no eccentricity exists and the full Euler elastic buckling load would be required before cracks propagate.

6.4.4 Flexural failure

Flexural failure of ice, as opposed to buckling failure, may potentially occur as a result of any one of three basic mechanisms. They are:

(a) eccentric loading, in which the contact load is non-uniform and not aligned with the applied load;
(b) an irregular thickness profile;
(c) interaction with a sloping-sided structure.

We shall analyse these in turn. The analyses are approximate, but illustrate the general principles.

6.4.4.1 Eccentric loading

Consider the interaction geometry shown in Fig. 6.21(a). An initial surface crack of depth a is positioned on the upper surface of an ice plate, and the plate is loaded against a structure by an axial load P. Now suppose that the ice makes uneven contact with the structure and that the reaction force is off-axis by an eccentric distance, e. If the loading is sufficiently eccentric then a tensile crack-opening force may develop and propagate the surface crack a. The problem has been analysed by Wierzbicki (1985) and Tomin *et al.* (1986), and proceeds as follows.

The ice beam is effectively subjected to two loads: a pure bending moment of $M = Pe$, which tends to open the crack, and pure compression by the load P, which tends to close the crack. The resulting stress intensity factor K_I can then be written as

$$K_I = K_{IM} + K_{IP} \tag{6.29}$$

where K_{IM} is the stress intensity factor corresponding to the bending moment, and K_{IP} is the stress intensity factor corresponding to the compressive

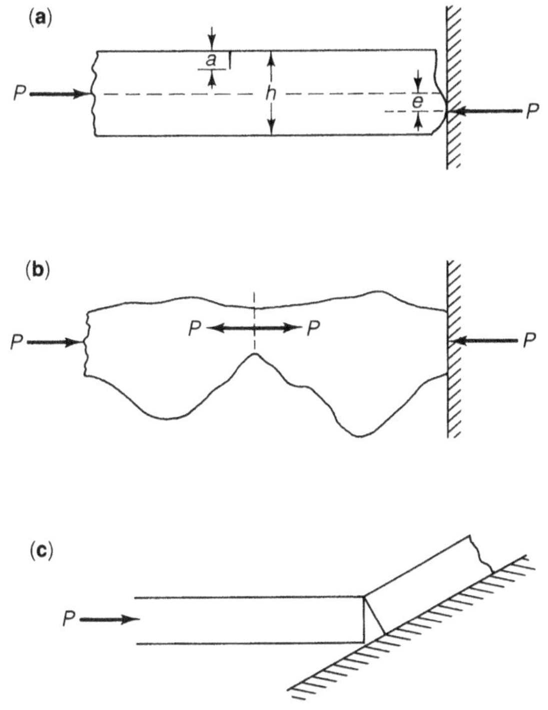

Fig. 6.21 Flexural failure mechanisms: (a) eccentric loading; (b) irregular thickness profile; (c) interaction with a sloping-sided structure.

force P. For a beam of uniform width, these factors are given by Collins (1981) as:

$$K_{IM} = \frac{6M\sqrt{\pi a}}{h^2} C_M \tag{6.30}$$

and

$$K_{IP} = -\frac{P\sqrt{\pi a}}{h} C_P \tag{6.31}$$

where C_M and C_P are factors depending on the ratio of crack depth to thickness, a/h, and are given approximately by

$$C_M = \frac{1}{\sqrt{\pi}}\left(1.99 - 2.47\frac{a}{h} + \ldots\right) \tag{6.32}$$

$$C_P = \frac{1}{\sqrt{\pi}}\left(1.99 - 0.41\frac{a}{h} + \ldots\right) \tag{6.33}$$

Propagation of a crack will occur if $K_I > K_{IC}$, so we can derive the force per unit width, P_f, required for flexural failure, from equations (6.29)–(6.31) as:

$$K_{IC} < \frac{P_f\sqrt{\pi a}}{h}\left(\frac{6eC_M}{h} - C_P\right)$$

That is:

$$P_f > \frac{K_{IC}h}{\sqrt{\pi a}}\left(\frac{6eC_M}{h} - C_P\right)^{-1} \tag{6.34}$$

Notice that flexure cannot occur unless we have:

$$\frac{6eC_M}{h} - C_P > 0 \tag{6.35}$$

since an infinite (or tensile) load would be required. Now, from equations (6.32) and (6.33) we see that under the reasonable approximation $a \ll h$ we can write $C_M = C_P = 1.99/\sqrt{\pi} = 1.12$. Thus, from equation (6.35), it is clear that we require $e > h/6$ for flexural failure to occur. If the load eccentricity is less than one sixth of the ice thickness then no tensile forces are present and flexural failure is not possible. In general, we might expect the average eccentricity to lie somewhere in the region $e = h/4$, and in this case the failure stress $\sigma = P_f/h$ is given by:

$$\sigma = \frac{K_{IC}}{\sqrt{\pi a}}(0.56) = 0.03/\sqrt{a} \quad \text{(MPa)} \tag{6.36}$$

If the initial crack size is, say, 10 mm, this corresponds to a stress of just 0.3 MPa.

This has been a slightly simplistic analysis for

two reasons. Firstly, it is two-dimensional, and therefore is taking for granted the existence of a crack a of effectively infinite extent in the transverse direction. Secondly, it likewise assumes that the eccentrically applied load is uniformly (and eccentrically) applied over the whole width of contact: in practice we might well expect local areas of eccentric contact, but not that they would be highly extensive laterally. To make the analysis more rigorous and applicable, statistics are needed on the characteristics of eccentric contact and of surface cracks available for propagation. Then, of course, a three-dimensional analysis should be performed. For these reasons it is likely that equation (6.36) is an under-estimate of the stresses actually required.

6.4.4.2. Irregular thickness profile

Ice is almost invariably modelled as a parallel-sided flat slab, but of course in practice it may be highly irregular in thickness. Examples of multi-year ice thickness profiles have been shown in Fig. 2.18. The irregularities are generally more pronounced at the lower surface, and this asymmetry can lead to the onset of flexural failure. The geometry of the problem is shown schematically in Fig. 6.21(b), and a two-dimensional analysis of it has been presented by Hallam *et al.* (1987), using the idealised geometry shown in Fig. 6.22. All quantities here refer to unit width of ice.

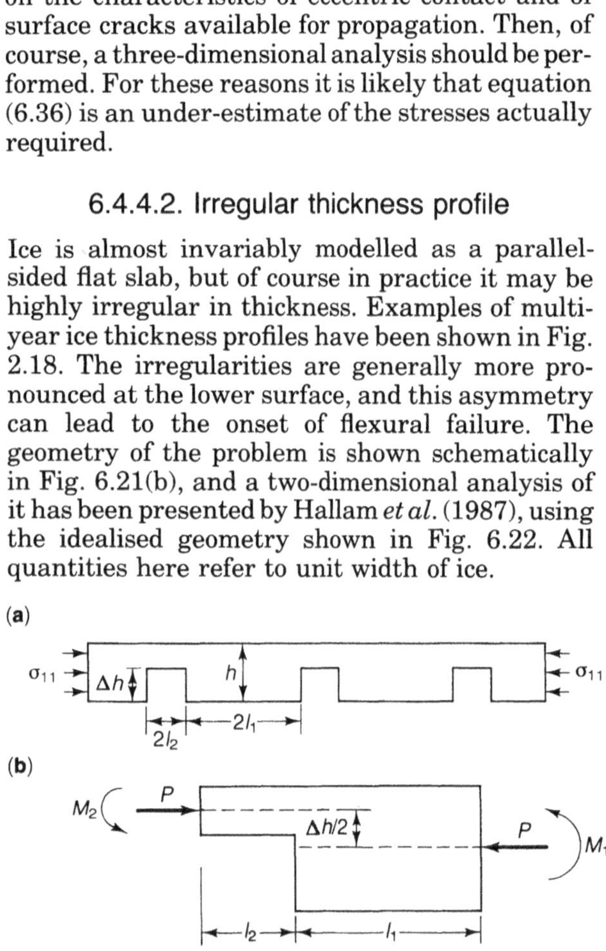

Fig. 6.22 Simplified geometry for analysis of the problem of irregular beams of ice under compression (after Hallam *et al.*, 1987).

The basis of the effect lies principally with the fact that the bottom-surface irregularity is more pronounced than the top-surface irregularity. It is not enough that the profile be simply uneven: if the top and bottom surface were irregular in an equal and opposite sense (that is, mirror images of each other), the stress field too would remain symmetrical and no flexural stresses would develop. However, because of the asymmetrical effect of gravitational versus buoyancy forces, and the effects of erosion by water currents, the undersurface is in fact considerably more uneven than the upper surface. This is reflected by the statistic that keel-to-sail depth ratios are generally of the order 3.2:1. In Fig. 6.22 the geometry is exaggerated, and the upper surface is assumed to be flat by comparison with the lower surface. This is a reasonable approximation and does not seriously impair the result.

Flexural failure can occur in much the same way as in the case of eccentric loading studied above: once again, an eccentric load is effectively being applied, since transmission of the load P through the reduced thinner section of the ice is not occurring along the same axis as the general applied load (Fig. 6.21(b)). Referring to the idealised geometry of Fig. 6.22, we consider slots of half-width l_2 and depth Δh in ice of general thickness h. The slots are separated by distance $2l_1$, and an overall uniform compressive stress σ_{11} is applied to the ends of the ice plate. No external bending moment is applied.

Considering the two beam sections of the ice plate, of thickness h and $(h-\Delta h)$, the load on each can be broken down into a compressive force P acting along the centroidal axis (dashed lines in Fig. 6.22) and bending moments M_1 and M_2 induced by the misalignment of the axes. The separation of the centroidal axes is $\Delta h/2$. Taking moments about any point we have, for static equilibrium:

$$P\Delta h/2 = \sigma_{11}h\Delta h/2 = M_1+M_2$$

Now, the ice sheet is not undergoing overall bending, so as small deflections occur the surface gradients at the centre of each section (the ends of the half-sections shown in Fig. 6.22(b)) must be zero, by symmetry, and furthermore the gradients at the junction of the two sections must be equal. Applying simple beam theory to each section we find

$$\frac{M_1l_1}{I_1} = \frac{M_2l_2}{I_2}$$

where I_1 and I_2 are the second moments of area per unit width, given by

$$I_1 = h^3/12$$

and

$$I_2 = (h-\Delta h)^3/12$$

Solving for M_2, we find

$$M_2 = \frac{\Delta h}{2}\sigma_{11}\frac{h}{(1+I_s/L_s)} \qquad (6.37)$$

where we have introduced the quantities

$$I_s = I_1/I_2$$

and

$$L_s = l_1/l_2$$

This allows us to calculate the bending stress σ_b in the outer fibre of the thinner section as

$$\sigma_b = M_2(h-\Delta h)/2I_2 \qquad (6.38)$$

The total stress at the surface is then given by the sum of this tensile stress σ_b and the superimposed compressive stress, that is:

$$\sigma = \sigma_b - \sigma_{11}\frac{h}{(h-\Delta h)} \qquad (6.39)$$

Consider now the tensile stress σ_t required to induce failure. If the ice is flaw-free and of large grain size, this will be the stress required for crack nucleation, as shown in Fig. 4.16; if the grain size is small, or significant flaws already exist in the material, it will be simply

$$\sigma_t = K_{IC}/\sqrt{\pi a}$$

where a is the flaw length. The condition for flexural failure is then that $\sigma > \sigma_t$. Rearranging equations (6.37)–(6.39), this means that the overall compressive stress σ_{11} must be great enough that

$$\sigma_{11} > \frac{(1-\lambda_s)[L_s(1-\lambda_s)^3+1]}{3\lambda_s L_s(1-\lambda_s)^2-[L_s(1-\lambda_s)^3+1]}\sigma_t \qquad (6.40)$$

where we have written relative depth of the bottom surface notches as

$$\lambda_s = \Delta h/h$$

The most interesting special case of this is the case where $l_1 \gg l_2$: that is, the bottom surface notches are very widely spaced. In this case, L_s tends to infinity, and we can write the condition as

$$\sigma_{11} > \frac{(1-\lambda_s)^2}{(4\lambda_s-1)}\sigma_t \qquad (6.41)$$

This condition for failure is plotted in Fig. 6.23 as a function of the fractional thickness reduction λ_s. It is seen that for $\lambda_s < 0.25$ no bending failure occurs, since no tensile stress is developed at the surface. This means that bottom surface irregularities must constitute at least a quarter of the total

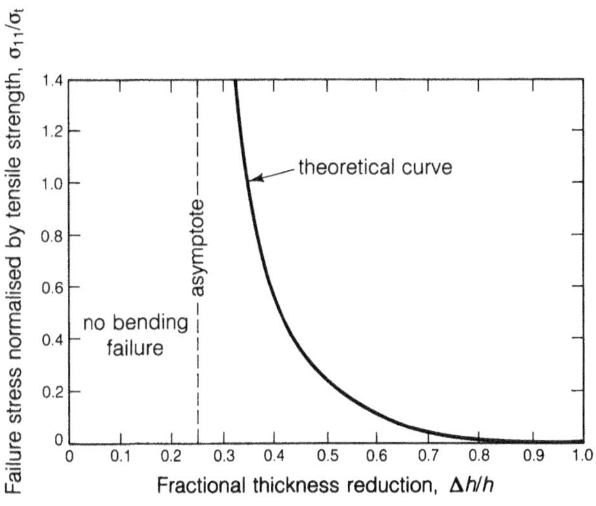

Fig. 6.23 Theoretical flexural failure curve for an isolated notch in a uniform plate under compression. Failure stress σ_{11} is normalised by tensile strength σ_t. Flexural failure can occur only if $\Delta h/h > 1/4$ (after Hallam *et al.*, 1987).

thickness. For values of λ_s exceeding about 0.35, however, the theoretical strength *in compression* is seen to be substantially less than the normal value of strength *in tension*. This is potentially a very great reduction, since the strength of ice in tension is of the order only about 1 MPa.

Note, however, that this is again a highly simplified model: most seriously of all, it is a purely two-dimensional model, and implicitly requires subsurface notches of infinite extent perpendicular to the loading direction. Although the subsurface of ice may be irregular to the required degree ($\lambda_s > 0.25$) it is not likely to be so in a predominantly linear fashion. In practice, it is more likely to resemble an egg-box than a series of linear corrugations. This can be seen from some full three-dimensional measurements of multi-year floe thickness recently presented by Poplin *et al.* (1987). The stress required to induce flexural failure in such a medium would be very much higher than in the simple two-dimensional case. Two things are required:

(1) Reliable statistics on the irregularity of ice thickness, on a full three-dimensional basis, rather than in simple two-dimensional profiles.
(2) An analysis of such irregular bodies under compression, probably using finite element stress analysis.

Such an approach would no doubt support a reduction of compressive strength due to flexural failure, but the reduction would not be as extreme as that shown by equation (6.40).

6.4.4.3 Sloping-sided structures

A satisfactory fracture analysis of ice interaction with sloping or conical structures has not, as far as I know, yet been performed. The analyses which are available are of two types: elastic representations, assigning the ice cover a single flexural strength σ_f (e.g. Nevel, 1972; Frederking, 1980); or plasticity representations, assigning the ice cover a yield strength σ_Y (e.g. Ralston, 1977a). Both of these approaches have shortcomings, since they tend to over-estimate the forces required to break ice in flexure at full scale. "Flexural strength" has been found to show pronounced scale-dependent behaviour (Gold, 1977), as might be expected, since it is a form of tensile fracture, and is thus susceptible to the propagation of single controlling flaws. Plasticity analysis, on the other hand, is essentially a continuum analysis of the problem, and represents an approximation to the force required to fail ice in flexure by creep. For these reasons, both these types of analyses supply upper bounds to forces to be expected at full scale.

Here we shall look at the two-dimensional analyses of the problem using the concept of flexural strength. Flexural strength is not a basic material property at all, but simply an index to the behaviour of beams of ice under flexural loading conditions. It has been measured by numerous experiments, generally using simply supported beams of ice or, more commonly, cantilever beams (Fig. 6.24). The flexural strength σ_f, as derived from a cantilever beam test, is normally defined as

$$\sigma_f = \frac{6M}{Wh^2} = \frac{6Pl}{Wh^2} \tag{6.42}$$

where P is the load required for failure of a beam of length l, thickness h and width W. M is the bending moment capacity of the beam. Experimental values for σ_f are generally in the range

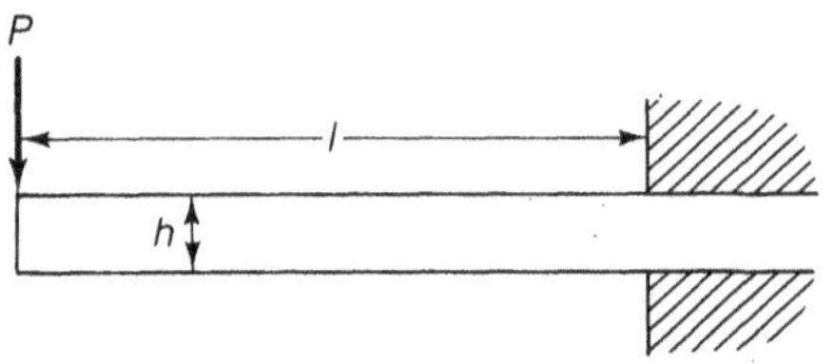

Fig. 6.24 Geometry for measurement of "flexural strength" of ice. This is not a true material property, but is useful in the analysis of bending failure against sloping structures.

0.5–1.0 MPa for freshwater ice and 0.1–0.8 MPa for sea ice (Schwarz and Weeks, 1977).

Consider now the geometry of interaction shown in Fig. 6.25, in which an ice sheet of thickness h advances onto a stucture sloping at angle α from the horizontal. The following analysis is based on Croasdale (1980).

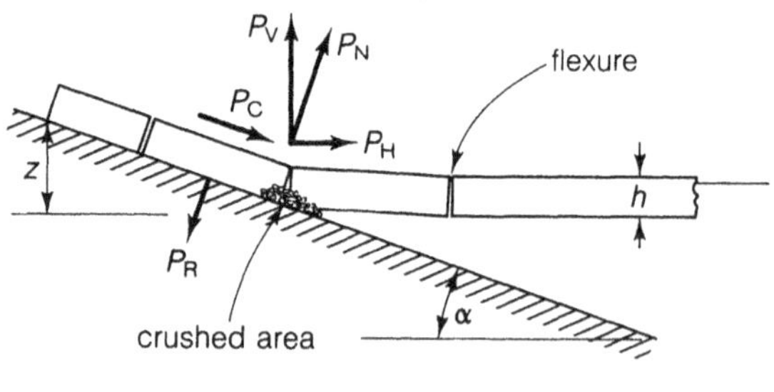

Fig. 6.25 Geometry for two-dimensional analysis of forces on a sloping structure.

Two processes have to be carried out during the interaction: failure of the advancing ice in flexure and pushing of failed blocks up the sides of the structure. At the contact of ice with the structure we can resolve the forces acting into two components, P_V acting vertically and P_H acting horizontally. If the coefficient of friction between the ice and the structure is μ, then these components may be written in terms of the normal force, P_N, as:

$$P_H = P_N \sin\alpha + \mu P_N \cos\alpha$$
$$P_V = P_N \cos\alpha - \mu P_N \sin\alpha$$

which supplies:

$$P_H = \left(\frac{\sin\alpha + \mu\cos\alpha}{\cos\alpha - \mu\sin\alpha}\right) P_V \tag{6.43}$$

Now if an edge load P_V is applied to a semi-infinite elastic beam floating on an elastic foundation, the maximum bending moment M in the beam is given by (Hetenyi, 1946):

$$M = \frac{P_V}{\beta_c e^{\pi/4}} \sin(\pi/4) \tag{6.44}$$

where $1/\beta_c$ is the characteristic length, defined by

$$\beta_c = [\rho_w g W/4E\mathcal{I}]^{1/4} \tag{6.45}$$

Here, ρ_w is the density of the sea water, E is Young's modulus and $\mathcal{I}$ is the second moment of area of the beam cross section ($= Wh^3/12$). Combining equations (6.42), (6.44) and (6.45), we find that the vertical force P_V on the structure is given by

$$P_V/W = 0.68\sigma_f(\rho_w g h^5/E)^{1/4} \tag{6.46}$$

Hence, the initial horizontal force per unit width required for flexural failure against the sloping structure is:

$$P_H/W = 0.68\,\sigma_f(\rho_w g h^5/E)^{1/4}\left(\frac{\sin\alpha+\mu\cos\alpha}{\cos\alpha-\mu\sin\alpha}\right) \quad (6.47)$$

Following failure of the ice in this manner, force still has to be applied in order to push broken blocks up the sloping surface (Fig. 6.25). Let us suppose that ice is pushed up the slope a vertical distance z (after which it is assumed to clear by some other mechanism – for instance, it falls back on itself or clears around the sides of a conical structure). The force P_C, parallel to the slope, required to do this is

$$P_C = \frac{z}{\sin\alpha} hW\rho_i g(\sin\alpha+\mu\cos\alpha) \quad (6.48)$$

where ρ_i is the density of the ice (which is assumed to be wholly removed from the water).

The total horizontal force P_H required both to break and lift the ice is then given by:

$$P_H = (P_V+P_C\sin\alpha)\left(\frac{\sin\alpha+\mu\cos\alpha}{\cos\alpha-\mu\sin\alpha}\right)+P_C\cos\alpha$$

Substituting equations (6.46) and (6.48), we can express the horizontal force per unit width as

$$P_H/W = \sigma_f\left(\frac{\rho_w g h^5}{E}\right)^{1/4} C_1+zh\rho_i g C_2 \quad (6.49)$$

where C_1 and C_2 are two functions depending only on slope angle α and coefficient of friction μ. They are given by:

$$C_1 = 0.68\left(\frac{\sin\alpha+\mu\cos\alpha}{\cos\alpha-\mu\sin\alpha}\right)$$

$$C_2 = (\sin\alpha+\mu\cos\alpha)\left(\frac{\sin\alpha+\mu\cos\alpha}{\cos\alpha-\mu\sin\alpha}+\frac{\cos\alpha}{\sin\alpha}\right)$$

These two functions are plotted in Fig. 6.26 for typical values of μ and α. In general, C_1 is of the order of about 1 and C_2 is of the order of 2–3. The first term in equation (6.49) represents the ice breaking force; the second term represents the ice ride-up force.

A sample calculation, for a conical structure of slope angle $\alpha = 45°$ and coefficient of friction $\mu = 0.3$, is shown in Fig. 6.27. The ice is assumed to have a flexural strength of 0.7 MPa and is allowed to slide a vertical distance $z = 5$ m up the structure.

The theoretical forces and pressures are quite low. For ice of thickness 1 m the force per unit width is approximately 0.13 MN m^{-1}, while for ice of thickness 3 m it is approximately 0.47 MN m^{-1}. The force is dominated by the ice ride-up force for this range of ice thicknesses, and it is seen from equation (6.49) that this depends linearly on ice thickness h. This means that average failure pressure against a sloping-sided structure is approximately independent of ice thickness. For the range of ice thickness h = 1–3 m, the average failure pressure predicted by equation (6.49) and Fig. 6.27 is only about 0.15 MPa.

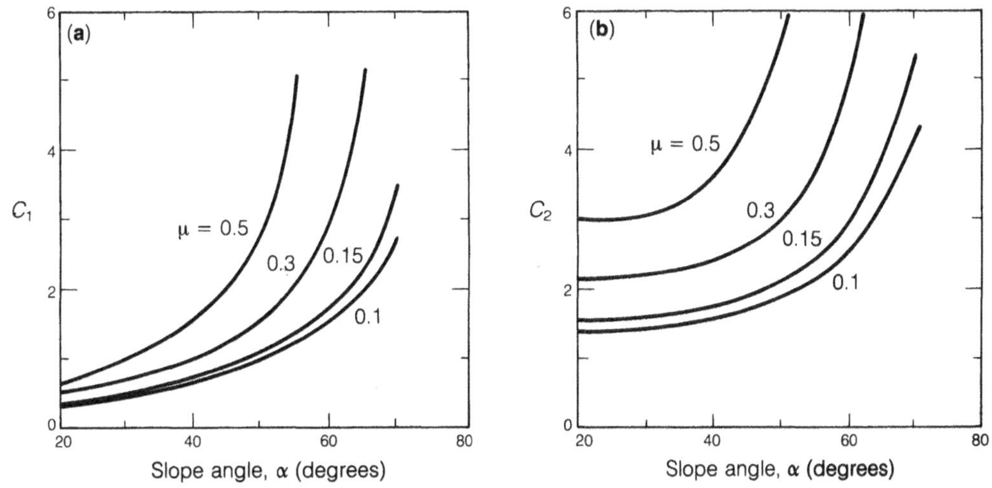

Fig. 6.26 The coefficients C_1 and C_2 used in equation (6.45), as a function of slope angle α and coefficient of friction μ (after Croasdale, 1980): (a) the coefficient C_1; (b) the coefficient C_2.

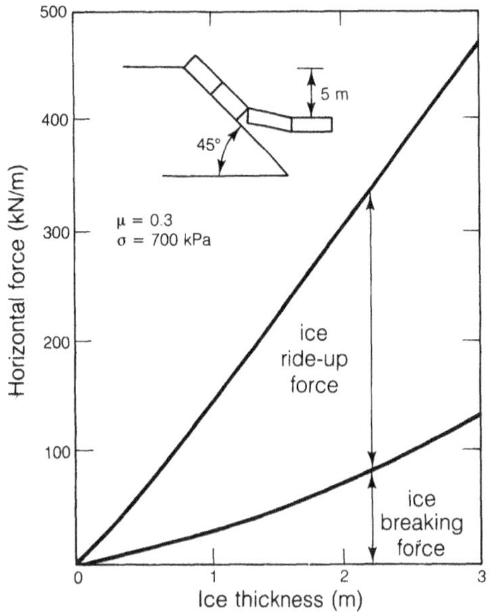

Fig. 6.27 Horizontal ice force against a sloping structure as a function of ice thickness, using equation (6.45) (after Croasdale, 1980).

The analysis given here is purely static. A further analysis by Frederking (1980) showed that hydrodynamic and inertial forces are generally negligible for the range of velocities normally encountered.

There are, however, some serious caveats. A sloping structure depends critically on efficient clearing of ice from its sloping sides. If for any reason ice begins to adhere to the structure, or if a rubble field, floating or grounded, develops around the structure, then all the advantages of the sloping sides are immediately nullified. Under such circumstances (Fig. 6.28), the structure presents an essentially vertical face to advancing ice, and the analysis expounded in this section no longer applies. For this reason, most of this book concentrates on the analysis of vertical-sided structures. I do not propose here to deal with the three-dimensional plastic limit analysis of ice interaction with conical structures. For details of this treatment, which constitutes an upper bound to expected ice forces, the reader should refer to Ashton (1986).

6.4.5 Non-simultaneous failure

We come now to the subject which I believe to be the most important to the understanding of full-scale forces on structures: non-simultaneous failure.

Once fracture has initiated close to a structure, of if the initial leading edge of ice is highly irregular, then failure may occur non-simultaneously in discrete local zones. Although crushing may appear to be occurring over the full width D of gross contact, at any one time it is likely to be concentrated only in a few zones of width d_i (Fig. 6.29). A real-life example of the onset of such a process is shown in Fig. 6.30, from the Hans Island programme of 1980. Initial contact is irregular, and zones appear to fail in such a way that contact remains irregular throughout the impact process. At no time is all the ice simultaneously failing, so the average total load is reduced in a statistical fashion.

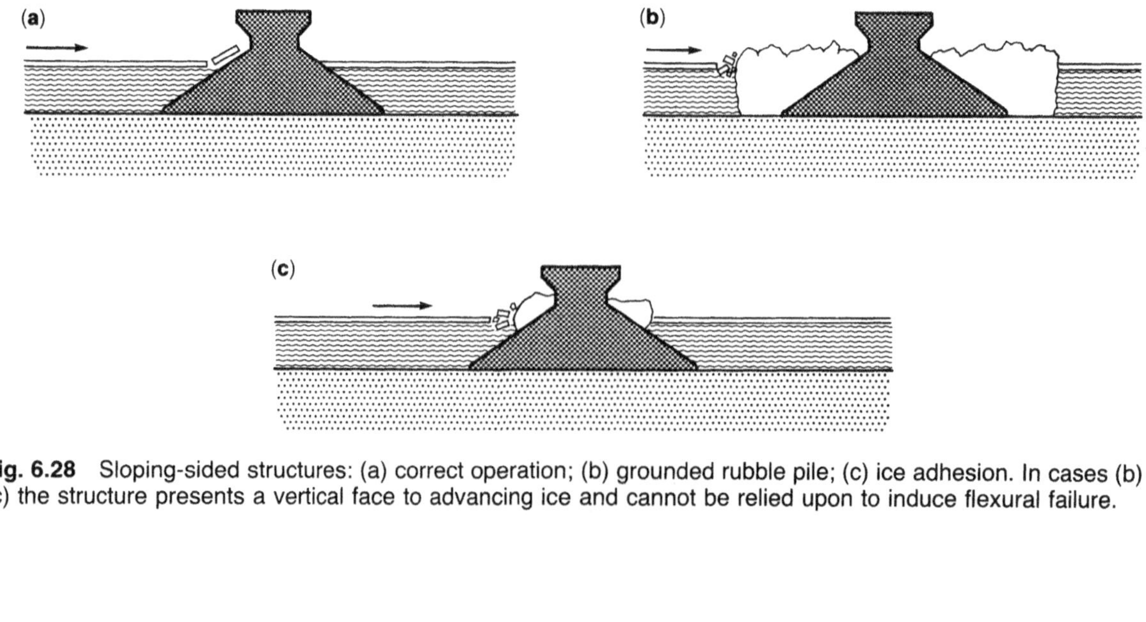

Fig. 6.28 Sloping-sided structures: (a) correct operation; (b) grounded rubble pile; (c) ice adhesion. In cases (b) and (c) the structure presents a vertical face to advancing ice and cannot be relied upon to induce flexural failure.

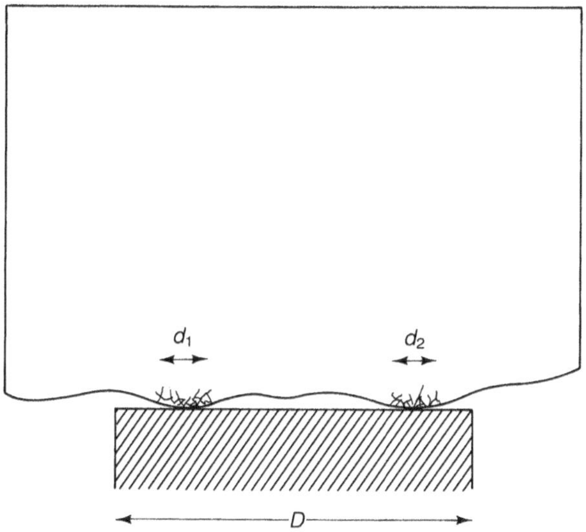

Fig. 6.29 A schematic representation of non-simultaneous failure.

6.4.5.1 A mechanical model

Several authors have presented analyses of the statistics of non-simultaneous failure (Kry, 1980b; Slomski and Vivatrat, 1983). These analyses make use of the hypothesis that large-scale failure occurs by successive fracture of independent zones (Fig. 6.31). The treatment is probabilistic and looks at the statistical sum of individual stress time series for a large number of zones. The statistical sum results in a "smearing" of peak stresses and leads to the conclusion that peak stresses over a large multi-zone area should be lower than over the area of a single zone. The manner of the reduction depends on the details of the statistics used.

More recently, Ashby *et al.* (1986) have presented an alternative approach to non-simultaneous failure which is based on a simple mechanical argument and backed up by experiments with brittle foams. As we shall see, it gives a realistic dependence on contact area. I emphasise, however,

Fig. 6.30 A small multi-year floe colliding with the north-west corner of Hans Island, August, 1980. Failure occurred by crushing, and was initially concentrated in discrete zones of irregular contact (photograph courtesy of Dome Petroleum).

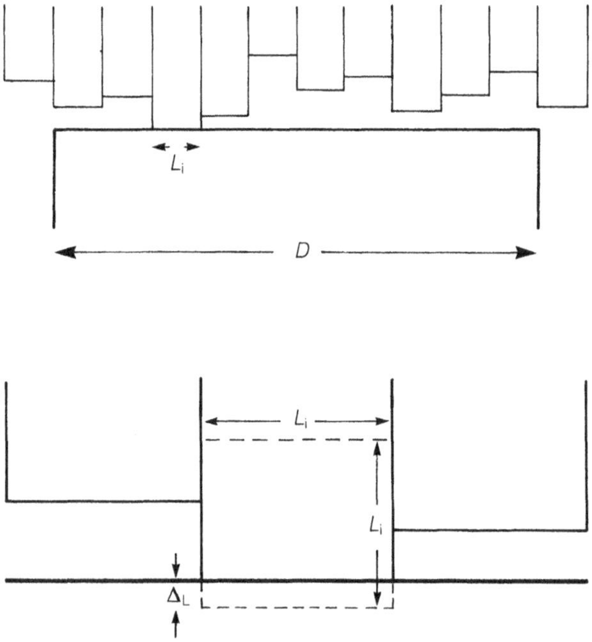

Fig. 6.31 (a) Non-simultaneous failure. (b) Detailed geometrical assumptions and definitions for analysis by Ashby *et al.* (1986).

that the argument is simplistic, and makes assumptions which require experimental support. It proceeds as follows.

Consider an irregular block of thickness h in contact with an indentor of width D (Fig. 6.31). Assume that it can be idealised as a set of cubical independent cells of dimension L_i. As each cell comes into contact with the indentor, assume that a distance Δ_L can be moved before the cell fails (this might correspond to critical displacement before flexural failure or to critical strain before crushing failure – it does not matter). This assumption about load history is illustrated in Fig. 6.32. For simplicity, we assume uniform cell size.

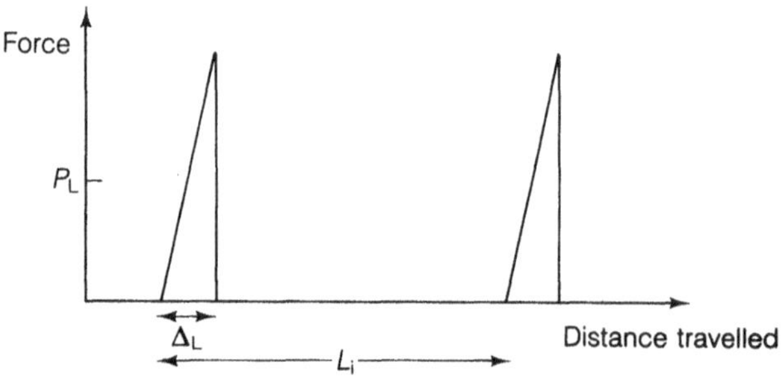

Fig. 6.32 Diagram of model assumptions for the force history over a particular zone. Force rises to $2P_L$ as the ice advances by distance Δ_L and falls to zero after failure. The ice has to advance by a distance L_i before a new zone of ice makes contact.

Let the average force during this period of contact be P_L over the area L_i^2 of the cell concerned. Now, the probability, p, of a particular cell area being in contact is

$$p = \Delta_L/L_i \tag{6.50}$$

(where Δ_L is very much less than L_i so that p is small). In the total area of contact $A = Dh$, the number of cell areas $\mathcal{N}$ is

$$\mathcal{N} = A/L_i^2 \tag{6.51}$$

Now, simple probability theory (Ashby *et al.*, 1986) can be used to show that the expected number of contacting cells within the area A is $\mathcal{N}p$, and the standard deviation is approximately $(\mathcal{N}p)^{1/2}$.

For illustration, if we take the three-standard-deviation level as a reasonable extreme case, the maximum number of areas simultaneously in active contact is approximately:

$$\mathcal{N}p+3(\mathcal{N}p)^{1/2}$$

or, using equations (6.50) and (6.51),

$$\frac{A\Delta_L}{L_i^3}\left[1+3\left(\frac{L_i^3}{A\Delta_L}\right)^{1/2}\right]$$

The total force P resulting is then

$$P = \frac{P_L A\Delta_L}{L_i^3}\left[1+3\left(\frac{L_i^3}{A\Delta_L}\right)^{1/2}\right]$$

and the indentation pressure σ is

$$\sigma = \frac{P}{A} = \frac{P_L\Delta_L}{L_i^3}\left[1+3\left(\frac{L_i^3}{A\Delta_L}\right)^{1/2}\right] \tag{6.52}$$

Notice that the second term of the equation demonstrates the functional dependence on the inverse square root of area, as observed in the pressure–area curve plots of Fig. 5.36 and 5.38.

We can, for illustration, now assume reasonable values for the constants of equation (6.52) and fit it approximately to the observed data. Firstly, let us suppose that $L_i \sim 1$ m, which is the order of magnitude of ice fragment size typically observed in the Arctic after a failure event. Secondly, assume that P_L, the force required to fail a single cell area, is about 15 MN (a pressure of 15 MPa, as observed for data groups A and B of Fig. 5.36(b)). Now, we know that for contact areas of $A \sim 100$ m^2 the peak stresses measured in the field are in the range approximately 1–2 MPa (group C of Fig.

5.36). Here we shall use a pressure of 1 MPa to fit our theoretical curve, since the higher pressures of 2 MPa were obtained from the Molikpaq structure: as discussed in Section 5.4.1.12, these may be unrealistically high owing to dynamic amplification effects. Substituting into equation (6.48), we can infer that $\Delta_L = 0.02$ m, or 2 cm. The resulting formula would then be:

$$\sigma = 0.33\left(1+\frac{20.3}{\sqrt{A}}\right) \qquad (6.53)$$

which is valid for $A \gg L_i^2$. For areas less than or equal to L_i^2 we expect pressure of the order of a single zone failure pressure, or 15 MPa. The resulting theoretical curve is shown in Fig. 6.33, plotted over the data of Fig. 5.36. Note that the treatment is not strictly valid for $A \sim L_i^2$.

It will be seen that it agrees fairly well with the observed data (which is not very surprising, since it is to a large degree been empirically fitted to them). Nonetheless, this does seem to provide a plausible model of observed behaviour. Note that for very large contact areas the stress predicted by equation (6.53) levels off to 0.33 MPa, which is rather higher than that actually required by mesoscale models (data group D in Fig. 5.36).

However, the model has many shortcomings and should not be regarded as a quantitative explanation. Its main defects are:

(1) No satisfactory physical description of the zonal failure process is given; the model is purely mechanical, and leaves open the question of what is happening during failure. There are, however, many likely candidates for failure mode in an individual zone: crushing, spalling, radial cracking followed by flexural failure, etc.
(2) It has been assumed (Fig. 6.32) that a failed zone clears perfectly and effortlessly, allowing the stress to drop to zero. In the next section we shall elaborate on this.
(3) The values of Δ_L and L_i have been arbitrarily chosen or fitted: work is still needed in order to derive them explicitly as a function of a physical mode of failure.
(4) The statistical approach to peak stresses is arbitrary. The selection of three standard deviations from the mean requires justification and refinement, and the curve of Fig. 6.33 would need to be put into a more formal context of risk analysis before it could be practically applied.

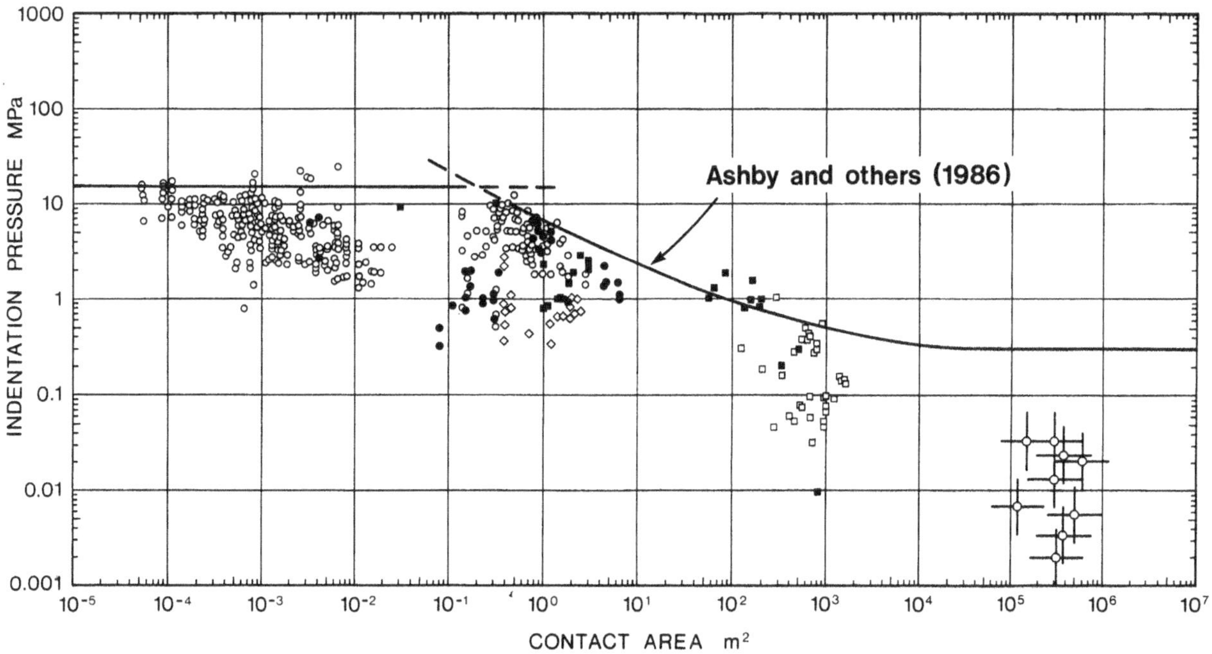

Fig. 6.33 Theoretical curve of Ashby *et al.* (1986) constrained to the observed data. The curve uses $L_i = 1$, $P_L = 15$ MN and $\Delta_L = 0.02$ m.

For these reasons the approach using non-simultaneous failure has serious limitations for the practising engineer. Although it provides a plausible explanation of trends in existing data it does not readily yield design values for loading on structures. This is because the probabilistic aspects of ice failure are insufficiently well defined, and this, in turn, is because insufficient data exist.

In the following sections we shall refine the model a little further, and illustrate the way in which it can be used to give a detailed simulation of an impact event. Then, in Chapter 7 we shall provide guidance on the way in which the model and the data which it describes can be used in a quantitative way to derive loads and associated risk levels.

6.4.5.2 Clearing forces

Firstly, let us include clearing forces. The simple model of Fig. 6.32 treats only the force required to fracture successive impacting zones, and it ignores the forces required to clear the debris formed during failure. We shall derive here a simple expression giving an approximation to these forces.

Consider the process illustrated in cross section by Fig. 6.34(a). As ice advances onto a wide structure and fails by fracture (crushing, flexure, spalling, etc.), the debris is unable to clear laterally and has to clear by movement up or down out of plane. (Note that during creep indentation this is not necessarily so, and that ice may undergo significant lateral creep deformation.) To clear ice out of plane, gravitational forces must be overcome, and work is done against them. Also, frictional resistance is offered by the ice rubble as it extrudes.

To quantify this, consider the model shown in Fig. 6.34(b). As a zone of the ice sheet of width W fails and moves forward a distance x it must displace a volume xhW of ice. For simplicity, let us suppose that ice debris is cleared in roughly equal proportions up and down and that the debris is raised or lowered by an average distance of about $h/2$. We shall ignore frictional resistance. Consider now the work done against gravity: to raise the top half of the ice clear of the water the work done is approximately

$$\frac{xhW}{2}\rho_i g\frac{h}{2} \qquad (6.54)$$

where ρ_i is the density of ice, and I have neglected the buoyancy which may be provided by water during the initial stages of lifting. The work done in submerging the lower half of the failed ice is, similarly,

$$\frac{xhW}{2}(\rho_w-\rho_i)g\frac{h}{2} \qquad (6.55)$$

where ρ_w is the density of sea water. Now let us equate the work done in raising and lowering the ice debris to the work done in moving the ice forward. If the applied force is P, the work done is Px. Thus, from the sum of equations (6.54) and (6.55),

$$Px = \frac{xWh^2}{4}\rho_w g$$

and the average stress to perform clearance of the debris, σ_d, is:

$$\sigma_d = \frac{P}{Wh} = \rho_w g\frac{h}{4} \qquad (6.56)$$

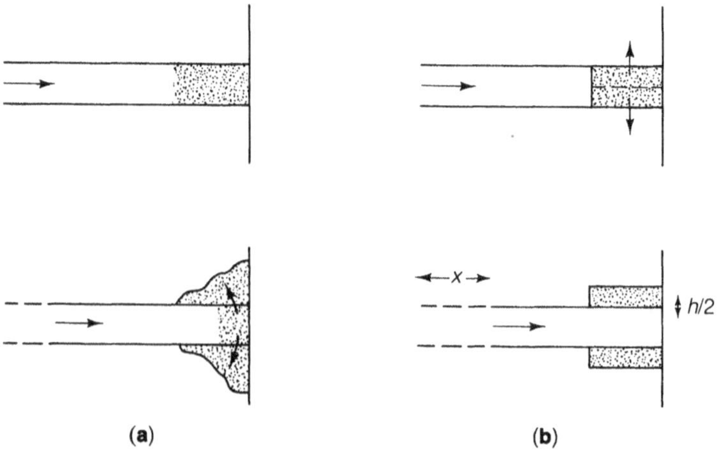

Fig. 6.34 Clearing processes: (a) upward and downward clearance of crushed debris; (b) a simple model of the process.

Note that this theoretical clearing force is proportional to thickness h, but independent of movement rate. Inserting the values

$\rho_w = 1.028 \text{ Mg m}^{-3}$

and

$g = 9.81 \text{ m s}^{-2}$

we find

$\sigma_d \sim 2500h$

For ice of thickness $h = 10$ m this implies a clearing force of $\sigma_d = 0.025$ MPa.

In practice, several effects tend to increase the likely stress required to clear ice, and these make equation (6.56) a lower bound:

(a) We have ignored frictional resistance of the failed material, which may be appreciable. This could be included explicitly if required (see Fransson and Sandkvist, 1985; Hallam, 1986). It would, however, contribute a force which is dependent on movement rate, which is an unnecesary complication at this stage.
(b) Near to the structure at the contact zone it is likely that much of the ice may need initially to be raised a greater distance than $h/2$, perhaps as much as h (see Fig. 6.34(a)).

For these reasons, and as a simple rule of thumb, I suggest that the clearing force can be approximated by doubling the formula, I have derived. This gives the following formula, which we shall use for illustration of an impact process:

$$\sigma_d \sim 5000h \quad \text{(Pa)} \tag{6.57}$$

or

$\sigma_d \sim 0.005h \quad \text{(MPa)}$

For 10 m thick ice this implies a clearing force of about 0.05 MPa.

6.4.5.3 Simulation of an impact

Let us now look at how well a mechanical model of non-simultaneous failure can cope with simulation of a real impact event.

A real impact event involves the arrival at a structure of a large mass of ice travelling with a large amount of kinetic energy. During the collision process a widening area of contact develops between the structure and the ice floe, and its kinetic energy is gradually absorbed until the floe comes to rest or passes the structure. The total force exerted on the structure varies with time and can be written $P(t)$, where t is elapsed time. The velocity of the floe may be denoted $U(t)$. Let the floe initially impact at time t_0 and come to rest at time t_1. Then we can simply equate initial kinetic energy to total work done, and write:

$$\tfrac{1}{2}m[U(t_0)]^2 = \int_{t_0}^{t_1} P(t)U(t)\,dt$$

where m is the effective mass of the impacting ice floe (including any allowance for added mass). It is instructive now to look at an example case of a progressive impact, using the methodology of non-simultaneous failure.

Consider the impact geometry shown in Fig. 6.35. A multi-year floe of radius R and thickness

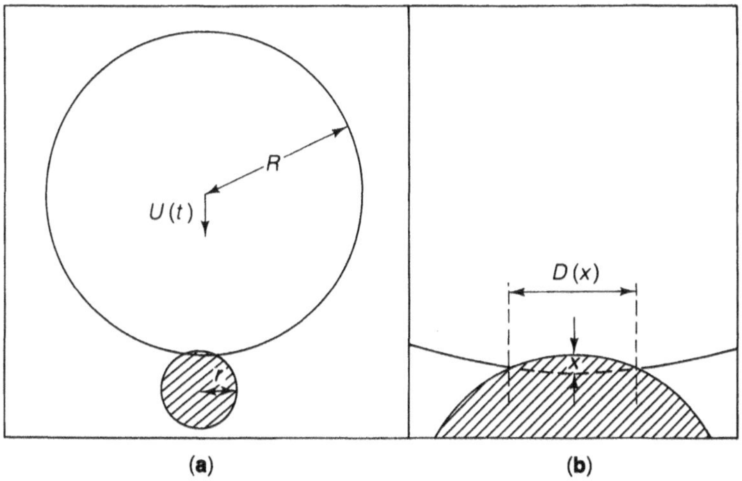

Fig. 6.35 Geometry for impact of a floe of radius R with a structure of radius r: as the floe is penetrated to a distance x, it develops a contact width $D(x)$.

h travels at initial velocity $U(t_0)$ towards a vertical-sided structure of radius r_s. The relative separation of the structure and the floe is denoted by x and we define $x = 0$ at the moment of first contact, $t = t_0$.

As the floe impacts with the structure, its contact zone progressively widens, and at penetration distance x the width $D(x)$ of the contact zone is given, by simple geometry, as:

$$D(x) = 2\sqrt{x_0(2r_s - x_0)} \quad (6.58)$$

where

$$x_0 = \frac{x(2R-x)}{2(r_s+R-x)}$$

Let us now conceptualise the impacting floe, along the lines of Section 6.4.5.1, as an ensemble of independent zones of area $A = L_i^2$ For the purposes of simple illustration of the construction of a numerical impact model, we divide the floe in the manner shown in Fig. 6.36. The thickness of the floe is divided into three layers, giving zones which are approximately square.

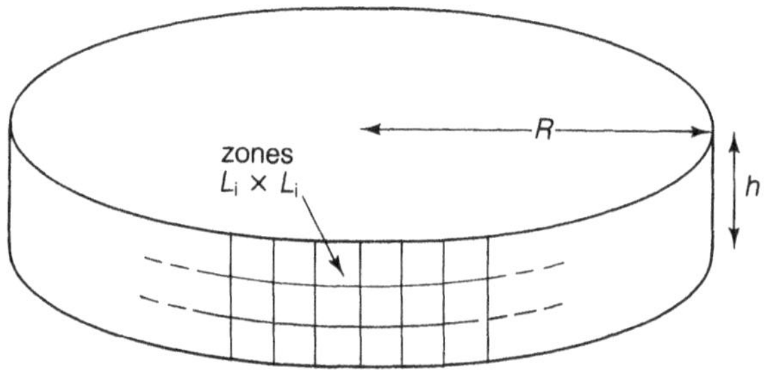

Fig. 6.36 Conceptual division of a model floe into zones of size $L_i \times L_i$.

During the impact process we shall suppose that each zone behaves independently of all the others and that it goes through the stress–time history shown in Fig. 6.37. This consists of:

(a) a stress σ_c to cause failure (for instance, crushing), followed by;
(b) a stress σ_d to clear the debris caused by failure.

Note that this picture differs slightly from the form used in Fig. 6.32: the peaks are square waveform, for computational ease, and the stress required for clearance is explicitly included. We now assume that successive failure peaks for a particular position on the contact zone are separated on average by a distance L_i, but show stochastic variation. For the purposes of the present model we shall assume that the distance between peaks is a random variable, with an equal chance of lying anywhere in the range approximately 0 to $2L_i$. We assume also that the initial impact of a zone (first contact) corresponds to a failure peak. Each failure peak lasts a distance Δ_L of penetration.

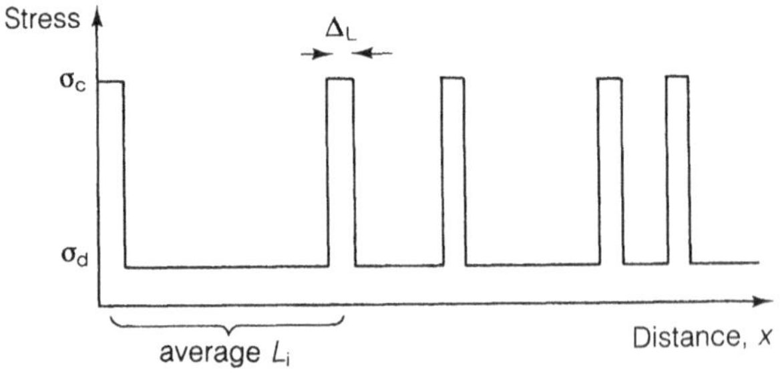

Fig. 6.37 Stress history used for simulation of impact: failure stress σ_c and clearing stress σ_d.

We are now in a position to give an approximate simulation of the progressive impact of a multi-year floe and produce a synthetic force–time history for the overall impact process. To do this we proceed as follows:

(1) Beginning at the moment of first contact, $t = t_0$ and $x = 0$, we progress the floe forwards in distance steps of $x = \Delta_L$.
(2) At each stage of penetration, calculate from equation (6.58) the total contact diameter D.
(3) Calculate the number of zones, $\mathcal{N}$, which are potentially in contact (either crushing or clearing). This is given by

$$\mathcal{N} = D(x)h/L_i^2$$

(4) As each zone comes into contact we prepare a stochastic time series of local stress, along the lines of Fig. 6.37, with separation of peaks randomised between Δ_L and $(2L_i - \Delta_L)$.
(5) Add together all the forces on individual zones, to give a total force P.

This procedure supplies us with total force $P(x)$ as a function of penetration distance x for one particular random realisation. To obtain this as a function of time instead of distance we can now calculate loss of kinetic energy, to give kinetic energy as a function of distance:

$$\tfrac{1}{2}m[U(x)]^2 = \tfrac{1}{2}m[U(x_0)]^2 - \int_0^x P(x)\,dx$$

In practice, of course, the integral is replaced by a summation, and at the rth displacement step we can calculate the velocity by:

$$U(x)^2 = U(0)^2 - \frac{2}{m}\sum_{0}^{r} P(x)\Delta_L$$

where $x = r\Delta_L$. From this, it is a short step to calculate time elapsed at the rth displacement increment as:

$$t = \sum_{0}^{r} \frac{\Delta_L}{U(x)}$$

and this enables us to give velocity $U(t)$ and force $P(t)$ as a function of time elapsed, t.

We now take a specific example. Let us suppose that a multi-year floe has a diameter of 2 km and thickness 10 m. Its initial velocity is 0.5 m s^{-1} and it impacts with a structure of diameter 100 m. We have:

$R = 1000$ m
$h = 10$ m
$U = 0.5$ m s^{-1}
$r_s = 50$ m

and the mass m of the floe is given by $m = \rho_i \pi R^2 h = 2.8\times10^{10}$ kg, where $\rho_i = 0.9$ Mg m^{-3} is the density of ice. As it impacts with the structure its contact width grows with penetration x in the manner tabulated in Table 6.5. Notice how rapidly the contact width grows.

TABLE 6.5 Contact width as a function of penetration distance for a floe of diameter 2 km colliding with a structure of diameter 100 m

Penetration distance, x (m)	*Gross contact width, D(x) (m)*
0	0
1	19.4
2	27.4
3	33.3
4	38.3
5	42.6
10	58.8
20	78.8
30	90.7
40	97.5
50	100.0
> 50	100.0

Each zone is modelled as having dimensions $L_i \sim 3$ m. (In fact, for ease of calculation I have taken the zones to be 3.33 m in height and 3 m in width; in this way there are precisely three layers of zones in the thickness of 10 m, and at full penetration there are 100 zones in contact.) At any penetration distance x the total number of zones potentially in contact is then given simply by the penetration width (in metres) as in Table 6.5. That is, if $D = 9$ m, nine zones are at that moment interacting with the structure. As each zone comes into contact it initially remains at failure stress σ_c for a penetration distance, Δ_L.

For the present purpose, we shall arbitrarily take $\Delta_L = 0.15$ m; that is, an individual zone is in contact at failure stress σ_c for roughly 5% of its history. During this time we shall assume that it fails at stress $\sigma_c = 3$ MPa, since this is the value which would be indicated by the pressure–area curve for such an area of 10 m^2. (Note that these values differ from those used in Section 6.4.5.1. This is because we are using zones with $L_i \sim 3$ m rather than $L_i \sim 1$ m. The analysis can be done with $L_i \sim 1$ m if required, but the calculation is more tedious.) After failure, while ice is merely clearing, it sustains a stress $\sigma_d = 0.05$ MPa, in line with equation (6.53). It remains at this stress for an average distance L_i.

A simulation along these lines has been performed using a random number generator, and it yields a total force history as shown in Fig. 6.38(a). This, obviously, is only one stochastic realisation of the model, and running the model a second time would yield a different result. For this realisation, the impact process lasts about 90 s and a peak force of 340 MN is generated. This corresponds to 10 individual zones simultaneously in contact at full failure stress σ_c, at elapsed time $t = 60$ s. At this moment the modelled contact area was 825 m^2 (Fig. 6.38 (b)) and a total of 82 zones were nominally in the contact area.

This synthetic force history of an impact should be compared with actual impact histories recorded at Hans Island (Figs 5.22, 5.24 and 5.26). It will be seen that the impact histories are plausibly similar in form, and this lends support to the conceptual model of non-simultaneous failure. However, there are at present insufficient high-resolution full-scale data in the public domain to allow realistic specification of the parameters L_i, Δ_L, σ_c and σ_d. The values used here should be regarded as an illustration not a prescription.

It is, of course, a straightforward matter to run the random number generator model repeatedly,

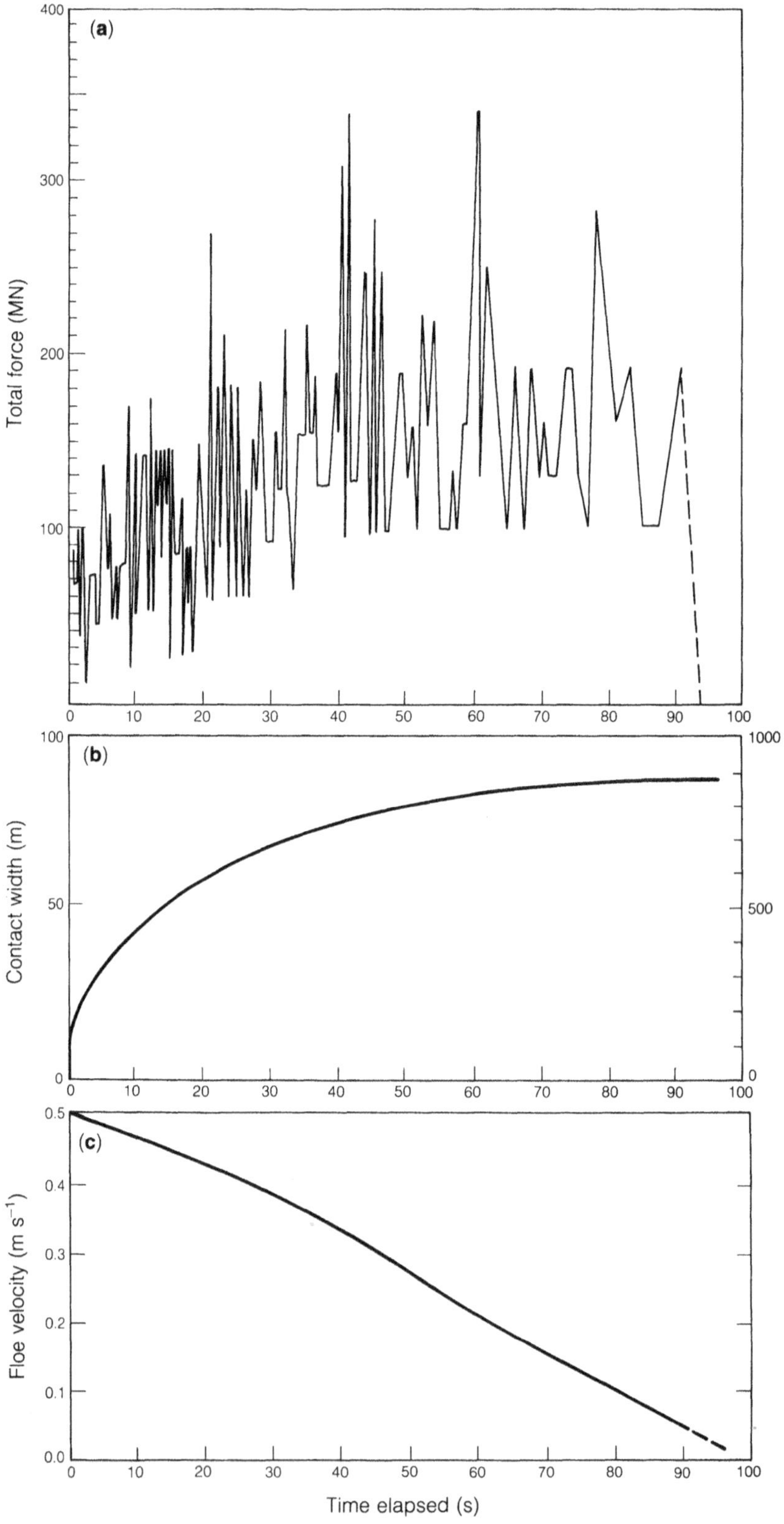

Fig. 6.38 Results of a numerical simulation of a multi-year floe impact: (a) total load; (b) contact width and area; (c) velocity. See text for details of input parameters and methodology.

to generate a series of impact histories for this particular scenario and allow calculation of the probability of occurrence of various peak loads. Although this is the correct procedure in principle, it is not justified at this stage, since the model input parameters are too arbitrarily chosen.

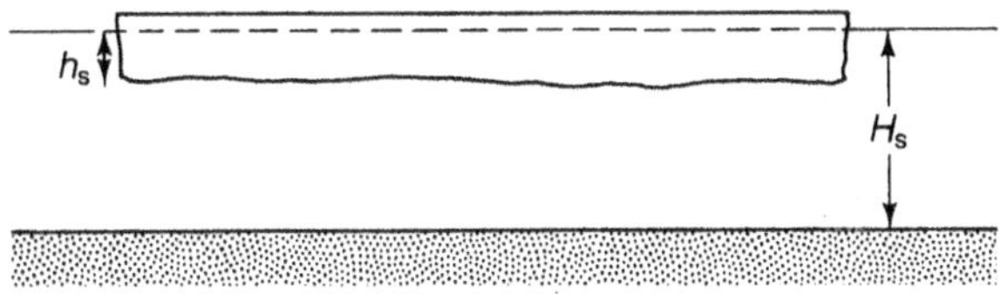

Fig. 6.39 Geometry for added mass calculation.

6.4.5.4 Refinements to the model

This has been quite a simple model, but it illustrates the important principles involved and appears to give a broadly reasonable simulation of an impact. Bear in mind that we have made a number of simplifications and that improvements may easily be made, though only if justified by the availability of data to show that the improvements are real ones. The principal simplifying assumptions are:

(a) *Wind and current drag.* We have neglected wind and current drag: these may easily be added, if known, using the methodology to be explained in Section 7.5, and would imply that the floe continues to exert some force even after it comes to rest. In most practical applications this force is negligible compared to the dynamic impact force.

(b) *Added mass.* We have ignored effects of hydrodynamic added mass. Added mass is an effective mass associated with a body accelerating or decelerating in a fluid, and is caused by the changes that are necessary to the surrounding flow field as the body's motion alters. It can be expressed simply as an additional mass which is added to the body's inertial mass. For a large, flat, smooth floe of submerged thickness h_s in water depth H_s (Fig. 6.39) the total effective mass of the floe may be written as (Marcellus and Morrison, 1982):

$$\left(1+\frac{h_s}{2H_s-h_s}\right)m$$

where m is the mass of the floe alone. Notice that this depends quite sensitively on the water depth H_s. If $H_s \gg h_s$ then there is no added mass effect. If, however, $H_s = h_s$ (that

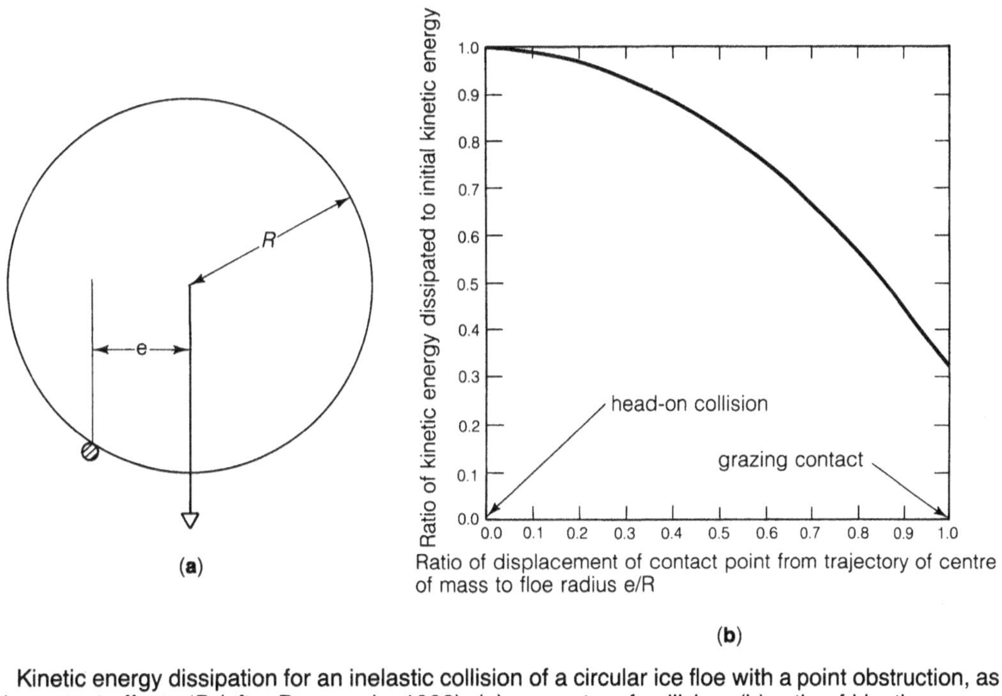

Fig. 6.40 Kinetic energy dissipation for an inelastic collision of a circular ice floe with a point obstruction, as a function of eccentric contact offset e/R (after Dunwoody, 1982): (a) geometry of collision; (b) ratio of kinetic energy dissipated to the initial kinetic energy.

is, the floe is almost touching the sea bed) then the added mass term is equal to 1, and doubles the effective total mass of the floe. For a more typical geometry, suppose that a 6 m thick floe impacts in water depth 30 m. In this case the submerged thickness is about 5 m and the added mass factor is 9%. This means that the total effective mass is 1.09 m. For most purposes the added mass term is thus quite small, but it is more appreciable in shallow waters where the thickness of the ice is comparable with the water depth.

(c) *Eccentric impact.* We have assumed (Fig. 6.35) that the impact is directly head-on. Rotational effects are thus not involved. More realistically, if a floe impacts eccentrically (Fig. 6.40(a)) then not all of its kinetic energy will be dissipated: a significant proportion will be converted into rotational energy. A calculation of the proportion of initial kinetic energy which will be dissipated by crushing under such circumstances has been performed by Dunwoody (1982), and is shown in Fig. 6.40(b).

(d) *Failure stresses.* The contact forces as a single zone fails have been assumed to be constant during the fracture process and to be equal for all failure events. A more realistic zonal failure history would perhaps involve some statistical variation of peak stress achieved during failure of a single zone. This is illustrated conceptually in Fig. 6.41, where failure stresses show variance about a mean stress $\bar{\sigma}_c$.

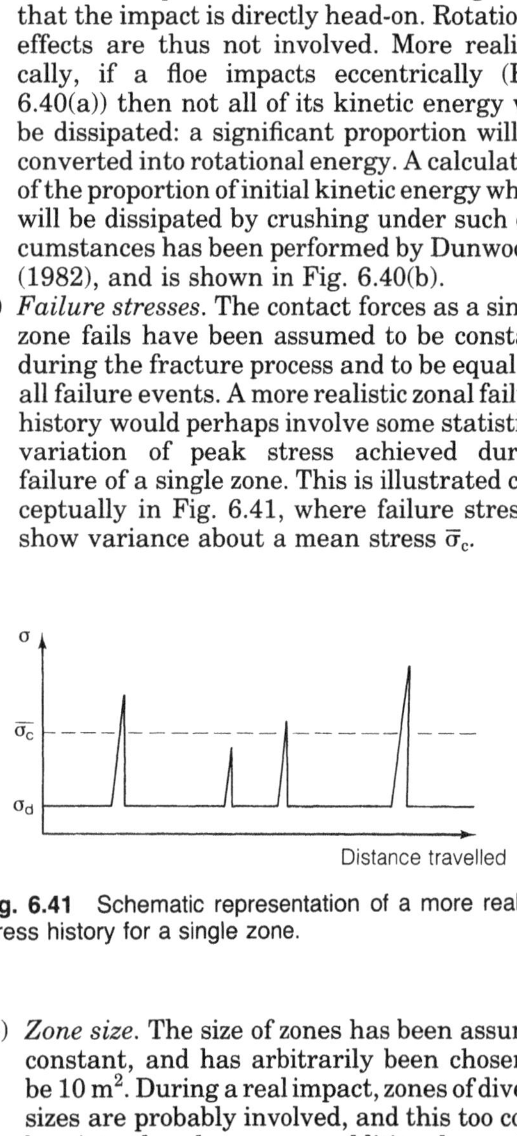

Fig. 6.41 Schematic representation of a more realistic stress history for a single zone.

(e) *Zone size.* The size of zones has been assumed constant, and has arbitrarily been chosen to be 10 m^2. During a real impact, zones of diverse sizes are probably involved, and this too could be introduced as an additional random variable.

6.5 KORZHAVIN'S EQUATION

Finally, to round off this chapter on theoretical analysis, let us look at the best known formula in the ice literature for calculating ice forces on structures. This is Korzhavin's equation (Korzhavin, 1962). We look at this outside the headings of "creep" or "fracture" behaviour, since the equation is intended to deal with both: this, indeed, is the main problem with its interpretation and application.

Many variants of the equation have been proposed, but they all boil down to the form:

$$P = IkDh\sigma \qquad (6.59)$$

where P is total load on the structure, I is an indentation factor, k is a "contact factor", D is the structure diameter, h is the ice thickness, and σ is the uniaxial compressive "strength" of ice. A "shape factor" is also sometimes included, again as a simple multiplier. However, on the basis of experiments it is generally ascribed the value 1.0 for a flat indentor and 0.9 for a circular indentor, and we shall ignore it in the following discussion.

Equation (6.59) is an empirical equation, and there is widespread confusion and disagreement about the interpretation of the individual terms and the functional dependence of the various terms on such factors as aspect ratio, strain-rate, etc. For instance, in Korzhavin's original analysis (1962) the indentation factor I is identified as being a function of W/D, where W is the overall gross width of the impacting ice sheet; however, most subsequent investigations (for instance, Afanasev *et al.*, 1971) identify I as being a function of aspect ratio D/h instead. Following the widespread application of plasticity theory to the indentation problem, I is now normally identified as being equal to the plasticity indentation factor already discussed in Section 6.2 and tabulated in Table 6.2.

The equation is nonetheless useful and shows the correct dimensional dependence. There are, however, serious problems with its practical application, since the meaning of the various terms has become conceptually confused. There are essentially three paradigm cases in which its application can give acceptable and sensible results (Fig. 6.42). They are:

(a) creep indentation where the indentor is frozen-in and perfect contact is achieved;
(b) creep indentation with partial contact between ice and indentor;
(c) fast fracture.

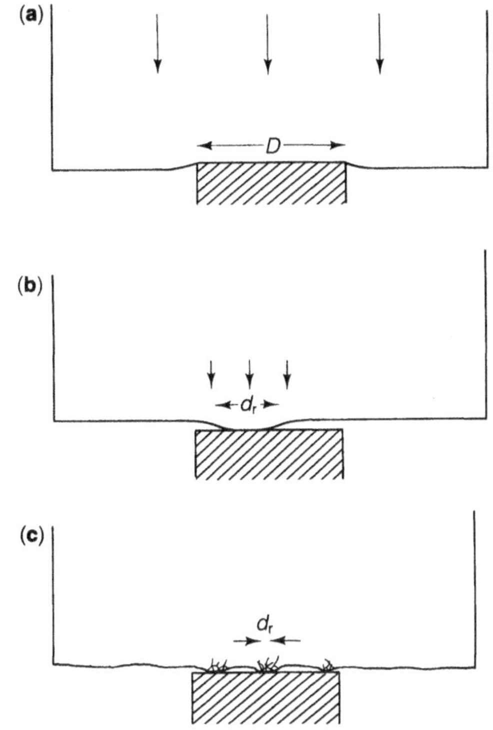

Fig. 6.42 Three paradigm cases for indentation analysis using Korzhavin's equation: (a) creep indentation with perfect contact over diameter D (e.g. frozen-in condition); (b) creep indentation with contact only over width d_r; (c) fracture with contact over n zones of width d_r ($r = 1,\ldots,n$).

The results are sensible only if a self-consistent set of choices is made for the various terms of equation (6.59). Let us look at these three cases in turn.

6.5.1 Creep indentation with perfect contact

We have already treated the formal analysis of creep indentation, by the techniques of plastic limit analysis and the reference stress method (Section 6.2). The similarity between Korzhavin's equation (6.59) and equations (6.17) or (6.21) is quite obvious. For creep indentation with perfect contact, it will be seen that Korzhavin's equation is a good representation if we simply calculate "compressive strength" σ as a function of indentation rate U/D, set I equal to the plastic indentation factors of Table 6.2 and set k, the contact factor, equal to 1.

We then have

$$P = IDh\sigma$$

where

$$\sigma = F_\sigma\{U/I\Psi D\} \tag{6.60}$$

using the reference method of Section 6.2.3.2. Note that we are assuming perfect contact, which will occur, for instance, in the case that ice is frozen-in around a structure. This implies that the relevant contact width in the indentation analysis is D and that the contact factor is unity.

6.5.2 Creep indentation with partial contact

Consider now creep indentation in which the contact is imperfect and is only achieved over a width d_r of the structure (Fig. 6.42(b)). (We have already considered this in Section 6.2.3.4.) In this case the methodology of reference stress (or plastic limit analysis) still applies, but we have to use the width d_r throughout all equations instead of D. Thus:

$$P = Id_rh\sigma$$

where

$$\sigma = F_\sigma\{U/I\Psi d_r\} \tag{6.61}$$

This can then readily be written

$$P = IkDh\sigma$$

where k, the contact factor, is literally the proportion of the indentor in contact, and is given by

$$k = d_r/D$$

Note, however – and this is vitally important – that the stress σ is calculated from equation (6.61), with d_r in the denominator, not equation (6.60) with D in the denominator. In addition, the parameters I and Ψ must be calculated for the aspect ratio d_r/h, not D/h. It is very common, and incorrect, practice to use a contact factor of less than 1 (imperfect contact) in conjunction with an ice stress and an indentation factor derived on the basis of perfect contact. This is an error.

6.5.3 Fast fracture

The same confusion afflicts the analysis of indentation fracture using Korzhavin's equation. During fast fracture (Fig. 6.42(c)) the physical process occurring is that of non-simultaneous failure: a large number, n_d, of small zones contribute statistically to a global average force, as described above in Section 6.4.5. Now this process can to some extent be described by use of the contact factor k in Korzhavin's equation to quantify the number of individual zones typically in contact (although the concept of a single numerical factor is obviously lacking in statistical sophistication). In this case, the quantity $I\sigma$ can be identified with the indentation pressure of a single zone at failure, and we can write

$$P = n_d I d_r h \sigma$$

This can then be rewritten as

$$P = IkDh\sigma \qquad (6.62)$$

where the factor

$$k = n_d d_r / D$$

expresses the relative degree of contact. Again, however, it is nonsense to identify σ as the creep strength during indentation over the total width D.

Now, if Korzhavin's equation is to be used in this way for realistic description of fast fracture indentation, then work needs to concentrate on determining appropriate values of the contact factor k as a function of size and aspect ratio. Experiments by Korzhavin (1962) indicated values of k in the range 0.4–0.7 for indentors of width 3–10 m, and measurements on Cook Inlet structures (Blenkarn, 1970) are consistent with a value of $k = 0.5$. Unfortunately, these values appear to have passed into the literature as reasonable values of general applicability, despite the limited data set from which they are derived. There is no indication that they are appropriate values to use, for instance, for structures of much greater diameter than about 10 m.

Consider, for instance, how Korzhavin's equation might cope with the most severe impact available from the Hans Island programmes (4 August 1981, as described in Section 5.4.1.11, Table 5.2). We have:

$P = 269$ MN
$D = 84$ m
$h = 8$ m

Let us take, for sake of argument, a value of $I = 1.15$ for isotropic ice under plane stress, and assume that the "uniaxial strength" of the sea ice is $\sigma = 5$ MPa. We can then substitute into equation (6.62) and find what value of contact factor is apparently appropriate. We find:

$$k = P/IDh\sigma = 0.07$$

As might be expected, considering the scale-dependent behaviour of non-simultaneous failure, this is very much less (10 times less) than the values of k derived from experiments very much smaller in scale. Similar retrospective calculations of k,

TABLE 6.6 Calculation of apparent contact factor k for various full-scale fracture events. A value of $\sigma = 5$ MPa is used in Korzhavin's equation (6.55)

Event	*Load, P (MN)*	*Diameter, D (m)*	*Ice thickness, h (m)*	*Indentation factor, I*	*Contact factor, k*
Hans Island 1980					
Event A	214	279	5.6	1.15	0.02
Hans Island 1981					
Event 2	269	84	8.0	1.15	0.07
Tarsiut P-45					
10 Dec. 1984	154	95	1.6	3.0	0.07
15 Dec. 1984	86	95	0.7	3.0	0.09
26 Jan. 1985	55	95	0.6	3.0	0.06
21 Feb. 1985	207	95	2.1	3.0	0.07
26 May 1986	165	95	0.9	3.0	0.13
Amauligak I-65					
7 Mar. 1986	270	100	6.5	1.15	0.07
12 April 1986					
Event A	500	90	10.0	1.15	0.10
Event B	210	100	5.5	1.15	0.07

based on the data available for full-scale islands and structures undergoing indentation fracture, are tabulated in Table 6.6. To draw up this table I have, for simplicity, assumed throughout that $\sigma = 5$ MPa; I have used a value of $I = 1.15$ for (isotropic) multi-year ice, and a value of $I = 3$ for (anisotropic) first-year ice. It would be a simple matter to recalculate these factors in the light of more sophisticated assumptions about temperature, salinity and geometry. However, the message is clear: the contact factor k for full-scale interactions appears generally to be less than 0.1, and is typically of the order 0.07.

I do not recommend the use of the contact factor methodology, since it is physically and statistically naive. But if it is to be used it must be used on an empirical basis with reference to experiments carried out under appropriate conditions. For a full-scale indentation event this means that we should be using $k \sim 0.07$ and not $k \sim 0.5$.

Chapter 7

Calculation of Design Ice Loads

7.1 INTRODUCTION

It will be clear from the last chapter, on theoretical analysis, that there are many plausible and quite highly developed methods for calculating ice loads under a variety of loading conditions. They are, however, all hampered by an inability to predict the onset of fracture and quantify the stress at which it occurs. Our understanding of continuum behaviour is good, but we need to know at what point fracture begins: until we know that, the considerable battery of sophisticated continuum treatments available is largely useless. We can speculate about flaw sizes, thickness irregularities and pre-existing cracks, but to date there are insufficient data on such matters to allow us to predict loads with any great reliability.

Similarly, we have developed quite a good mechanical understanding of non-simultaneous fracture, but in the final analysis it has to be calibrated closely to full-scale empirical data, and it is a tool for explaining rather than predicting.

For these reasons, I shall in this chapter follow the course of the dedicated large-scale empiricist. First and foremost come the large-scale data, which we have examined in Chapter 5. From these data, and using the theoretical understanding we have developed in Chapter 6, I shall propose quantitative practical guidelines for calculation of ice loads on structures.

To a pure scientist this may seem unsatisfactory. It would indeed be much more pleasant to be able to state an analytic equation from which ice loads could be calculated under any circumstances, following simple and well understood laws of physics. However, this is not yet possible. Instead, we have to view the full-scale data set available, interpret it in the light of our mechanical understanding and then use it as a predictive tool.

First, however, we must look at the general philosophy of design load calculation and provide some introduction to the art of risk analysis.

7.2 DESIGN LOAD PHILOSOPHY: RISK ANALYSIS

The calculation of design loads for offshore structures subject to ice hazards passes through a series of logical steps, which are shown schematically in Fig. 7.1. The steps are as follows.

(A) *Ice-loading conditions.* The first step is to define the probability of certain ice-loading conditions occurring: that is, during the course of a year (or during the design life of 100 years, or 10 000 years) what is the likely range of impact events a structure will experience? Ideally we should aim to collect statistics which allow a probability density function (p.d.f.) to be constructed for each relevant ice parameter, such as thickness, velocity, temperature, etc. Some such probability functions have already been described in Chapter 3. These can then be combined, either analytically (or more likely by Monte Carlo randomising techniques), to provide an ensemble of ice-loading scenarios which may be expected to arrive at a structure. As we shall see, in practice we need to do two things: firstly calculate the arrival rate of events, and then secondly calculate the likely range of ice parameters given that the event has arrived.

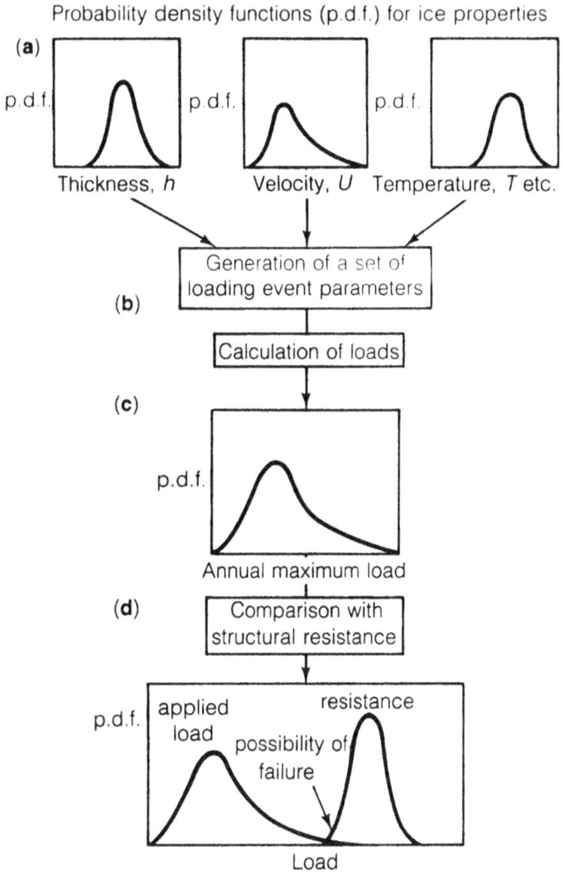

Fig. 7.1 Conceptual steps in calculating an ice design load.

(B) *Ice-load calculation for loading events.* Given a certain particular ice-loading condition arising from step (A) above, we now have to calculate a peak load for that event. That is; supposing a multi-year floe of 5 km diameter, 8 m thickness and velocity 1 m s^{-1} strikes a structure, what load is generated? The load calculation is, of course, independent of the probability of that set of ice-loading parameters being realised. The calculation itself may, however, be probabilistic: our lack of knowledge is such that we cannot say deterministically exactly what load will be exerted by a particular-sized floe striking a structure at a particular velocity, and we may wish to introduce uncertainty into the load calculation method.

(C) *Probabilistic ice-load generation.* Steps (A) and (B) now need to be combined: given the range of ice event conditions generated by step (A), what range of ice forces is generated from them if we apply the load calculation methodology of step (B)? A large number of calculations of load may then be combined to yield a probability density function plot for applied load (Fig. 7.1). For purposes of risk analysis, the most useful function to plot is the p.d.f. for maximum load experienced during one year. That is, over a very long trial period, what are the statistical characteristics describing the annual maximum load?

(D) *Risk of failure.* The final step is then to look at structural resistance afforded by the structure (for instance, foundation resistance or buckling strength of structural members) and calculate the probability that the maximum ice load during a year exceeds this resistance. If it does, of course, failure of the structure occurs. In the simplest case the strength of the structure may be well enough defined that it can be idealised as a single number: that is, the ultimate load it can withstand is P_0, and it will certainly withstand loads below P_0 and certainly fail if the load exceeds P_0. The risk of failure per year is then simply the area under the annual load p.d.f. lying to the right of the load value P_0 (Fig. 7.2(a)). This is simply the probability that the maximum load during a year exceeds P_0. More realistically, it is better practice to associate a probability density function with structural resistance also. Uncertainties invariably exist in many areas: soil properties are generally poorly defined, and the strength of engineering materials is always subject to statistical scatter. In addition, controls on quality of construction are such that the final structure as-built may not have the resistance calculated by the designer. As a result, the resistance of

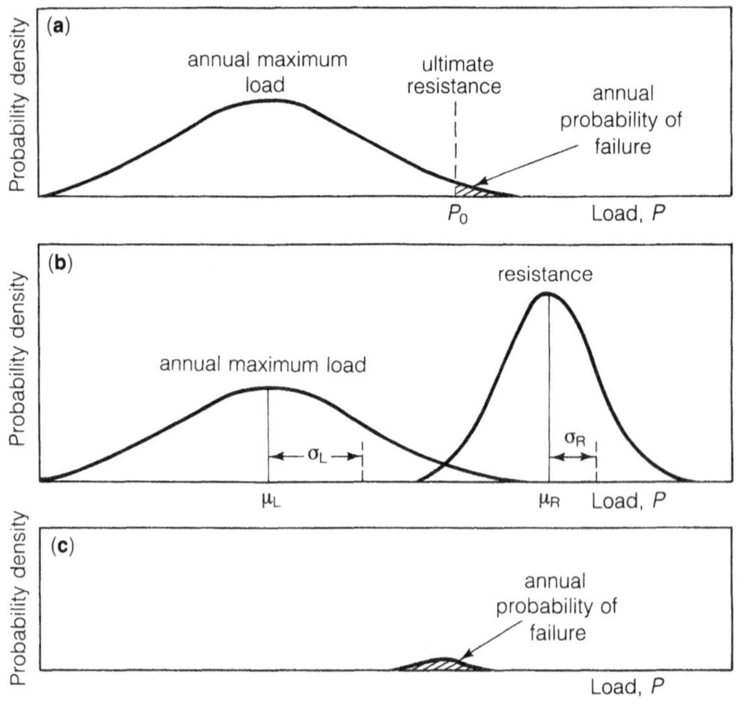

Fig. 7.2 Risk of failure. (a) If ultimate resistance is a simple well defined value, P_0, the probability of failure is the area under the p.d.f. exceeding P_0. (b) If ultimate resistance is also a probability distribution, then the probability of failure is determined from the overlap of the two p.d.f. curves. Note that it is *not* simply the area under the two overlapping curves: it is the area under a separate p.d.f., as shown in (c) and calculated in the text.

the structure has a probability density function associated with it, as shown in Fig. 7.2(b). The expected resistance should of course substantially exceed the expected ice loads, and the probability of failure of the structure may be calculated from the range of overlap of the two p.d.f. curves. Specifically, the probability density function for failure (Fig. 7.2(c)) can be constructed by calculating, as a function of load P and the probability that the load applied exceeds P *and* that the resistance is less than P.

If both the probability distributions (of load and of resistance) are normal, then some quite simple relationships can be written. (The notation and definition of probability distribution is summarised in Fig. 7.3.)

Consider the population of annual maximum ice loads P over a long trial period. If the ice loads P are normally distributed then we can represent the population by the probability density function $f_I(P)$, defined by:

$$f_I(P) = \frac{1}{\sigma_I\sqrt{2\pi}} \exp\left[-\tfrac{1}{2}\left(\frac{P-\mu_I}{\sigma_I}\right)^2\right] \qquad (7.1)$$

which is the standard normal distribution function, with a mean annual maximum ice load μ_I and a standard deviation σ_I.

On the other hand, let the resistance of the structure be represented by a similar p.d.f. denoted $f_R(P)$ and defined by:

$$f_R(P) = \frac{1}{\sigma_R\sqrt{2\pi}} \exp\left[-\tfrac{1}{2}\left(\frac{P-\mu_R}{\sigma_R}\right)^2\right] \qquad (7.2)$$

where μ_R is the mean resistance and σ_R is the standard deviation. The annual probability of failure of the structure is then given by the sum of all joint probabilities of the resistance being P but the annual maximum load exceeding P. The probability that the annual maximum ice load exceeds P is

$$F_I(P) = \int_P^\infty f_I(l)\, dl$$

(where l is a dummy variable) and thus the probability of failure, p_f, is given by the integral:

$$p_f = \int_{-\infty}^\infty F_I(P) f_R(P)\, dP \qquad (7.3)$$

where $f_R(P)$ is the probability density of the resistance to load being P. If both the load and the resist-

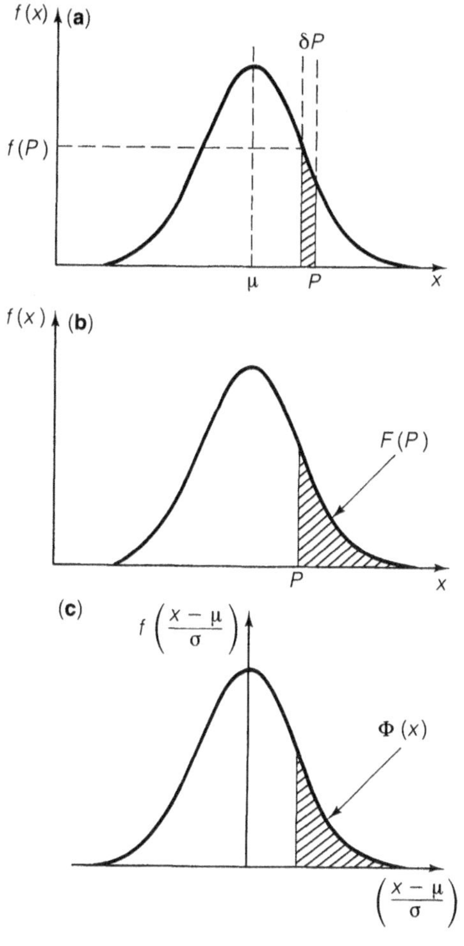

Fig. 7.3 Summary of probability distributions. (a) Probability density function $f(x)$. The probability of the variable x lying between P and $(P+\delta P)$ is given by $f(P)$. δP. (b) The probability that x exceeds P is denoted $F(P)$ and defined as

$F(P) = \int_P^{\infty} f(x)\,dx$

It is the area under the curve $f(x)$ lying to the right of $x = P$. (c) The function $\Phi(x)$ is essentially the same in (b) above, but normalised with respect to the mean μ and the standard deviation σ of the distribution. It is defined as:

$\Phi(x) = F\left(\frac{x-\mu}{\sigma}\right)$

ance are indeed normally distributed (equations (7.1) and (7.2)), then it can be shown that the probability of failure (equation (7.3)) reduces simply to

$$p_f = \Phi\left[\frac{\mu_R - \mu_I}{\sqrt{\sigma_R^2 + \sigma_I^2}}\right] \qquad (7.4)$$

where Φ is the normalised exceedance function shown in Fig. 7.3(c), which is defined by the equation

$$\Phi(x) = \frac{1}{\sqrt{2\pi}} \int_x^{\infty} e^{-l^2/2}\,dl$$

Equation (7.4) effectively represents the probability that a variate having a standard deviation of $(\sigma_R^2 + \sigma_I^2)^{1/2}$ exceeds its mean by more than $\mu_R - \mu_I$. Obviously, the larger the value of mean resistance μ_R relative to mean load μ_I, the greater the argument of equation (7.4) and the lower the probability of failure.

It is worth illustrating this with a simple worked example. Let us suppose that the annual maximum ice load has a mean of

$\mu_I = 300$ MN

and a standard deviation of

$\sigma_I = 100$ MN

(This means that, in 95% of years, the peak load measured experienced during the year should lie within two standard deviations of the mean, that is, between 100 and 500 MN. This is intended to correspond approximately to the experience of the Molikpaq structure, as described in Section 5.4.1.12, but the precise quantities should not be taken literally.) Assume now that the resistance afforded by the structure is nominally a mean of

$\mu_R = 700$ MN

but with an uncertainty of

$\sigma_R = 50$ MN

(Conceptually, this means that, if a large batch of similar Molikpaq-type structures were constructed, we should expect their structural resistances to have a mean of 700 MN and a standard deviation of 50 MN. 95% of the structures so constructed would have resistances lying in the range 600–800 MN. Again, the precise quantities should not be taken literally.)

From Equation (7.4) we can now simply calculate the annual probability of failure. For these parameters we have:

$p_f = \Phi(3.6) = 0.0002$

where Φ is evaluated from standard normal probability tables. This would indicate an annual probability of failure of about 1 in 5000.

It is the business of the designer to ensure that his design resistance characteristics lead to an acceptable probability of failure when combined with the probability distribution of applied loads. This level of "acceptable" probability of failure varies from industry to industry and from country to country, and is, of course, critically dependent on the costs and consequences of failure occurring. If "failure" merely means superficial and safe structural damage, we may be prepared to put up with an annual failure probability of as much as 0.1. If, however, "failure" entails a major disaster involving loss of human life and widespread contamination of the environment, the acceptable annual failure probability may be set at a very low level indeed. In the nuclear power industry, the design level for annual failure is intended to lie somewhere in the region of 10^{-6} to 10^{-7} per annum – that is less than 1 in a million. In the oil industry it is typically in the range 10^{-4} to 10^{-6} per annum. Whether these design levels of risk are in practice achieved is, of course, another question, which can be answered only by examining real statistics of incidence of failure.

For practical design purposes it is in fact quite rare, even now, for rigorous probabilistic calculations of load levels and resistance levels to be carried out. It is much more common practice to work, instead, in terms of design load levels calculated for a nominal "design life" of the structure and then to build the structure with a certain conventional "safety factor" over and above the calculated design load. This is the procedure generally enshrined in civil engineering codes and standards, and it is not rigorously associated with a specified annual risk of failure. Instead, the task of design proceeds in two stages. Firstly, a calculation is made of the maximum load to be expected during a design life of, say 100 years. This is defined as the load which has an annual probability of 1% of occurring. Next, this load is multiplied by a conventional safety factor, typically between 1.3 and 1.5, and the structure is designed to withstand this load.

This has much the same final result as the more formally correct probabilistic procedure illustrated by Fig. 7.2, and in practice gives rise to annual probabilities of failure of the order of 10^{-4} to 10^{-6} per annum. Note that the specification of a design load which occurs about once every 100 years (strictly, with a probability of 63% every 100 years) does *not* mean that the structure will fail once every 100 years. Because a safety factor has been applied to the design load, the structure should easily withstand the 100-year load, and would be at serious risk only if a load with a return period of several thousand years were to arrive.

Specifications of design-life and safety factors vary from code to code, and also depend on the type of construction and purpose for which the structure is built. I do not propose to treat these variations in detail here. For most offshore applications the relevant codes require ice design loads to be calculated which have an annual exceedance probability of between 1% and 4%. These correspond to a nominal return-period of 100 years and 25 years respectively.

7.3 TYPES OF ICE LOADING

For the purposes of calculation of design loads it is convenient to divide ice-loading scenarios into two broad categories: static and dynamic.

(a) *Static loading.* The loading state is static if ice exists in stationary contact with a structure and then experiences an increasing load applied to it by natural driving forces. Two examples are as follows:

   (i) *Winter landfast ice conditions,* in which a structure is typically surrounded by more or less uniform first-year ice. Wind and current drag then may act on a large expanse of ice cover and load it slowly against the structure, which acts as a single isolated pinning point resisting the applied driving force (Fig. 7.4(a)). The contact between ice and the structure may or may not be initially perfect. The event at Netserk F-40 on 17 December 1975 is an example of static loading (see Section 5.4.1.4).

   (ii) *"Winter-blocking" condition:* the possibility also exists in some regions that a massive ice feature such as a multi-year floe, a hummock field or a multi-year ridge may become lodged against a structure and concentrate load onto it. The load may be generated by a combination of wind, current and the action of surrounding slowly moving thinner ice (Fig. 7.4(b)). Under such circumstances, a calculation has to be made to determine which is greater: the applied driving force or the capacity of the massive ice feature to sustain it. If the strength of the ice feature is very great, the force may be limited by the driving forces available. Under most circumstances, however, the force is limited by failure of the blocking ice feature. The event at Amauligak I-65 on 12 April 1986

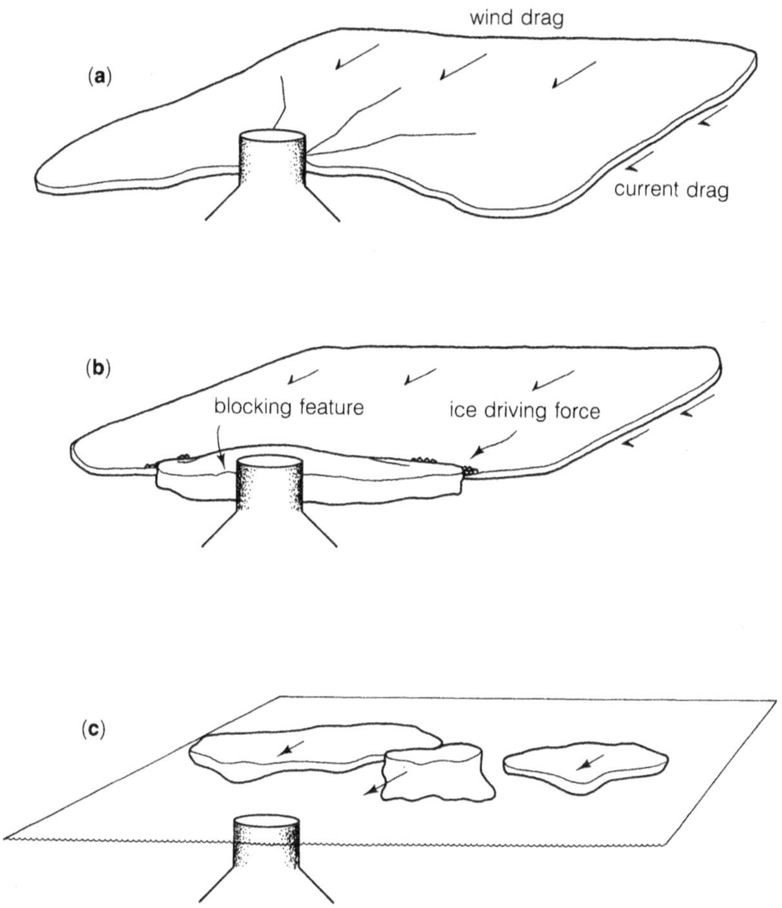

Fig. 7.4 Principal loading scenarios: (a) static winter loading (first-year ice); (b) winter-blocking; (c) dynamic impact.

is an example of the winter-blocking scenario (see Section 5.4.1.12).

(b) *Dynamic loading*. The loading state is dynamic if an ice feature is not initially in contact with a structure, but arrives and strikes it with appreciable velocity (Fig. 7.4(c)). The loading state differs significantly from static loading in two respects: firstly, the initial contact conditions are invariably irregular and non-uniform; secondly, the duration of the impact is generally determined by the kinetic energy of the impacting ice feature, which may come to rest during the impact process. The most important examples are:

(i) impact by multi-year floes (for instance, as at Hans Island, see Section 5.4.1.11);
(ii) impact by iceberg or ice island (of which no instances have yet been monitored).

In addition to these two broad categories of loading, static and dynamic, we need also to consider the magnitude of area over which the loading is important: are we looking at total forces on the whole structure or at local forces over a limited area? This question needs to be answered by reference to the expected failure mode of the structure under consideration, and in general the following two categories are relevant:

(1) *Total, or "global" load*. The total load sustained by a structure is important for considerations of foundation sliding resistance, foundation bearing capacity and overturning moment. The ice contact areas normally involved for such load considerations are in the range 100–1000 m^2.
(2) *Local loads*. Depending on the details of the structure design in question, it is common for

stresses over areas of the order of 10 m^2 to be important. Stresses over such an area determine the design and spacing of internal structure members and the dimensions of internal cell units. Loads over smaller areas, in the range 0.01–10 m^2, are also important to the design of the external skin of a structure.

7.4 TYPES OF STRUCTURE

All types of loading, static or dynamic, global or local, may depend substantially on the design and shape of the structure concerned. This book does not aim to provide a detailed treatment of structure design, but the principal respects in which structure design is important are: surface slope, degree of structural dynamic response, and formation of a surrounding rubble pile. We also have to consider how to deal with multi-legged structures.

7.4.1 Surface slope

Some conceptual designs of Arctic offshore structures are shown in Fig. 7.5. Figures 7.5(a)–(e) represent sloping-sided structures which are designed to fail ice in flexure. Under the correct operating conditions this results in significant reductions in load, especially for relatively thin ice, and the technique has already been successfully used by the floating *Kulluk* structure (Section 5.5.2.2). A two-dimensional analysis of the interaction process was presented in Section 6.4.4.3. I do not, however, consider that sloping-sided structures can be relied upon to function in the manner for which they are designed. In each case, both for upward-breaking and downward-breaking cones, the possibility exists that ice will fail to clear sufficiently quickly around the structure and that, as a result, a vertical contact face will be presented to advancing ice. If this happens, the global force on the structure will approach that for a simple vertical-walled structure.

Additional disadvantages associated with sloping-sided structures include the risk of serious ride-up and damage of surface facilities, and also the unnecessary introduction of significant vertical loads which may prove critical to design in poor soil conditions. Here I shall concentrate on design loads for vertical-sided structures, with the judgement that design forces for conical structures under severe ice conditions will in practice have to be similar.

7.4.2 Structural response

Many structures, especially those constructed of steel, show significant dynamic response and vibration when ice interaction occurs. This phenomenon has been observed at Cook Inlet, on lighthouses and on the Molikpaq structure, and has already been discussed in Chapter 5. It can lead to short-lived high peak loads being measured on a structure and high cyclic loads being transmitted to the foundations. This can be especially serious in poor soil conditions where partial liquefaction may occur. Vibration response is not well understood, and is also generally specific to the structure concerned, and in what follows we shall simply consider perfectly rigid fixed structures.

7.4.3 Rubble pile formation

Certain types of structure, such as the artificial island or the shallow caisson structure placed on a sand/gravel berm, are designed to encourage formation of a rubble pile of grounded ice. Indeed, some more recent Arctic drilling structures, the spray-ice islands, consist of nothing else. Schematic examples are shown in Figs 7.5(g) and (h). A rubble pile generally presents an effectively vertical face to advancing ice: continuing ice action tends to erode the exposed face of the rubble pile, and the opposite forces of gravity and buoyancy prevent the formation of a consistent sloping surface.

When a rubble pile forms around a structure it has two effects:

(i) It increases the effective diameter of the structure, thus presenting a greater area to advancing ice. This increases the expected load on the structure/rubble-pile system.
(ii) It increases the effective area of frictional contact with the sea bed, thus increasing structural resistance.

The effect has been discussed by Kry (1977, 1980a). Although there was some doubt in early days over the advisability of encouraging a rubble pile it is now fairly clear that rubble piles do afford protection. The advantages of increased structural resistance generally scale with approximately D^2, where D is the rubble-pile diameter, since the resistance depends principally on the area of rubble pile which is grounded ($\pi D^2/4$). The disadvantages of increasing width, however, derive from the increased diameter of structure being presented to advancing ice; the diameter scales, of

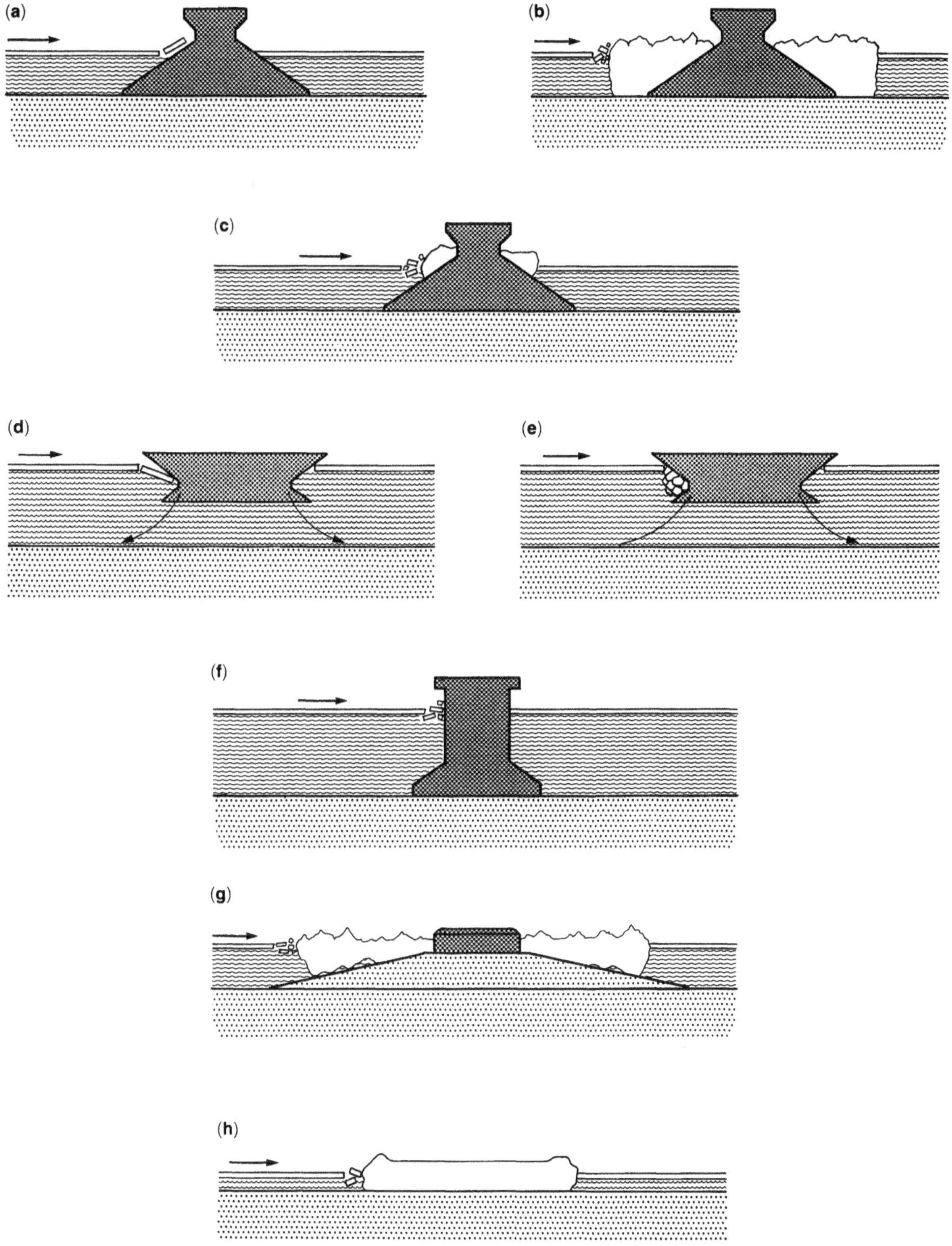

Fig. 7.5 Principal types of Arctic structure: (a) upward-breaking conical structure; (b) conical structure surrounded by grounded rubble pile; (c) conical structure with accreted ice rubble; (d) downward-breaking conical structure, floating but moored; (e) conical structure with ice jammed against it; (f) vertical-walled structure; (g) caisson-retained island with grounded rubble pile; (h) grounded spray-ice island.

course, with D, and the load (if the pressure–area curve is correct) scales probably with $\sqrt{D}$. The advantages therefore easily outweigh the disadvantages.

7.4.4 Multi-legged structures

Little attention has been paid in the literature to the analysis of ice forces on structures with several legs (for instance, the four-legged jacket structures used in Cook Inlet, Section 5.3.1).

This is an area in which scale-model experiments carried out in ice model tanks are of great assistance. I have chosen not to include here a detailed introduction to ice modelling; the technology is reviewed by Timco (1984), and its specific application to multi-legged structures is assessed by Timco (1986b). The principal conclusions are as follows.

Consider ice interaction with a structure which consists of two legs of diameter D separated by a distance W (Fig. 7.6). Model experiments then indicate that:

(1) If the spacing between the legs, W, is greater than about 6 times the individual leg diameter D then the legs may be treated entirely independently, with no interference effects occurring between them (Fig. 7.6(a)). If the load on each leg is P the total load on the structure is $2P$.
(2) As the leg spacing is reduced (W/D less than 6) then the horizontal force on each leg in the direction of ice motion decreases marginally below that for single isolated leg. This is because the force vectors on each leg tend to rotate inwards towards each other, by up to about 12° (Fig. 7.6(b)). The individual total force on each leg remains the same as in (1) above, but the aggregate force is marginally lower than $2P$.

Experiments with three- and four-legged model structures (Evers and Wessels, 1986) show similar behaviour, but indicate that the ratio W/D must exceed 7 or more before the legs may be considered independent.

Naturally, the loading geometry on a multi-legged structure changes radically if rubble accretes between the structure's legs (Fig. 7.6(c)). If this occurs then the effective total structure width is W, not two legs of width D. This sudden increase in structure width can, of course, have serious consequences, and caused problems with early structures in the Bohai Gulf of China (Wang, 1983). The load in such cases greatly exceeds $2P$.

7.5 DRIVING FORCES

Before considering the provision of design loads for the various categories of loading, we must also investigate the available natural driving forces. This is an essential initial ingredient in the logic of calculating maximum forces on structures, since the actual force exerted is in general the lesser of the following two:

(a) the force required to fail ice against the structure;
(b) the force available from natural driving forces.

The motion of floating ice is caused by three primary driving forces: wind, current and thermal expansion. In addition, motion and stresses may be caused by secondary driving forces; that is, contact with other ice masses which themselves have been driven by primary driving forces.

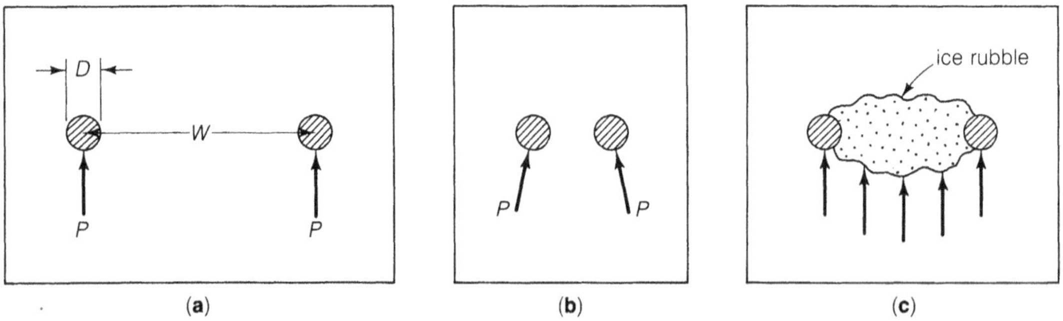

Fig. 7.6 Ice failure on a two-legged structure. (a) W/D greater than 6, where the legs behave independently. (b) W/D less than 6, where interference occurs; the resultant load decreases since the component loads on the two legs deviate towards each other. (c) If rubble accretes between the legs, the structure width is effectively increased to W.

7.5.1 Wind and current forces

Wind and current driving forces develop from drag applied by flow over the rough top and bottom surfaces of the ice cover. The uniform flow of fluid is perturbed by the presence of the surface and shows a sharp velocity gradient in its vicinity (Fig. 7.7(a)). Since the far-field unperturbed velocity a long way from the surface cannot generally be measured, the drag of wind and current is conventionally expressed in terms of the fluid velocity u_z a certain standard distance z from the drag surface. Drag forces are proportional to the square of velocity, and the general expression for the shear stress τ_f due to fluid flow is

$$\tau_f = \rho_f C_z u_z^2 \tag{7.5}$$

where ρ_f is fluid density and C_z is known as the drag coefficient. The value of C_z depends on the surface roughness of the ice in question and also on the reference height z at which velocity measurements are taken.

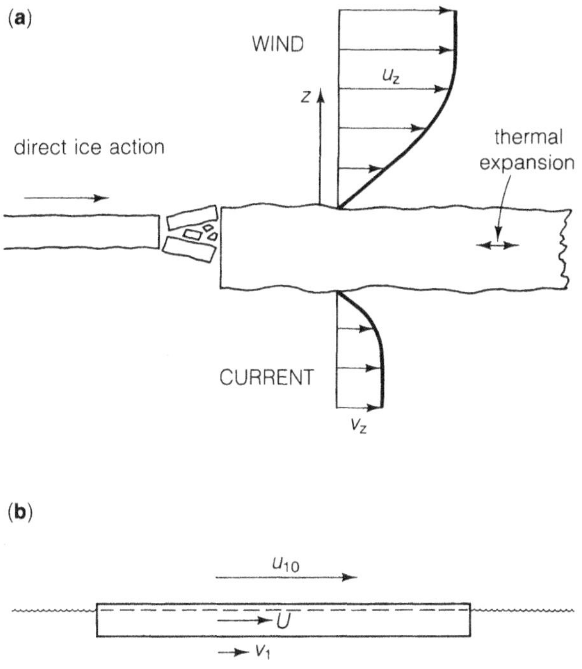

Fig. 7.7 (a) Four driving forces for ice action: wind, current, thermal expansion and direct ice action. (b) Geometry for analysis of floe velocity U as a function of wind speed u_{10} (at height 10 m) and current speed v_1 (at depth 1 m).

7.5.1.1. Wind drag

Wind velocity measurements are conventionally taken at a height of 10 m above the surface, and the equation of wind drag is then expressed in terms of the drag coefficient C_{10} corresponding to this reference height. It is:

$$\tau_a = \rho_a C_{10} u_{10}^2 \tag{7.6}$$

where u_{10} is the velocity of air at height 10 m above the surface. The density of air, ρ_a, is approximately 1.3 kg m^{-3} at low temperatures, and is not appreciably dependent on humidity under normal Arctic conditions. A large number of experiments have been performed to determine the drag coefficient C_{10} for ice and air (Banke and Smith, 1973), and it is found, as might be expected, to depend on ice type and surface roughness. Values for C_{10} found in Arctic regions range between 1.0×10^{-2} and 3.1×10^{-3}, and a value of $(1.9\pm0.5)\times10^{-3}$ is good guidance for generally smooth Arctic sea ice without major ridges. It has been found that fresh snowfall tends slightly to increase surface roughness (and hence C_{10}). A correlation is also apparent between ridge elevation and drag coefficient: drag coefficients increase as a function of surface undulations, particularly those of wavelength less than about 10–15 m.

Table 7.1 shows some sample calculations of typical wind drag forces on large expanses of ice. It shows forces on an expanse of ice 50×50 km (2500 km^2) and on a floe of area 1 km^2 for wind speeds up to 30 m s^{-1}. Bearing in mind the forces we know to have been measured as ice fails against offshore structures (that is, of the order of 500 MN), Table 7.1 makes two things clear. Firstly,

TABLE 7.1 Wind-induced driving forces for (a) a 50×50 km (2 500 km^2) expanse of landfast ice and (b) a multi-year floe of area 1 km^2

	Driving force (MN)	
Wind velocity ($m\,s^{-1}$)	*Landfast ice of area 2 500 km^2*	*Floe of area 1 km^2*
5	154	0.1
10	618	0.2
15	1 389	0.6
20	2 470	1.0
25	3 859	1.5
30	5 558	2.2
40	9 880	4.0

under landfast ice conditions the expanse of ice is so vast that only quite modest wind speeds (about 10 m s^{-1} or 20 knots) are required in order to develop enough force to fail ice against a structure – there appears to be little prospect of driving forces being insufficient. Secondly, if a medium-sized multi-year floe impacts a structure it is reasonable to neglect the effect of wind drag.

7.5.1.2 Current drag

Drag by subsurface water current obeys the same law as wind drag (equation (7.5)), but with different constants. Owing to its higher viscosity, the boundary layer of shear is much more limited for water than for air, and velocity measurements are usually referred to a depth 1 m below the bottom surface. We then have

$$\tau_w = \rho_w C_1 v_1^2$$

where v_1 is the current velocity at depth 1 m. The density of sea water, ρ_w, may be taken as 1.028 Mg m^{-3}. A series of measurements of C_1 (Langleben, 1980) have yielded values in the range 2.2–8.3×10^{-3} for multi-year ice and of about 1×10^{-3} for first-year ice.

For general purposes it is reasonable to take a value of 4×10^{-3}. Typical current drag forces for currents in the range up to 1 m s^{-1} are tabulated in Table 7.2. Again drag forces on a single floe are usually negligible, while on a major expanse of ice they may be very high indeed.

TABLE 7.2 Current-induced driving forces

Current velocity ($m\,s^{-1}$)	Driving force (MN): Landfast ice of area 2 500 km^2	Driving force (MN): Floe of area 1 km^2
0.1	103	0.04
0.2	411	0.2
0.4	1645	0.7
0.6	3701	1.5
0.8	6579	2.6
1.0	10280	4.1

Comparing wind and current drag forces, we see that the force F_a due to wind on a static area $\mathcal{A}$ of ice is given by

$$F_a = (2.5\times10^{-3})\mathcal{A}u_{10}^2$$

while the force due to current, F_w, is given by

$$F_w = 4.1\mathcal{A}v_1^2$$

This means that a current of a given velocity exerts approximately 1600 times as much drag as a wind of the same velocity. Putting this another way, a water current of 1 m s^{-1} is as effective in terms of drag as a wind of 40 m s^{-1}.

7.5.1.3 Terminal velocity of a floe

Drag considerations are useful also in calculating the theoretical terminal velocity of a freely drifting floe.

Considering an isolated floe acted on by a wind of velocity u_{10} and a current v_1 (Fig. 7.7(b)). For the purpose of this simple two-dimensional calculation, we assume that the wind and current lie in the same direction, and we ignore the Coriolis effect. This means that the floe will accelerate in the direction of the wind and current, attaining a velocity U. It is clear that once the floe velocity U exceeds the current velocity v_1 it will begin to experience restraining drag due to relative motion through the water, and that it will continue to accelerate only until the wind drag forces and current drag forces exactly balance each other. The net drag on a floe travelling at velocity U (where $v_1 < U < u_{10}$) is:

$$\rho_a C_{10}(u_{10}-U)^2-\rho_w C_1(U-v_1)^2$$

and to find the terminal velocity we must set this equal to zero. We find

$$U = v_1+(u_{10}-U)\left(\frac{\rho_a C_{10}}{\rho_w C_1}\right)^{1/2}$$

Now, in general, we expect the floe velocity U to be negligible by comparison with the wind velocity u_{10}, and we can therefore write approximately,

$$U = v_1+u_{10}\left(\frac{\rho_a C_{10}}{\rho_w C_1}\right)^{1/2} = v_1+0.025u_{10}$$

using appropriate values for constants. This means that the terminal velocity of a floe is approximately current speed plus 2.5% of the wind speed. The value of 2.5% is rather lower than the rule-of-thumb value of 3.5–4% found in Section 3.2.3 (Fig. 3.14) but is in reasonably good agreement. It differs for two principal reasons:

(1) Sea currents, which are ignored in formulating the rule-of-thumb value, are often correlated with wind, which tends to exaggerate the direct dependence on wind speed.

(2) Because of the dependence of drag on the square of velocity, floes are influenced disproportionately by gusts of high wind. The root-mean-square (r.m.s.) wind velocity would be a better measure than the arithmetic mean.

Notice that all floes apparently have the same terminal velocity, irrespective of their diameter. This is because current and wind drag forces both depend linearly on floe area. Interestingly enough, it would appear that velocities during the initial acceleration phase, before terminal velocity is reached, are also independent of floe area. This is because the floe mass m is also a function of floe area. The mass of a floe of area $\mathcal{A}$ and thickness h is given by

$$m = \rho_i \mathcal{A} h$$

where ρ_i is the density of ice. Consequently, its acceleration at any moment is given by

$$\frac{[\rho_a C_{10}(u_{10}-U)^2 - \rho_w C_1(U-v_1)^2]\mathcal{A}}{\rho_i h \mathcal{A}}$$

which is clearly independent of floe area $\mathcal{A}$.

We do, however, expect that *thicker* floes will accelerate less rapidly as a result of their greater inertia. The acceleration a is inversely proportional to thickness h. This means that during build-up of high winds, thicker floes should move more slowly; as winds die away, however, they will take longer to slow down.

7.5.2 Thermal expansion

Thermal driving forces occur when ice undergoes a temperature change and tries to expand. It may or may not be prevented from expanding. These are two quite distinct cases, corresponding to different boundary conditions, and have to be treated separately. The cases are:

Case A: restricted expansion. Ice is confined in a limited space such as a reservoir or canal and tries to expand (Fig. 7.8). Because the ice is restricted it cannot physically expand in the horizontal direction and so develops compressive stresses of just sufficient magnitude to prevent it from expanding. These stresses have to be imposed by the immovable boundaries of the reservoir structure, and the problem is then one of uniform uniaxial or biaxial compression with a strain-rate defined by the thermal rate of strain.

Case B: unrestricted expansion. Ice is fixed along one boundary but may freely expand away from the boundary. If a structure is placed some distance L from the boundary and has a diameter D that is small compared with the length of the boundary, then the expanding ice moves past the structure (Fig. 7.8). This corresponds, for instance, to an island 100 m wide lying 50 km offshore from an extensive coastline. The structure does not significantly restrain the ice cover from expansion but experiences an indentation force as the ice moves inexorably forward. The restraining force exerted by the island in this way is negligible compared to the thermal forces available to drive the ice cover. The boundary condition is one of ice velocity due to thermal strain-rate over a fetch L.

Thermal expansion is the result of temperature changes in an ice cover, which occur according to three principal mechanisms.

(1) *Surface air temperature changes.* Any change of temperature at the surface will propagate slowly downwards through the ice cover, following the Fourier theory of heat conduction. The presence of thermal boundary layers of air makes the efficiency of this process dependent on wind speed. Propagation also depends on the thermal diffusivity of the ice and may be impeded by a surface layer of insulating snow.
(2) *Transport mechanism.* If the ice is permeable, as snow or firn may be, then percolation of air, rain or meltwater may cause rapid temperature changes. This is unusual, especially for sea ice.
(3) *Radiation.* Short-wave and long-wave radiation may play important roles in temperature change. Incoming solar radiation and long-wave radiation from the atmosphere result in temperature rise; and long-wave radiation emitted back by the ice may result in temperature decrease, especially at night in clear weather. The effects of radiation are all strongly dependent on weather conditions, snow cover and absorption characteristics of the ice. Temperature changes are effectively instantaneous but attenuate with depth.

All these factors are important and any may dominate under particular conditions. A comprehensive analysis of their net effect has been made by Bergdahl and Wernersson (1977, 1978) for particular Swedish lakes, and an analysis for any other environmental conditions could be made following their analysis. In many conditions, however, thermal changes are dominated by surface

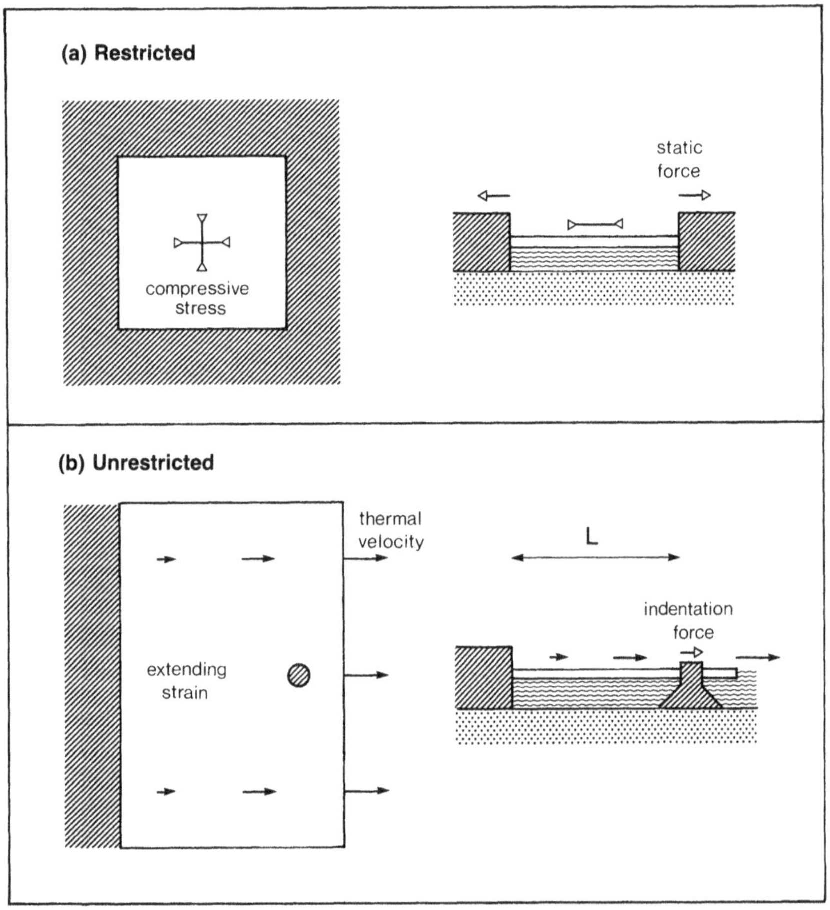

Fig. 7.8 Ways in which a thermal ice force is generated. (a) Restricted expansion (case A), where no movement occurs and compressive stresses develop. (b) Unrestricted expansion (case B), where movement occurs leading to indentation forces on a structure.

temperature changes and their propagation characteristics. Where the ice involved is highly dense this may be treated purely by heat conduction analysis with negligible convection.

Temperature changes at the surface of an ice cover propagate downwards according to Fourier theory of heat conduction (Carslaw and Jaeger, 1959). Any surface temperature variation may be analysed into a Fourier series of sinusoidal terms of different frequency and phase. We then need simply an analysis of how a surface temperature wave of frequency ω and amplitude T behaves with depth z. It is readily found that the temperature wave becomes attenuated according to:

$$\exp[-z(\omega/2k_i)^{1/2}] \quad (7.7)$$

where k_i is the thermal diffusivity of ice. The presence of ω in the negative exponent means that rapid temperature fluctuations are attenuated more quickly with depth than slower ones. Only slow cyclic temperature changes propagate deeply through an ice cover. For pure ice the constant k_i is about 1.15×10^{-6} m^2 s^{-1} (Van Dusen, 1929) and equation (7.7) then gives the depth

$$z_0 = (2k_i/\omega)^{1/2}$$

at which temperature the wave is attenuated to 1/e of its surface amplitude. This represents a "skin depth" for propagation of a wave frequency ω. It is tabulated in Table 7.3 and represented schematically in Fig. 7.9 for three characteristic frequencies 6 hours (a warm afternoon), 24 hours (diurnal cycle) and 14 days (a warm spell).

In analysis of thermal stresses we are particularly interested in maximum rate of change of temperature at depth. There is a neat answer to this. If a series of surface temperature waves of different frequency but the same amplitude are applied to the ice we can calculate by differentiation the wave frequency which causes the maximum rate of temperature dT/dt. At depth z,

TABLE 7.3 Depth of penetration of surface temperature waves through an ice cover. The "skin depth" (depth at which the wave is attenuated to 1/e of its surface amplitude) is shown as a function of wave period and frequency ω. Direct radiative heating would increase this depth substantially

Period	ω *(rads s⁻¹)*	*Skin depth (m)*
1 h	1.7×10^{-3}	0.04
2 h	8.7×10^{-4}	0.05
6 h	2.9×10^{-4}	0.09
12 h	1.5×10^{-4}	0.13
24 h	7.3×10^{-5}	0.18
7 days	1.0×10^{-5}	0.47
14 days	5.2×10^{-6}	0.67
30 days	2.4×10^{-6}	0.97
1 year	2.0×10^{-7}	3.40

the wave frequency ω which dominates is (Sanderson, 1978):

$$\omega = 8k_i/z^2$$

This is tabulated in Table 7.4 and shows why, for most applications, temperature changes are most

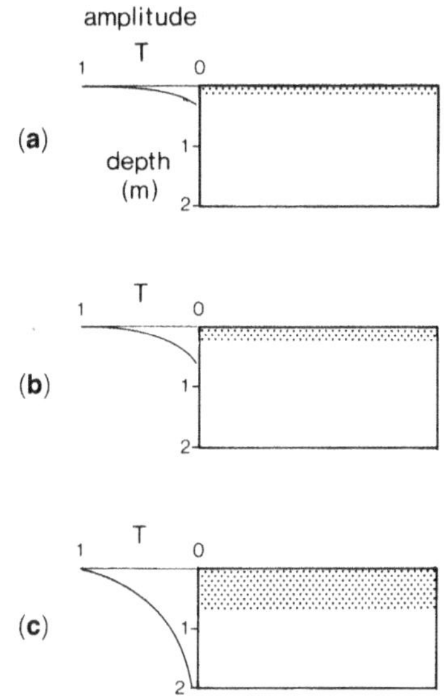

Fig. 7.9 Attenuation of a surface temperature wave propagating by conduction, for three wave periods: (a) 6 hours (a warm afternoon); (b) 24 hours (diurnal cycle); (c) 14 days (a warm spell). Amplitude of wave is shown as a function of depth, and skin depth is shaded. The ice is drawn as 2 m thick.

TABLE 7.4 Period of surface temperature wave which at a particular depth gives rise to maximum rate of change of temperature (all waves are assumed to be of the same amplitude)

Depth (m)	*Dominant period*
0.1	1.9 h
0.2	7.6 h
0.3	17 h
0.4	30 h
0.5	2 days
1.0	8 days
2.0	30 days

significant only in the top 0.3–0.4 m. The important temperature cycles of period one day and less dominate only at depths less than 0.36 m.

Now, where expansion is restricted (Case A, above) the restraint conditions are usually biaxial, and the stress associated with the thermal activity can be calculated using appropriate application of the equations of Section 4.2.5. To do this, we assume that by being confined, the ice is effectively straining in compression at a rate equal and opposite to the rate at which it would expand if it were unconfined and stress-free. This procedure is explained in more detail in Sanderson (1984b).

The case of more interest here is the unconfined case (Case B). Where an ice cover is unrestricted it actually expands in response to any temperature change. In this case the process is dependent on ice thickness, since the presence of a thick underlying layer of ice undergoing no temperature change will restrain the tendency of the active thermal surface layer to expand.

A thick ice cover undergoing a temperature change effectively consists of two coupled layers (active and passive), which together expand at a slower net rate than an ice cover would if it were undergoing a uniform temperature change (Fig. 7.10(a)). The ice cover supports balanced tensile and compressive stresses as the surface active layer tries to stretch the passive lower layer, and the passive lower layer tries to restrain the upper layer. If the stresses due to these thermal effects exceed the restoring forces due to gravity then the entire sheet may arch and fracture (Fig. 7.10(b)). This may happen especially easily with thinner ice.

The rate of expansion of a thick ice cover is difficult to calculate exactly since it is a coupled problem. To first order, for an ice cover many times

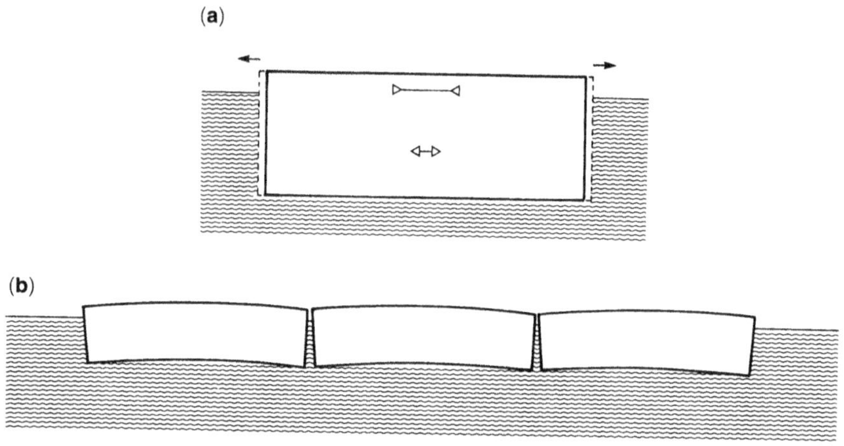

Fig. 7.10 Expansion of an ice cover shown in section. (a) Thick ice, in which an active surface layer is partially restrained by the passive lower layer. (b) Bending and fracture of a thinner ice cover when thermal stresses overcome the gravitational restoring forces.

thicker than the active surface layer, it may be calculated as follows:

(a) calculate thermal stress σ in the active layer of thickness h_T (Fig. 7.9), on the assumption that it is perfectly restrained owing to its attachment to the unstraining lower layers;
(b) distribute this stress over the total thickness h, so that the average stress is $\sigma(h_T/h)$;
(c) calculate gross strain-rate due to this distributed stress.

In general, this means that thick ice deforms very much more slowly than thin ice. A rigorous calculation of typical deformation rates is quite involved, especially in the presence of a vertical temperature gradient and variations of ice crystallography with depth. However, example cases treated by Sanderson (1984b) show that an island lying 50 km offshore surrounded by ice 2 m thick might experience ice velocities of a few metres per day due to thermal expansion. This would correspond to thermal strain-rates in the order of 10^{-10} to 10^{-9} s^{-1}. This is consistent with measurements of thermal movement rates which have been made in the Arctic ocean around artificial islands. At Netserk F-40 in winter 1975–6 steady movements of 10^{-5} m s^{-1} (0.9 m per day) were observed and attributed to thermal strain (Strilchuk, 1977).

7.5.3 Pack-ice action

Lastly, we shall look briefly at indirect driving forces due to distant pack-ice action (Fig. 7.11). Action by pack-ice may occur even if winds and currents are negligible, since the pack-ice of the Arctic Ocean is constantly in a state of motion and stores a vast amount of kinetic energy. The force which can be exerted by pack-ice is limited by the force required to build ridges. Unfortunately, this is not a well known quantity at all, so it is difficult to set limits to the forces which may be developed by pack-ice action. Most estimates of ridge building forces lie in the range 0.01–0.1 MN m^{-1} (Croasdale, 1984), for ice of thickness 1–2 m. These estimates are based on energy considerations (e.g.

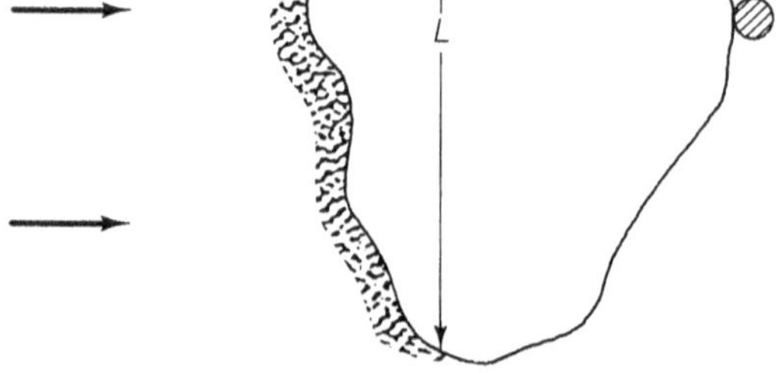

Fig. 7.11 Pack-ice action on a blocked ice feature of width L.

Parmerter and Coon, 1973), or on fitting of data to meso-scale modelling as described already in Section 5.6 (Pritchard, 1977, 1980; Hibler, 1980a,b). However, it is very likely that the ridge-building force is scale-dependent in some manner and so it is inadvisable to adopt a single value independently of the distance or area over which it is thought to act. Over smaller areas it is likely to be higher; over larger areas, lower. In calculating potential pack-ice driving forces I suggest that Fig. 5.36, the pressure–area curve, should be used in conjunction with the scenario in question. As an illustration, suppose that a floe of diameter $L = 5$ km is blocked against a structure and that pack-ice ridging activity begins behind it. If the pack-ice is of thickness 2 m we have a gross contact area of 10^4 m^2 and a maximum average failure stress, from Fig. 5.36, of about 0.2 MPa. This constitutes a driving force of 2000 MN. A similar calculation for a more typical floe of width $L = 1$ km suggests a potential driving force of about 800 MN, while for a vast floe of width $L = 10$ km we would have a driving force of about 2500 MN. Clearly, the driving forces from pack-ice are considerable and under most circumstances exceed the forces which have actually been measured during failure of large features against Arctic structures (Chapter 5). These forces were 100–500 MN. It will be readily seen from Tables 7.1 and 7.2 that wind and current forces, too, may easily exceed the forces apparently required for ice failure against typical structures. For this reason, I consider that the concept of a "limited driving force" as a constraint on maximum ice forces is something of a red herring, at least in normal applications. Under typical Arctic conditions, forces developed from wind, current and pack-ice action are almost invariably enough to fail ice against a structure. This accords with the observations at, for instance, Amauligak I-65 (Section 5.4.1.12), that practically whatever the size of the ice feature presented to the structure it nonetheless underwent failure at the structure, and did not remain in place as a rigid blocking floe while ice failed behind it.

This is not to dismiss the concept entirely. It is an important logical check which should routinely be carried out, and there are certainly scenarios in which it is likely that driving force would limit the load applied by ice to a structure. Such a scenario would, however, likely have to involve an exceptionally thick feature such as an iceberg or ice island.

7.6 STATIC ICE LOADING

7.6.1 General

Conceptually, "static" loading of ice against a structure can be divided into three separate stages, shown schematically in Fig. 7.12.

Consider a structure surrounded by static first-year or multi-year ice, which is initially in more or less close contact with the structure. If the ice cover is initially adfrozen to the structure the contact may be essentially perfect; it if is not adfrozen it may be quite irregular. Now let us introduce a slowly increasing driving force, such as wind, which causes the ice cover to move forward onto the structure at a steadily increasing velocity. The interaction typically passes through the following three distinct phases:

(1) *Pure creep.* To begin with, the ice will deform by pure creep, in a continuum mode (Fig. 7.12), and during this period the load may be calculated as a function of the ice velocity, if it is known, using the reference stress method. The calculation will differ according to whether the ice is in perfect contact (solid line) or in imperfect contact (dashed line).

(2) *Transition to fracture.* At a certain limiting velocity or limiting stress (at present rather ill-defined), the ice undergoes a transition between creep behaviour and fracture behaviour. During this phase, the load is at or near its maximum and velocity may subsequently increase without a corresponding increase in load.

(3) *Non-simultaneous fracture.* Following the onset of fracture, ice fails in a truly brittle fashion, and gives rise to a spikey and highly variable load trace (Fig. 7.12). In general, it is to be expected that the loads experienced during this phase will be lower than those experienced at the critical peak of transition from creep to fracture behaviour. Certainly, a load peak arising from non-simultaneous fracture is most unlikely to exceed the transition-point peak for perfect contact conditions. However, if a sufficiently long period of fracture interaction is experienced then subsequent "non-simultaneous failure" peaks may well exceed the stress at the transition point for irregular contact.

For the sake of increasing our understanding, we shall look in turn at each of these stages. However, for purposes of practical calculation of loads, we

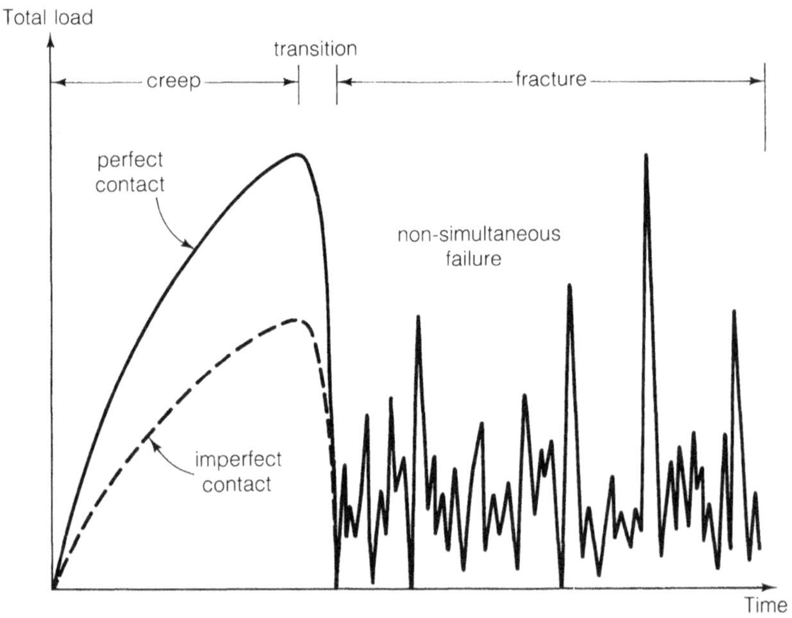

Fig. 7.12 Schematic diagram of load history during a static interaction event. During the initial stages, when deformation is by creep alone, the load rises smoothly. The load depends on whether the contact is perfect, for instance adfrozen (solid line) or imperfect (dashed line). Once fracture initiates, the load trace is irregular.

shall see in Section 7.6.2 that we can in general ignore the creep phase of behaviour, and carry out our analysis entirely in terms of peak loads measured at or after the onset of fracture.

7.6.1.1 Pure creep

It is very rare indeed for the ice interaction conditions around a structure to be so limited and well defined that we can confidently predict that only creep interaction will occur. In general, there is likely to be sufficient driving force available that fracture behaviour has to be considered and we cannot restrict ourselves to a simple continuum analysis. However, in certain circumstances, for instance, nearshore landfast ice conditions where ice velocities are dominated by thermal expansion effects, we may be able to place confident limits on the ice velocities to be expected during the design life of the structure. If these velocities are so low that the ice may be expected to deform exclusively by creep, then it is fairly straightforward to calculate an upper bound to the load associated with the interaction.

I shall not present here a full probabilistic risk analysis of the design load to be used under such circumstances, since it depends critically on the probability distribution to be adopted for ice velocity. We shall look instead at loads predicted by continuum mechanical methods, for specific examples of ice characteristics and a range of ice velocities. We use the reference stress method, as developed in Section 6.2.3.2 and we shall assume perfect contact conditions. The calculation provides an upper bound to ice loads at a given velocity, for two reasons:

(1) We make the assumption that secondary creep is occurring. This presents the highest load associated with a given velocity, and if primary creep is also occurring, the load may be expected to be lower.
(2) Under most circumstances, we do not expect adfrozen perfect-contact conditions. If contact is imperfect, then peak loads for a given velocity are in general lower. Methodology for treating this has been given in Section 6.2.3.3, but needs quantitative calibration before it can easily be applied.

Now, natural sea ice in contact with a structure is not made wholly of one type of ice. It is typically a combination of granular and columnar ice, the proportions varying according to growth conditions (Fig. 7.13). Near to the shore, in stable growth conditions, the ice may be largely columnar beneath an initial surface granular layer, and if significant sea currents exist its lower layers may be of oriented columnar ice (Fig. 7.13(a)). Further offshore, where rafting and wave action disturb growth, ice may typically be more as in Fig. 7.13(c).

As we have seen, granular ice is about 50% more creep-resistant than columnar ice under comparable conditions (Section 4.2.3). In addition, there is almost always an appreciable temperature gradient through the ice, meaning that surface layers are colder and harder than lower layers. The increasing brine volume with depth has the same effect. If blocks of ice of the type shown in Figs 7.13(a) and (c) are compressed at uniform strain-rate we expect the highly unbalanced stresses shown in Figs 7.13(b) and (d). Much of the load is borne by the granular layers because they are harder and colder.

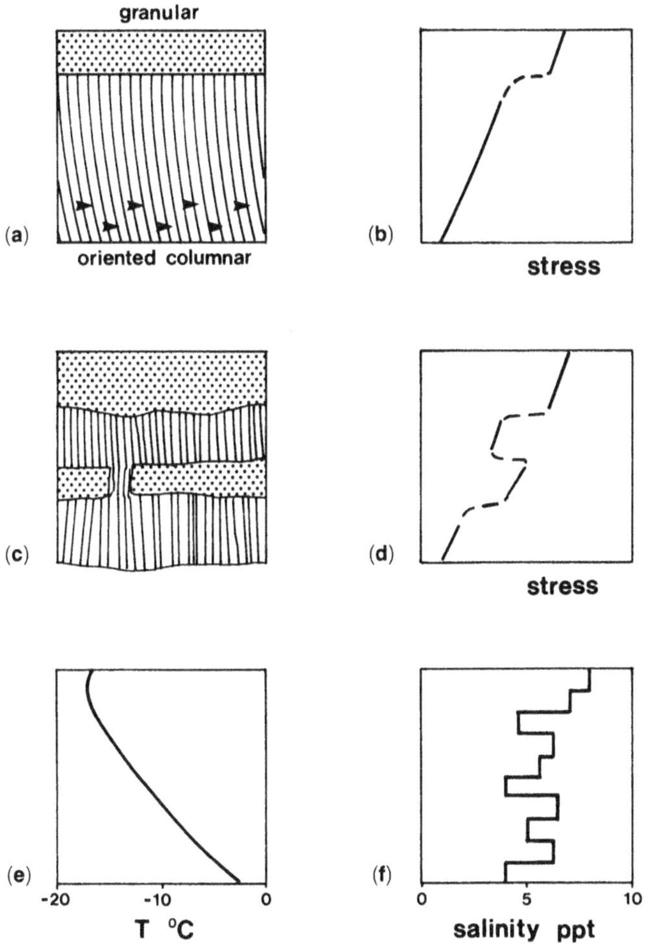

Fig. 7.13 Naturally occurring sea ice, shown with schematic theoretical stress profiles under uniform uniaxial compression: (a) typical landfast ice, showing thin granular surface layer, unoriented columnar ice and oriented columnar ice at the bottom; (b) stress profile for landfast ice, case (a); (c) typical disturbed ice, showing rafting and uneven texture; (d) stress profile for ice of case (c); (e) typical temperature profile; (f) typical salinity profile.

To a first approximation, the two layers can be treated independently. In fact, the layers couple awkwardly since under uniaxial compression columnar ice tries to expand more in the transverse direction than isotropic granular ice does. It is similar to an elastic sandwich of two materials of contrasting Poisson's ratio, and under stress the sandwich should tend to bend. In practice, the fact that the ice is floating and flat tends to ensure strain compatibility at the interface, and a compromise is reached. Here I shall ignore the interaction.

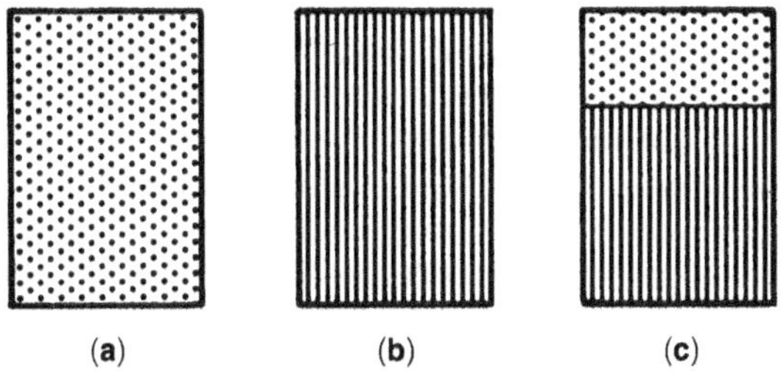

Fig. 7.14 Indentation cases considered: (a) entirely granular; (b) entirely columnar; (c) 1/3 granular and 2/3 columnar.

As an example, we will look at the case of first-year ice of thickness $h = 1.5$ m with a uniform temperature gradient running between -20°C at the surface to -2°C at the bottom (Fig. 7.13(e)). We assume uniform salinity of 6 ppt and look at three cases, as shown in Fig. 7.14. They are:

(a) entirely granular;
(b) entirely columnar;
(c) 1/3 granular and 2/3 columnar.

For the temperature of a particular layer we adopt simply the mean temperature through the layer as a "reference temperature". This is a good approximation.

We now assume that the ice moves at a velocity U past a structure of width $D = 100$ m. We have $D \gg h$, so we can apply plane stress conditions and calculate the effective strain-rate as $U/I\Psi D$, with I and Ψ defined, as appropriate, by Table 6.2. For the three cases of ice crystallography under consideration the results are obtained as follows, and are summarised in Table 7.5 as a function of ice velocity U.

TABLE 7.5 Creep indentation loads predicted for sea ice 1.5 m thick indenting a structure of diameter $D = 100$ m at velocity U. Loads are calculated by the reference stress method, and are very insensitive to the relative proportion of granular and columnar ice. See text for details. At velocities exceeding about 1 m h^{-1}, the creep treatment is probably inadequate

Ice velocity U (mh^{-1})	Indentation rate U/D (s^{-1})	Total load (MN)		
		(a) Granular	(b) Columnar	(c) Granular/columnar
0.001	3×10^{-9}	46	46	47
0.01	3×10^{-8}	99	99	101
0.1	3×10^{-7}	213	213	218
1.0	3×10^{-6}	458	458	469

(a) *Entirely granular.* We have $I = 1.15$ and $\Psi = 0.39$, and can use equation (6.21) to calculate the total load P as:

$$P = 1.15DhF_{\sigma}\{U/0.44D\} \qquad (7.8)$$

The stress may be calculated from equation (4.10), for granular sea ice, using a reference temperature of $T = 262$ K (-11°C). Brine volume is calculated from equation (2.1) using salinity of 6 ppt and temperature -11°C. It is 0.03 (3%). The total load calculated on this basis is tabulated in Table 7.5(a).

(b) *Entirely columnar.* We have $I = 3.13$ and $\Psi = 0.45$, and can write:

$$P = 3.13DhF_{\sigma}\{U/1.41D\} \qquad (7.9)$$

where stress is calculated using equation (4.9), for columnar sea ice, again using $T = 262$ K and a brine volume of 3%. The total load calculated on this basis is tabulated in Table 7.5(b). It will be seen that the results are indistinguishable from the loads calculated for case (a) above. That the results should be identical is fortuitous, but the closeness is due to compensation by the opposing factors in the indentation analysis: columnar ice is a softer material than granular ice under uniaxial compression, but its anisotropy means that it is more resistant to biaxial compression under indentation conditions. This is reflected in the higher indentation factor $I = 3.13$ used in equation (7.9).

(c) *1/3 granular and 2/3 columnar ice.* To cope with a mixture of ice types, we ignore coupling problems, and make a simple proportional sum of results from equations (7.8) and (7.9). For the granular layer we have $h = 0.5$ m and an average temperature of $T = 256$ K (-17°C); for the columnar layer we have $h = 1.0$ m and $T = 265$ K (-8°C). For both cases $D = 100$ m and salinity is 6 ppt. The results are shown in Table 7.5(c). As might be expected the result is, again, much the same (it is not identical, since the equations involved are non-linear). This fact is rather useful: under typical indentation conditions it really does not matter much what the relative proportions are of granular and columnar ice.

The table of loads shown (Table 7.5) extends only up to a velocity of 1 m h^{-1}. On the basis of observations at full-scale islands, it is very unlikely that indentation at rates exceeding this can occur without formation of cracks and a transition to brittle behaviour. In fact, to be confident that the creep-indentation analysis is applicable, movement rates need to be in the order of 0.1 m h^{-1} or less. This amounts to no more than a few metres per day.

7.6.1.2 Transition to fracture behaviour

At a certain point during the loading process a transition occurs from creep behaviour to fracture behaviour. We have seen that for small-scale laboratory indentation tests this transition occurs at an indentation rate U/D of about 10^{-2} s^{-1}, corresponding to indentation pressures of about 10–20 MPa. From the few large-scale indentation tests we have access to, it seems that at full scale the transition has occurred at indentation rates in the order of 10^{-7}–10^{-6} s^{-1}. This corresponds to an indentation pressure of 1–2 MPa.

The reasons for this are poorly understood, but are probably a combination of the following factors:

Loading-rate effects. As explained in Section 6.3.2, if a slowly increasing load is applied to ice then first cracks appear to nucleate as a function of delayed elastic strain. Under loading rates typically encountered in the Arctic, such cracks may nucleate at stresses below 1 MPa. It is interesting to speculate what is the maximum possible rate of stress increase (and peak stress attainable) for static wind-driven ice loading. Typical sustained high winds in the Beaufort Sea are 15 m s^{-1} ($\sim$ 30 knots) and it appears that wind drag typically acts over an area of several 100 km^2 if the ice is to fail against the structure (Croasdale, 1980). However, it is obvious that the wind cannot instantaneously jump from 0 to 15 m s^{-1} over this entire area. In practice, an approaching weather system moves towards the structure and gradually acts on a greater and greater area of ice in contact with the structure.

For illustration, suppose that an area of ice 20×20 km square (400 km^2) rests against a structure, and that weather conditions begin perfectly calm. Now suppose that a front of high wind, of speed 15 m s^{-1}, begins rapidly to move across the area of ice. In general, such a wind system will establish itself at a speed somewhat less than the speed of the wind itself, but here, for simplicity, assume that the system moves forward at 15 m s^{-1}. It thus takes 22 minutes to establish itself, and during this time, the theoretical wind drag load (using equation (7.6)) rises from 0 to 220 MN. For ice of thickness 1.5 m and a structure diameter of 100 m, this represents a loading rate of about 1 kPa s^{-1}, and from Fig. 6.15 we might therefore expect a stress of 0.5–1.0 MPa before formation of first cracks.

Imperfect contact. As explained in Section 6.3.1, if the contact between ice and structure is imperfect then local strain-rates may greatly exceed the gross strain-rate U/D. If, for instance, zones only of diameter of the order of the ice thickness h are in contact, then we would expect fracture when

$$U/h > 10^{-2}\ s^{-1}$$

If we have $h = 1.5$ m and $D = 100$ m then this corresponds to a gross strain-rate of

$$U/D = 1.5\times10^{-4}\ s^{-1}$$

Flaws and irregularities. If the ice is known to contain a large population of sufficiently large flaws or if its thickness is known to be highly irregular, then the analyses of Sections 6.4.1 and 6.4.4.2 indicate significant reductions in strength. Unfortunately, statistics on flaw populations and (three-dimensional) thickness irregularities are not available.

As general guidance, for a particular loading event, a transition to brittle behaviour will occur when any, any two or all three of these conditions is exceeded. As they stand, however, the criteria are not very quantitative, and are therefore impossible to apply. For this reason, we have to look at full-scale data on ice interaction to find out at what stresses such a transition occurs.

7.6.1.3 Fast fracture behaviour

Following the onset of brittle behaviour, the load history may be represented as the sum of loads in a series of independent zones failing non-simultaneously. This has been described in Section 6.4.5. Note, however, that the peak force during a particular interaction event depends on the duration of the event: the longer the event lasts, the greater the number of stress peaks which are sampled and the higher the expected peak load. This is simply because there is a higher probability of a larger number of independent zones failing simultaneously.

Attempts have been made to model this process: my own Fig. 6.38 is one such attempt (for the dynamic interaction problem, with progressive penetration) and Kry (1979a, 1980b) has developed predictions of stress statistics using log-normal distributions. These models do not, however, yet provide a convincing quantitative tool for calculating ice loads.

Again, the problem is that there is not a sufficiently extensive public data set available to permit calibration of conceptual models to reality. In the following section, therefore, we shall approach from the opposite direction; we shall look directly at the available data and use it to formulate design load criteria. We do this in the knowledge that although the data cannot yet be quantitatively predicted by theoretical models, it can at least be accounted for conceptually.

Note that once fast fracture has been initiated there is very little distinction between "static" loading and dynamic impact loading. The only important distinction lies with contact width: for static loading we can generally assume contact over the whole structure width; for dynamic impact the contact and penetration is progressive and dependent on initial kinetic energy. We shall look at this in more detail in Section 7.7.

7.6.2 Design load recommendation

Because of the uncertainties surrounding a rigorous detailed analysis of ice interaction events, I propose that design loads are calculated on a pragmatic basis, by introducing the concept of "normalised stress at peak load during an event", and examining the statistics of available full-scale measurements.

Consider the pressure–area curve of Fig. 5.36. The pressure data, despite the scatter, show a clear functional dependence on contact area. The data themselves suggest that indentation pressure $\sigma = P/Dh$ may depend on gross contact area $A = Dh$, with a functional dependence of the form

$$\sigma \propto A^{-\beta}$$

where β has a value lying somewhere between 1/3 and 1. As we have seen from our analysis of non-simultaneous failure (Section 6.4.5), there are theoretical grounds for believing that β should be approximately 1/2. This, also appears to describe quite well the upper bound of the full-scale data (group C in Fig. 5.36).

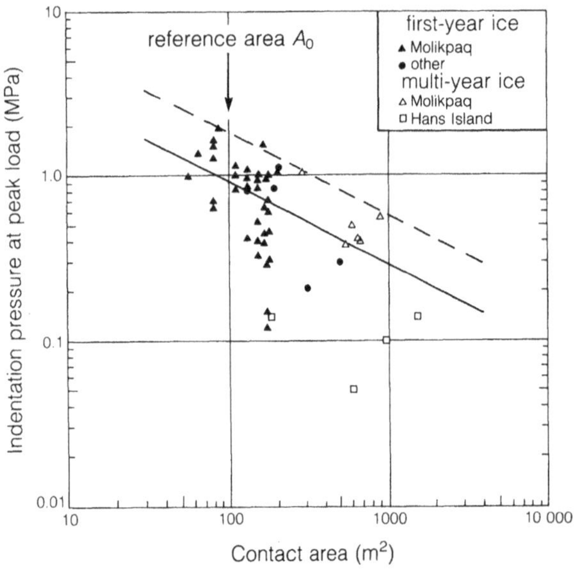

Fig. 7.15 Plot of full-scale data on indentation pressure at peak load for the events tabulated in Tables 7.6 and 7.7. Data are plotted as a function of contact area. If normalised to a reference area $A_0 = 100\ \text{m}^2$ using equation (7.10), the data have a mean normalised stress σ^*_m of 0.92 MPa and a standard deviation of $\sigma^*_s = 0.45$ MPa. The mean is shown by the solid line; the two standard deviation exceedance level is shown by the dashed line, and corresponds well to an upper bound.

The data for full-scale interactions are replotted in Fig. 7.15, including, for statistical purposes, a large body of data collected at Tarsiut P-45 in 1984–5, for which only incomplete data exist. These load data are digitised from Fig. 5.28, and the ice thickness has been assumed simply to approximate to the seasonal average shown for the Beaufort Sea in Fig. 3.11. These data are subject to change or revision – especially since Wright *et al.* (1987) have recently published an alternative analysis of the load data – but I do not believe the data are misleading in form. The full data set used for Fig. 7.15 is tabulated in Table 7.6 (first-year ice) and Table 7.7 (multi-year ice). Quantities which are poorly known and for which assumptions have been made are shown in brackets.

The observation that the upper bound of the data appears to depend on inverse square root of area ($A^{-1/2}$), and the fact that theory would lead us to suppose that indeed it should do so, suggests that a normalisation of these data can be carried out.

In particular, it suggests that we can normalise all pressure measurements by the inverse square root of contact area A, and express them relative to a single "reference contact area" A_0. This means that for any measurement of stress σ over an area A we can calculate a normalised stress σ^* over the reference area A_0, using the assumption of inverse square root dependence on area. Thus

$$\sigma^* = \sqrt{\frac{A}{A_0}}\,\sigma \qquad (7.10)$$

Effectively, this means shifting all the data points of Fig. 7.15 towards a reference area A_0, along lines parallel to a line of slope $-\frac{1}{2}$. This process yields a single histogram for the frequency distribution of normalised indentation pressure σ^* over an area of A_0. I have carried out this process on the available data set, using an arbitrary reference contact area of $A_0 = 100\ \text{m}^2$. The normalised data are tabulated in Tables 7.6 and 7.7 and are shown in the form of frequency histograms in Fig. 7.16.

The histogram of σ^* for 43 first-year events alone has a mean of 0.93 MPa and standard deviation of 0.42 MPa (Fig. 7.16 (a)). For the 10 multi-year events available the mean is 0.88 MPa and the standard deviation is 0.58 MPa (Fig. 7.16 (b)). If the two distributions are combined and analysed

TABLE 7.6 Summary of peak loads during first-year ice loading events at full scale. Assumed values are in brackets. Average stress is shown normalised to contact area $A_0 = 100$ m^2

Site	*Date*	*Peak load (MN)*	*Contact area (m^2)*	*Average stress (MPa)*	*Normalised stress σ^* ($A_0 = 100$ m^2) (MPa)*
Adgo P-25	Dec. 1974	105	130	0.81	0.92
Netserk F-40	Dec. 1975	160	190	0.84	1.16
Arnak L-30	Mar. 1977	227	202	1.12	1.60
Adams Island	1986	64	310	0.21	0.36
Kadluk	1984	150	(500)	0.30	0.67
Tarsiut P-45	Dec. 1984	50	(80)	0.63	0.56
		55	(80)	0.69	0.61
		154	152	1.01	1.25
		100	(80)	1.25	1.12
		120	(80)	1.50	1.34
		130	(80)	1.63	1.45
		86	64	1.34	1.08
	Jan. 1985	110	(110)	1.00	1.05
		125	(110)	1.14	1.19
		90	(110)	0.82	0.86
		55	56	0.98	0.73
		90	(110)	0.82	0.86
	Feb. 1985	140	(130)	1.08	1.23
		55	(130)	0.42	0.48
		140	(130)	1.08	1.23
		125	(130)	0.96	1.10
		207	199	1.04	1.47
		110	(130)	0.85	0.96
		140	(130)	1.08	1.23
	Mar. 1985	140	(150)	0.93	1.14
		60	(150)	0.40	0.49
		50	(150)	0.33	0.41
		125	(150)	0.83	1.02
		80	(150)	0.53	0.65
	Apr. 1985	65	(165)	0.39	0.51
		105	(165)	0.64	0.82
		75	(165)	0.45	0.58
	May 1985	55	(175)	0.31	0.42
		105	(175)	0.60	0.79
		125	(175)	0.71	0.94
		80	(175)	0.46	0.60
		175	(175)	1.00	1.32
		165	86	1.92	1.78
	June 1985	20	(170)	0.12	0.15
		50	(170)	0.29	0.38
		25	(170)	0.15	0.19
	July 1985	165	(170)	0.97	1.27
Amauligak I-65	May 1986	250	(162)	1.54	1.96

TABLE 7.7 Summary of peak loads during multi-year ice-loading events at full scale. Assumed values are in brackets. Average stress is shown normalised to contact area $A_0 = 100$ m^2

Site	*Date*	*Peak load (MN)*	*Contact area (m^2)*	*Average stress (MPa)*	*Normalised stress σ^* ($A_0 = 100$ m^2) (MPa)*
Hans Island	Aug. 1980	214	1520	0.14	0.55
	July 1981	95	970	0.10	0.31
	Aug. 1980	269	672	0.40	1.04
		26	180	0.14	0.19
		31	600	0.05	0.13
Amauligak I-65	Nov. 1985	295	(285)	1.04	1.75
	Mar. 1986	270	650	0.42	1.06
		300	(600)	0.50	1.22
	Apr. 1986	500	(900)	0.56	1.67
		210	550	0.38	0.90

together then they give a mean normalised indentation stress $\sigma_m^* = 0.92$ MPa and a standard deviation of $\sigma_s^* = 0.45$ MPa (Fig. 7.16(c)). Now, there are two important points to make about these distributions. Firstly, it is remarkable that the distributions for first-year ice and for multi-year ice have very similar statistics, despite referring to contact areas which are in general an order of magnitude different (Fig. 7.15). Secondly, it is clear that the aggregate frequency distribution (Fig. 7.16(c)) is approximately normal. For reference, a normal distribution with mean 0.92 MPa and standard deviation 0.45 MPa is plotted over the data. The data are clearly consistent with the normal distribution: a Kolmogorov-Smirnov test (Stephens, 1974) on the cumulative frequency distribution indicates a high degree of correspondence between the data and the normal distribution, and certainly indicates that the hypothesis that the distribution is normal cannot be rejected at the 90% confidence level. (This is in fact somewhat contrary to expectations. The sequence of load peaks during a single interaction event might be expected to be normally distributed, since the Central Limit theorem (see e.g. Cramér, 1955) indicates that the addition of independent stochastic

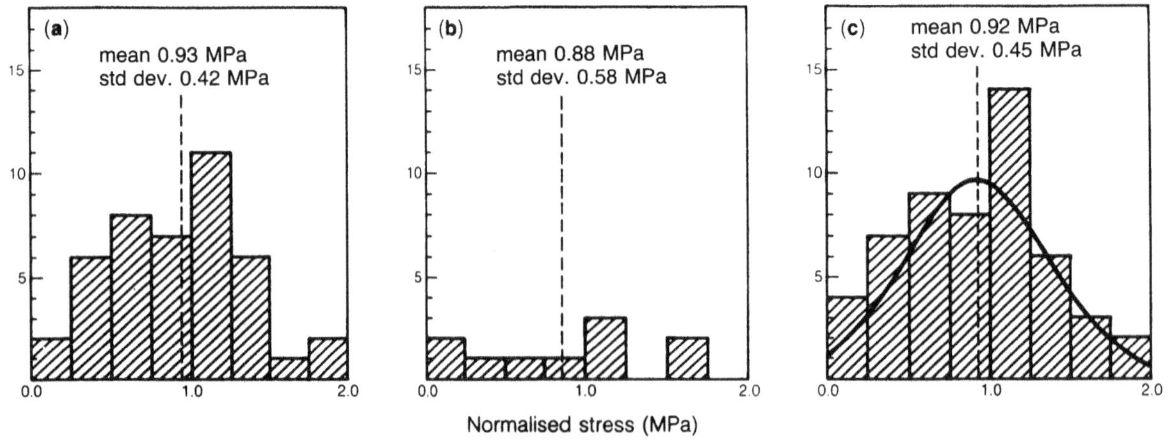

Fig. 7.16 Frequency histograms of normalised indentation stress, σ^*, for (a) first-year events (43 events), (b) multi-year events (10 events) and (c) all events (53 events). The distributions are all consistent with a normal distribution of mean 0.92 MPa and standard deviation 0.45 MPa. This distribution is shown in (c).

fracture events should be normally distributed. However, the statistics of the maximum loads during an ensemble of events might be expected to follow a log-normal or other extremal function (see, for instance, Gumbel, 1958). It would be easy here to waste time debating which extremal distribution is most appropriate but I shall instead restrict the analysis to simple normal distributions.)

A further remarkable feature of the distributions of Fig. 7.16 is this: not only are all three frequency histograms consistent with a normal distribution, they are in fact all consistent with the *same* normal distribution shown in Fig. 7.16(c). That is, the hypothesis that all these loading events are from the same statistical population cannot be rejected at the 90% level. For this reason, we can characterise all ice-loading events, first-year or multi-year, by means of one single normalised indentation stress distribution. On the basis of the data currently available, it has a mean of 0.92 MPa and a standard deviation of 0.45 MPa. This probability distribution for peak stress during an event forms the heart of the method I shall be proposing for design load calculation.

It leads us to a very simple method for generating probabilistic loads during static first-year (or multi-year) ice interaction events. Bear in mind that the data plotted in Fig. 7.15 refer specifically to indentation stress at "peak load during an event". Only one such stress value is plotted for each event, and the statistical distribution may thus be used to generate stress at peak load for a long hypothetical series of events during the design life of a structure.

Because the loading we are considering in this section is "static", we may assume that contact occurs over the full width of the structure. All we need then, to generate peak load from "stress at peak load", is the ice thickness during the event.

The definition of "an event" is somewhat loose, but is in effect quite easy to apply. Ice loading events are in general episodic and shortlived, and are quite easy to identify. A typical static loading event lasts for a period of the order of a few hours, is associated generally with high wind conditions and normally gives rise to ice movement of the order of 1–100 m. In order to generate design loads we simply need to know how many such events occur during a winter season, and we can then develop a probability distribution for peak loads over the design life of the structure. To do this, the only other parameter we need to know, and to assign a probability distribution to, is the ice thickness. This may be done on the basis of site-specific observations of monthly average and extreme ice thicknesses as provided, for instance, in Chapter 3. It may seem odd that effects of temperature, salinity and strain-rate are not explicitly treated in the methodology I am here proposing. This is not to say that they are ignored: their effects are implicitly included in the data sets of Figs 7.15 and 7.16, which contain a large statistical sample of various types and velocities of ice. Ideally, of course, the effects of temperature, salinity, crystallography and strain-rate should be separated out of the distribution, and each variable should then be re-introduced with its own probability distribution and an explicit calculation of its effect. However, it is as well at this stage for these secondary effects not to be explicitly included in the analysis, since their effects are really not well understood. For instance, Blenkarn (1970) found at Cook Inlet that the highest crushing forces were not necessarily associated with the coldest, hardest ice; contrary to expectations, warmer ice seemed to give rise to higher forces (Section 5.3.1 and Fig. 5.6). Also, as we have seen in Chapter 4 (Fig. 4.27) there is no evidence that high rates of indentation lead to higher forces, once brittle behaviour has begun. For these reasons, I suggest that the simple raw distribution of Fig. 7.16 is appropriate until further more detailed information becomes available.

The proposed method for calculating design loads is then as follows:

(A) Define the mean number, N, of static ice loading "events" which take place during a single year. For a typical Beaufort Sea structure near to the transition zone this might lie in the range 10–100, that is, several per month. For purposes of illustration, let us suppose that events occur at the rate of three per month during the November–June season. This is a total of 24 events per year.

(B) Generate, or specify, an ice thickness h_i to be associated with each event $i (i = 1, \ldots, N)$. This can be done using ice thickness data of Chapter 3 (for instance, Fig. 3.11 for the Beaufort Sea). The method of generation depends rather on the level of sophistication required and the quality of data set available. For illustration, we shall here assume a series of month-by-month normal probability distributions based on Fig. 3.11: in each month, the thickness has a mean defined by the solid line in the figure and the extreme thickness (the dashed line) will be assumed to represent two standard deviations (95% confidence level). The probability distributions assumed on this basis are tabulated in Table 7.8. During March, for instance, the mean ice thickness is 1.50 m and there is a 95% chance that it will lie between 1.22 and 1.78 m. A single value h_i can then be generated for each event using a Monte Carlo simulation process to generate random normally distributed values with the specified mean and standard deviation. It would be a simple matter, of course, to specify alternative statistics or forms of distribution if appropriate data are available. We have here ignored the fact that ice thickness in, say, March of a given year is strongly correlated with ice thickness in February of the same year. In the long run the effect of such correlation is negligible.

TABLE 7.8 First-year ice design loads: example calculation for the Beaufort Sea

Input data		
Structure diameter		100 m
No. of events per year		24
	Ice thickness (m)	
Month	*Mean*	*Standard deviation*
November	0.48	0.02
December	0.80	0.07
January	1.10	0.15
February	1.34	0.13
March	1.50	0.14
April	1.65	0.22
May	1.74	0.18
June	1.70	0.10
Stress statistics		
	Mean, σ_m^*	*Standard deviation* σ_s^*
Normalised indentation stress, σ^* (MPa) over $A_0 = 100\ m^2$	0.92	0.45
Design loads		
25-year design load		277 MN
50-year design load		291 MN
100-year design load		305 MN

(C) For each of the N events generated we now generate a normalised stress at peak load, σ_i^*. This may be done from Fig. 7.16 using, again, a Monte Carlo method of generation. Here we shall assume that σ^* has a mean 0.92 MPa and a standard deviation 0.45 MPa. Now, the gross contact width during a static loading event remains constant at D, so that the gross contact area A_i during event i is

$$A_i = Dh_i$$

The peak load P_i is then given simply by

$$P_i = \sqrt{A_i A_0}\,\sigma_i^* = \sqrt{Dh_i A_0}\,\sigma_i^* \tag{7.11}$$

where A_0 is the normalising contact area associated with σ^*, here taken to be 100 m^2. For example, suppose that an event occurs during March, and that it is associated with a thickness $h_i = 1.62$ m generated by the Monte Carlo process (slightly higher than average). We then generate a normalised stress σ_i^* for the event, of, say, 0.85 MPa (slightly lower than average). The peak load during this event, for a structure of width $D = 100$ m, would then be from equation (7.11),

$$P_i = \sqrt{100\times1.65\times100}\times0.85 = 108 \text{ MN}$$

For an extreme ice thickness of $h_i = 1.78$ m and an extreme normalised stress of 1.82 MPa the peak load, by a similar calculation, would be 243 MN.

(D) Now we run the Monte Carlo generation procedure repeatedly, to generate a series of maximum annual ice loads. Each year contains an average of N events, and we record, for each modelled year of the generation programme, what is the maximum load generated. (The occurrence of these events should, strictly, be modelled as a Poisson process with a mean number N of events per year. Here, I shall model it simply as a regular series of N events per year, for computational ease. This does not lead to appreciable errors.) The expected distribution of peak loads during a single year is shown in Fig. 7.17(a). The expected peak load during an event is 105 MN. Once a large number of hypothetical model years has been simulated it becomes possible to define an annual exceedance curve for ice loads. An annual exceedance curve of this type is shown in Fig. 7.17(b) plotted on logarithmic probability paper, for the static first-year ice loading parameters summarised in Table 7.8.

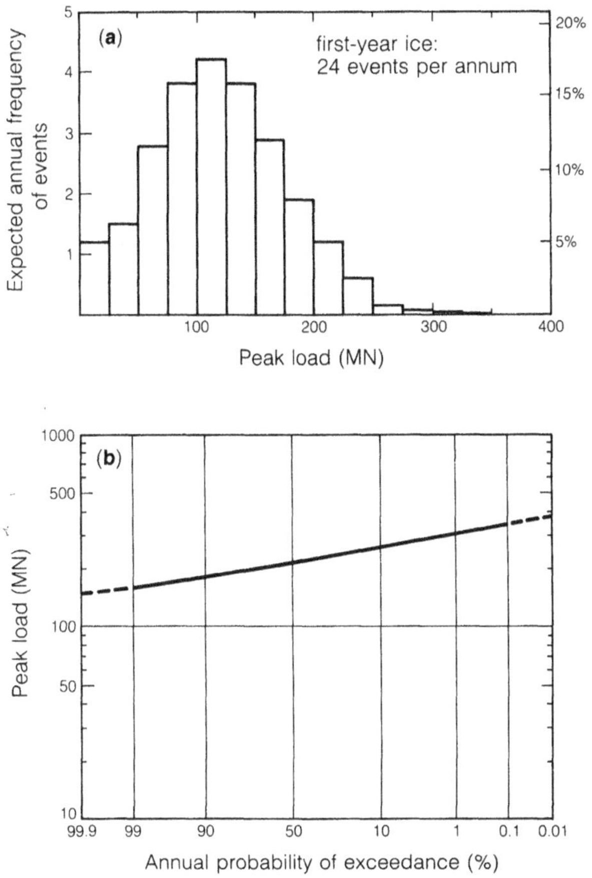

Fig. 7.17 (a) Histogram of expected first-year ice loading events experienced during a single year, generated on the basis of parameters in Table 7.8. (b) Annual exceedance probability for static first-year ice loads, generated on the basis of parameters in Table 7.8.

This curve defines the probability that a certain load will be exceeded during the year. The fact that the line is nearly straight indicates that the exceedance probability approximately follows a log-normal distribution.

(E) For design purposes it is common practice to design to a specific annual exceedance level. For instance, a 25-year design load specification means that the structure must be designed on the basis of a load with an annual exceedance probability of 0.04 (4%). A 100-year design load corresponds to an annual exceedance probability of 0.01 (1%). The exceedance level adopted depends on how long the structure is expected to operate (for instance, whether it is a seasonal exploration structure or a more permanent production structure), and also on the codes applying in the country where it is to be constructed. The design loads corresponding to these exceedance levels may be read directly from Fig. 7.17(b). For the example under discussion (Table 7.8), the 25-year design load is 277 MN and the 100-year design load is 305 MN.

I should emphasise again that this treatment has been quite a simple one. A full analysis would take account of the statistical variability of all input parameters such as temperature, salinity, crystallography, velocity, geometry, flaw-size, thickness irregularity, penetration distance, etc. It is, however, pointless to incorporate all these variables until their effects are properly understood. I have avoided including these more complex influences for several reasons: to avoid unjustified sophistication, to avoid appearing too definitive, and to aid clarity in exposition of the methodology.

7.7 DYNAMIC IMPACT: MULTI-YEAR FLOES

7.7.1 Introduction

The analysis of dynamic impact is more complex than that of static interaction, since it involves progressive penetration and a growing area of contact during the impact process. We have already looked at the simulation of this process in Section 6.4.5.3. There are two principal complications that we need to deal with.

(1) The maximum penetration of an impacting floe, and hence its maximum contact area, depends on its initial kinetic energy and the rate at which this energy is dissipated during the impact process. We cannot assume that contact initially occurs over the full width of the structure, and the kinetic energy of the floe may not be sufficient for full penetration to be achieved at any stage.
(2) We need to know the kinetic energy in order to model the impact process. However, the probability distribution for initial kinetic energy of impacting floes cannot be simply taken from the probability distributions presented in Chapter 3 for floe diameter, thickness and velocity. This is because the probability of impact itself depends on diameter and velocity. Hence, the probability distributions for diameter and velocity of *impacting floes* need to be

biased towards floes of larger diameter and higher velocity.

We shall look at these problems in turn: first, the analysis of penetration distance, and then the analysis of impact frequency and its associated floe velocity and diameter probability distributions. Finally, I shall propose some simple recommendations for dynamic ice-load calculations, and provide an example analysis.

7.7.2 Penetration distance

When a multi-year floe impacts with a structure it continues moving until all its kinetic energy is dissipated. If the floe has mass m and initial velocity U_0, then it will come to rest at total penetration X, when:

$$\tfrac{1}{2}mU_0^2 = \int_0^X P(x)\,dx \tag{7.12}$$

where $P(x)$ is the total force exerted when the floe has moved a distance x after first contact. The load $P(x)$ may, in turn, be written as

$$P(x) = \bar{\sigma}(A)A(x)$$

where $A(x)$ is the contact area achieved after penetration by a distance x and $\bar{\sigma}(A)$ is the mean stress associated with this contact area. The details of the function $A(x)$ depend entirely on geometrical considerations, and have already been given by equation (6.54) and Fig. 6.35 for the case of a circular floe and a circular indentor. As for the mean stress $\bar{\sigma}(A)$ associated at any stage with failure over an area A, we can derive this approximately from the data of Fig. 7.15. Now, the data of Fig. 7.15 show stress $\hat{\sigma}$ at peak load and the analysis has shown that the mean of these data, $\bar{\hat{\sigma}}$, can be represented by the solid line in the figure, which follows the form:

$$\bar{\hat{\sigma}}(A) = \sigma_m^* \sqrt{A_0/A} \tag{7.13}$$

where $\sigma_m^* = 0.92$ MPa, the mean normalised stress at peak load over area $A_0 = 100$ m^2. This is not quite what we need: equation (7.13) supplies the expected *peak* stress $\bar{\hat{\sigma}}(A)$ for a given contact area, and not the mean stress $\bar{\sigma}(A)$. The difference between these two concepts is shown schematically in Fig. 7.18, and real examples have been shown for Hans Island and Amauligak I-65 in Figs 5.22, 5.24, 5.26 and 5.30. A synthetic simulation of the process was shown in Fig. 6.38. During an impact the peak loads and peak stress are highly variable spikes, while the mean load and mean stresses are

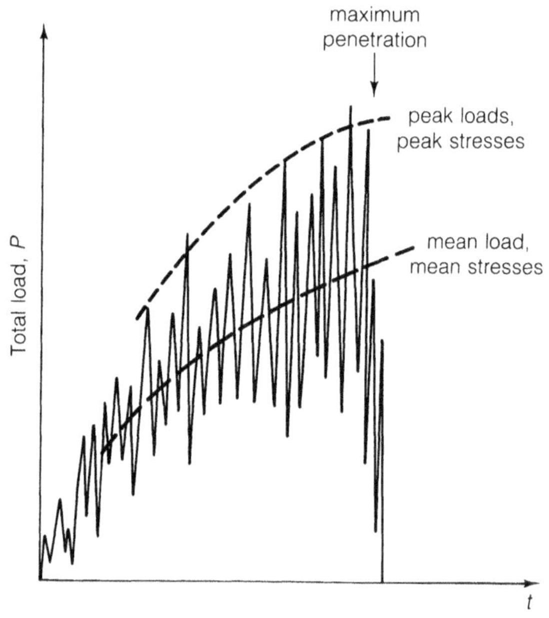

Fig. 7.18 Schematic diagram of a multi-year impact: peak loads and peak stresses are typically some 50% higher than mean loads and stresses.

considerably lower. Typically, the mean loads are some 60–75% of the peak loads: that is, peak loads exceed mean loads by about 50%. For the analysis of progressive penetration we require the mean stress $\bar{\sigma}(A)$ as a function of contact area, since it is according to this, not the peak stresses, that kinetic energy is dissipated. We can write, approximately, from equation (7.13):

$$\bar{\sigma}(A) = \lambda_m \sigma_m^* \sqrt{A_0/A} \tag{7.14}$$

where λ_m is a factor expressing the average ratio of mean load to peak loads. For this analysis we shall adopt a value of $\lambda_m = \frac{2}{3}$. The relationship between these two functions, mean peak stress $\bar{\hat{\sigma}}(A)$ (equation (7.13)) and mean stress $\bar{\sigma}(A)$ (equation (7.14)) is shown in Fig. 7.19.

Returning to equation (7.12) we can write

$$\tfrac{1}{2}mU_0^2 = \int_0^X \bar{\sigma}(A)A(x)\,dx$$

and substituting equation (7.14) we have

$$\tfrac{1}{2}mU_0^2 = \lambda_m \sigma_m^* \sqrt{A_0} \int_0^X \sqrt{A(x)}\,dx \tag{7.15}$$

Now, contact area $A(x)$ may be calculated by

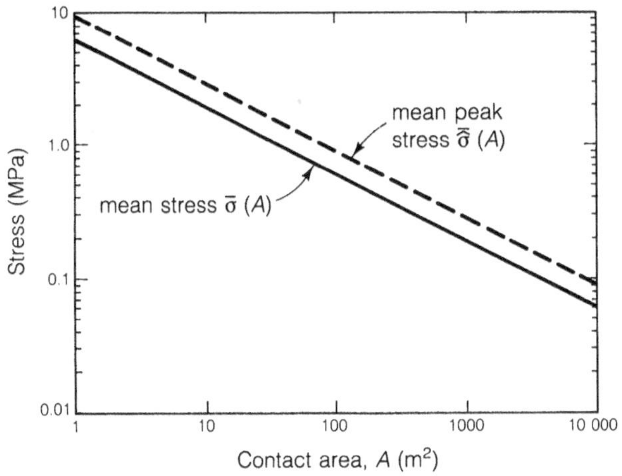

Fig. 7.19. Representation of equations (7.13) and (7.14) for mean peak stress during an event $\hat{\bar{\sigma}}$ (A) and mean stress $\bar{\sigma}$ (A), as a function of contact area A.

geometrical considerations. For a circular floe indenting with a circular structure we should, strictly, use equation (6.54) to give contact width as a function of penetration distance x. However, this is a clumsy expression and leads to unnecessary complications when we come to perform the integral in equation (7.15). For simplicity, and in order to obtain an analytic solution, we shall look therefore at a simplified interaction geometry, as shown in Fig. 7.20. Instead of a circular structure, we shall assume a square structure of diagonal width D, and we assume that it is loaded on its edge by an effectively straight-edged floe. This geometry differs somewhat from the circular

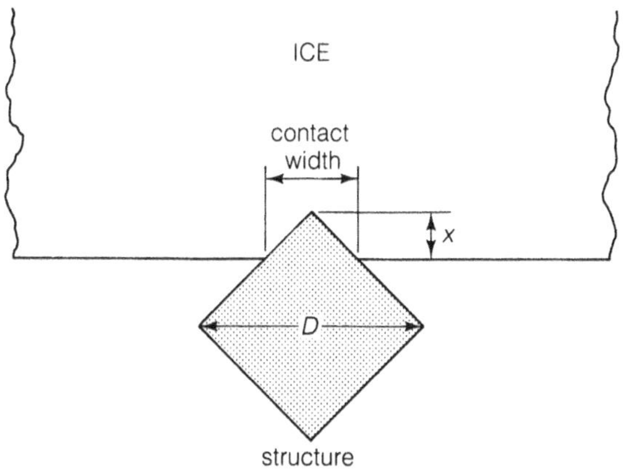

Fig. 7.20 Simplified interaction geometry for analytic model of progressive penetration.

geometry of Fig. 6.35, but is no less plausible as a representation of reality: floes are neither straight-edged nor circular and structures may be square, circular, octagonal or, indeed, quite irregular if they are surrounded by a natural or artificial rubble pile. The geometry of Fig. 7.20 has the great advantage that the expression describing projected contact area is appealingly simple. It is, for $x < D/2$,

$$A(x) = 2xh \tag{7.16}$$

We can now substitute equation (7.16) into equation (7.15) and integrate. We find:

$$\tfrac{1}{2}mU_0^2 = \tfrac{2}{3}\lambda_m\sigma_m^*\sqrt{2hA_0}\,X^{3/2}$$

or, solving for maximum penetration X:

$$X = \left[\frac{3mU_0^2}{4\lambda_m\sigma_m^*\sqrt{2hA_0}}\right]^{2/3} \tag{7.17}$$

It is interesting to put numbers into this equation, to get a feel for the degree of penetration achieved by typical floe impacts. In Table 7.9, equation (7.17) is used to calculate the maximum penetration achieved by two floes as a function of impact velocity. The floes are:

(A) a "typical" floe, of diameter 500 m and thickness 4 m (having a mass of 6×10^8 kg);
(B) an "extreme" floe, of diameter 5 km and thickness 8 m (having a mass of 1.2×10^{11} kg).

If the structure width D is 100 m, then total penetration is achieved when $X = 50$ m. Now, it will be seen from Table 7.9 that a "typical" floe is not expected ever to achieve full penetration. Even when travelling at an extreme velocity of 1.0 m s^{-1} ($\sim$ 2 knots), it has sufficient kinetic energy to penetrate only 8.8 m before coming to rest. At this point its contact zone would be some 18 m wide. To achieve full penetration of 50 m (contact width 100 m) it would, from equation (7.17), need to be travelling at 3.7 m s^{-1}. This is not possible. The "extreme" floe, on the other hand, may quite easily achieve full penetration. If it is travelling at a velocity exceeding 0.31 m s^{-1} (0.6 knots) then it will achieve the full contact width of 100 m. Note, however, that even this velocity is quite high by the standards of the Beaufort Sea, where the average floe velocity during summer is in the region of 0.2 m s^{-1} (Section 3.2.3). For a velocity of 0.2 m s^{-1}, total penetration by the extreme floe is about 28 m, and the contact zone is about 56 m wide. This illustrates that consideration of initial kinetic energy is important, and we cannot simply

TABLE 7.9 Penetration achieved by floes during imapct with a structure of diameter $D = 100$ m, for a range of initial velocities. Floe A (typical floe) is of diameter 500 m and thickness 4 m. Floe B (extreme floe) is of diameter 5 km and thickness 8 m. Penetration of more than 50 m is total

	Penetration (m)	
Initial velocity ($m\,s^{-1}$)	*Floe A ("typical")*	*Floe B ("extreme")*
0.1	0.4	11.0
0.2	1.0	27.8
0.3	1.8	47.8
0.4	2.6	total
0.5	3.5	"
0.6	4.4	"
0.7	5.4	"
0.8	6.5	"
0.9	7.6	"
1.0	8.8	"

assume that all floes achieve full penetration. A typical floe does not achieve full penetration, and even an extreme floe needs to be travelling at quite a high velocity before it achieves more than partial penetration. In the analysis which follows (Section 7.7.4), we shall be paying particular attention to the ice loads exerted at the point of maximum penetration (the end-point on Fig. 7.18). At this point the contact area A is, from equations (7.16) and (7.17):

$$A = 2h\left[\frac{3mU_0^2}{4\lambda_m\sigma_m^*\sqrt{2hA_0}}\right]^{2/3} \qquad (7.18)$$

Floe mass m may be calculated approximately in terms of ice density ρ_i, diameter L and thickness h by using equation (3.4), which gives

$$m = 0.66\rho_i hL^2$$

7.7.3 Impact probability

To perform a statistical analysis of multi-year floe impacts (or iceberg impacts) we need to analyse the frequency with which impacts occur. This analysis is important from two points of view:

(a) In order to generate extreme design loads, we need to know the expected number of impacts a structure will have to sustain during its design life.
(b) As already discussed, the probability of impact is not independent of floe diameter and velocity: the larger and faster a floe is, the more likely it is to impact with a structure, since on its travels it sweeps out a greater area. This means that the probability distributions we must use for diameter and velocity of *impacting floes* differ from the basic probability distributions for *all* floes.

7.7.3.1 Impact frequency

The geometry for the analysis is shown in Fig. 7.21. A population of floes of various diameters L and velocities U move across an area a_0, which contains a structure of width D. (The area a_0 is arbitrary, and ultimately disappears from the

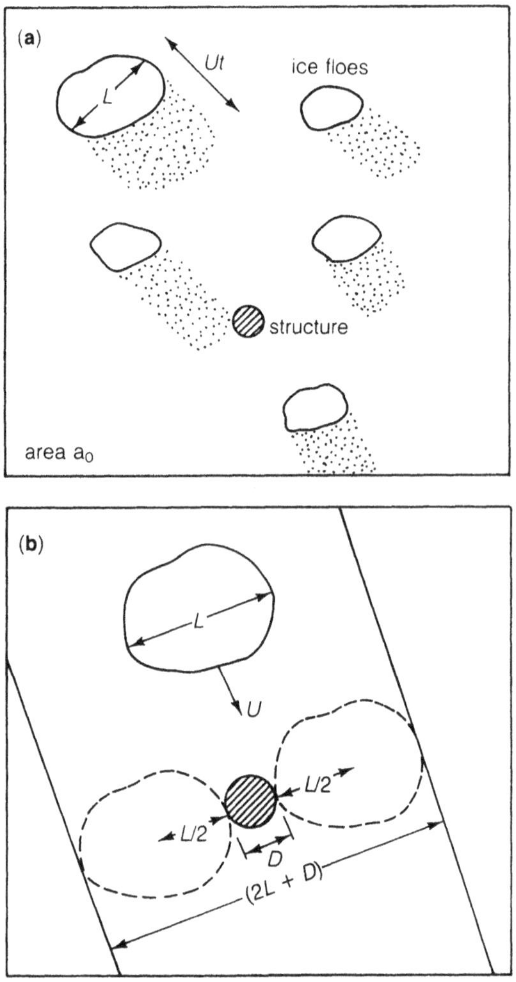

Fig. 7.21 Geometry for analysis of impact probability. (a) During time t each floe sweeps out an area LUt. (b) However, impact will occur if and only if the centre of a floe lies within a bandwidth of $(L + D)$ centred on the structure.

analysis.) The shaded areas in Fig. 7.21(a) represent the areas swept out by each floe during a time interval t (the paths of the floes have been drawn as approximately parallel to one another, since this is typically the case, but it is not necessary to the analysis).

Consider first a population of floes of uniform velocity U and diameter L. We shall develop the analysis for this uniform population and then generalise it to a variable population. Let there be a number N_0 of these floes of diameter L in the area a_0. This is an area density of $n_0 = N_0/a_0$ floes per unit area. Looking in detail at an impact path in the vicinity of the structure (Fig. 7.21(b)) it is clear that an impact of some description (head-on or glancing) will occur during a time interval t if and only if a floe lies entirely within a bandwidth of $(2L+D)$ upstream of the structure for a distance Ut. (Strictly this is only exactly true in the limit of long time interval t, so that $Ut \gg (L+D)$.) Now, for a floe to lie entirely within the bandwidth $(2L+D)$ its centre must lie within the bandwidth $(L/2+D+L/2)$, or $(L+D)$. Hence impact will occur if the centre of a floe lies within the area

$$(L+D)Ut$$

upstream of the structure. The probability that a particular floe within the area a_0 lies with its centre in the collision-course area is simply the ratio of areas:

$$(L+D)Ut/a_0$$

There are, however, N_0 floes in the area a_0, so the total probability of impact during a time t is

$$N_0(L+D)Ut/a_0 = n_0(L+D)\,Ut$$

and the probability $p(I)$ of impact per unit time interval is therefore

$$p(I) = n_0(L+D)U \tag{7.19}$$

where n_0 is the area density of floes. Note that, as expected, the probability of impact is proportional to velocity U and, for small structure width D, is approximately proportional to floe diameter L. If $L \gg D$ we have a probability of

$$n_0 LUt$$

which is simply the area swept out by the floe population during a time t (Fig. 7.21(a)).

The next step is to generalise the analysis to variable diameter and velocity, so that we can utilise statistical distributions of diameter and velocity as observed in nature. Now, to a good approximation, it can be assumed that the velocity distribution of floes is independent of the diameter distribution: by the driving force analysis of Section 7.5 it is clear that environmental forces tend to drive all floes at the same velocity, independently of diameter. Under transient conditions there is, as explained, some dependence of velocity on floe thickness, owing to inertial effects, but diameter and velocity are effectively independent. This means that we can integrate their probability distributions separately.

First, we define an area density $n'(L)$ of floes of diameter L. This represents the number of floes per unit area per diameter interval, and is defined by reference to the number of floes $N(L)$ in area a_0 having a diameter exceeding L. The definition is:

$$n'(L) = \frac{N(L)-N(L+\delta L)}{a_0\delta L} = -\frac{1}{a_0}\left[\frac{\partial N}{\partial L}\right] = -\frac{\partial n}{\partial L} \tag{7.20}$$

where n represents the cumulative number distribution of floes: that is, $n(L)$ means the number of floes per unit area having a diameter exceeding L. The area density may, for instance, be readily calculated from Fig. 3.13(a) by taking the derivative of the cumulative number n per km^2 with respect to mean caliper diameter L. We can also express this in terms of the probability density function $f_L(L)$ for floe diameters, by the relationship

$$n'(L) = \frac{N_0}{a_0} f_L(L) = n_0 f_L(L) \tag{7.21}$$

where N_0 is the total number of floes of any diameter in area a_0 and n_0 is the area density of all floes. The area density can also be expressed in terms of the total multi-year ice concentration c and the mean floe area $\bar{\mathcal{A}}$ by the relationship $n_0 = c/\bar{\mathcal{A}}$. Details on the definition of probability density functions have been given in Fig. 7.3.

Let us also define a probability density function for velocity of any floe, as $f_U(U)$. This has the meaning that the probability of a floe having a velocity between U and $U+\delta U$ is, in the limit of small δU,

$$f_U(U)\delta U$$

Again, in terms of exceedance plots of the type presented in Fig. 3.14, if $F_U(U)$ is the probability that a floe velocity exceeds U then the probability density is given by

$$f_U(U) = \frac{F_U(U)-F_U(U+\delta U)}{\delta U} = -\frac{\partial F_U(U)}{\partial U}$$

Consider now the joint probability distribution for velocity and diameter of an impacting floe. We denote this $f_i(U,L)$, where the subscript i means that the floe is an impacting floe. We can write, from equation (7.19):

$$f_i(U,L) = n'(L)(L+D)Utf_U(U)$$

or, by equation (7.20),

$$f_i(U,L) = -\left(\frac{\partial n}{\partial L}\right)(L+D)Utf_U(U) \quad (7.22)$$

This has the meaning, for instance, that the probability p of a floe impacting during a time t with a velocity between U and $(U+\delta U)$ *and* a diameter between L and $(L+\delta L)$ is

$$p = f_i(U,L)\delta U\delta L$$

or, from equation (7.22),

$$p = -\left(\frac{\partial n}{\partial L}\right)(L+D)Utf_U(U)\delta U\delta L \quad (7.23)$$

This allows us to write an expression for the overall probability $p(I)$ of impact of any sort occurring per unit time. It is the integral of equation (7.23) over all possible velocities U and diameter L, expressed per unit time interval, that is:

$$p(I) = -\int_L\int_U\left(\frac{\partial n}{\partial L}\right)(L+D)Uf_U(U)\,\mathrm{d}U\mathrm{d}L$$

Now, since U and L are independent variables, the integrals can be separated out and we have

$$p(I) = \left[-\int_L\left(\frac{\partial n}{\partial L}\right)(L+D)\,\mathrm{d}L\right]\left[\int_U Uf_U(U)\,\mathrm{d}U\right]$$

From equations (7.20) and (7.21), this can be rewritten as

$$p(I) = n_0\int_L f_L(L)(L+D)\,\mathrm{d}L\int_U Uf_U(U)\,\mathrm{d}U \quad (7.24)$$

This simplifies quite neatly in terms of mean floe diameter $\bar{L}$ and mean floe velocity $\bar{U}$. We have, by definition,

$$\bar{L} = \int_L Lf_L(L)\,\mathrm{d}L \quad (7.25)$$

and

$$\bar{U} = \int_U Uf_U(U)\,\mathrm{d}U \quad (7.26)$$

and also, of course, we have

$$\int_L f_L(L)\,\mathrm{d}L = 1$$

so we can write equation (7.24) as

$$p(I) = n_0(\bar{L}+D)\bar{U} \quad (7.27)$$

Notice that this is identical to equation (7.19), but with constant diameter L and velocity U replaced by their means $\bar{L}$ and $\bar{U}$. This is a remarkable result, since it is true for any probabilistic distribution of L and U so long as they are independent of each other. Equations (7.25) and (7.26) are fundamental definitions of the means $\bar{L}$ and $\bar{U}$ and are true for all types of distribution – normal, lognormal, exponential or whatever. Equation (7.27) is thus a general expression giving the impact probability per unit time as a function of $\bar{L}$, $\bar{U}$, structure diameter D and total area density n_0 of floes.

From equation (7.27) it is very simple to calculate the expected number of impacts during a time interval t. The expression $p(I)$ represents the probability of impact per unit time, which is , by definition, the long-term average frequency of impact per unit time. The expected number of impacts during time interval t is thus $p(I)t$, or

$$n_0(\bar{L}+D)\bar{U}t$$

7.7.3.2 Probability distributions for impacting floes: Bayes' Theorem

Bayes' Theorem is an important theorem in probability theory, and deals with conditional probabilities. It enables us to answer the question: given that a floe has collided, what is the probability distribution for its diameter and velocity? We have already, in Section 7.7.3.1 above, calculated the probability of impact, but we have not given details of the diameter and velocity of impacting floes. As we have already mentioned, faster bigger floes are more likely to be impacting floes, and we need to calculate conditional probability density functions *given that* impact has occurred.

Bayes' Theorem states the following:

$$p(H|F) = p(F|H)\frac{p(H)}{p(F)} \quad (7.28)$$

where $p(H|F)$ means the probability of an event H occurring, given that an event F (for instance, impact) has occurred. The overall probability of event H is simply $p(H)$. We can apply this theorem to the problem of conditional probability as follows.

The probability density function for diameter L, given that impact has occurred, may be denoted $f_L(L|I)$ and, by Bayes' Theorem, is given by

$$f_L(L|I) = \frac{p(I|L)}{p(I)}f_L(L) \quad (7.29)$$

That is, it is the same as the probability density function $f_L(L)$ for diameter L of *any* floe, but multiplied by the quotient of $p(I|L)$ (the probability of impact by floes given that they are of diameter L) and $p(I)$ (the probability of any impact at all).

Now, the probability of impact by floes of uniform diameter L, per unit time, would be

$$p(I|L) = n_0(L+D)\bar{U} \tag{7.30}$$

while the overall impact probability $p(I)$ per time is, from equation (7.27),

$$p(I) = n_0(\bar{L}+D)\bar{U} \tag{7.31}$$

for a population of floes of variable diameter.

Combining equations (7.29)–(7.31), we find that the probability density function for diameter of impacting floes is:

$$f_L(L|I) = \frac{L+D}{\bar{L}+D} f_L(L) \tag{7.32}$$

An entirely similar procedure may be followed in order to derive the probability distribution $f_U(U|I)$ for velocity of floes, given that they have impacted. We have, from Bayes' Theorem

$$f_U(U|I) = \frac{p(I|U)}{p(I)} f_U(U)$$

and we can show also that:

$$p(I|U) = n_0(\bar{L}+D)U$$

while

$$p(I) = n_0(\bar{L}+D)\bar{U}$$

as before, in equation (7.31). Hence we have

$$f_U(U|I) = \frac{U}{\bar{U}} f_U(U) \tag{7.33}$$

These two equations, (7.32) and (7.33), represent simple modifications which must be made to our basic probability density functions for diameter and velocity if we are to use them to analyse diameter and velocity of impacting floes. Their importance lies in the fact that they enable us to divide the calculation of impact probabilities and probability distributions into three discrete stages:

(a) Calculate frequency of occurrence of any impact. This is simply

$$n_0(\bar{L}+D)\bar{U}$$

per unit time, where $\bar{L}$ and $\bar{U}$ are mean diameter and mean velocity of *all* floes.

(b) Given that impact has occurred, generate the probability distribution for impacting floes. This is derived from the probability density function for all floes multiplied throughout by the quantity

$$\frac{L+D}{\bar{L}+D}$$

A resulting exceedance function for *impacting* floes, based on the data shown in Fig. 3.13(b) for *all* floes, is represented by the solid line in Fig. 7.22 (where we have assumed $L \gg D$ for simplicity, so that the multiplier is simply $L/\bar{L}$).

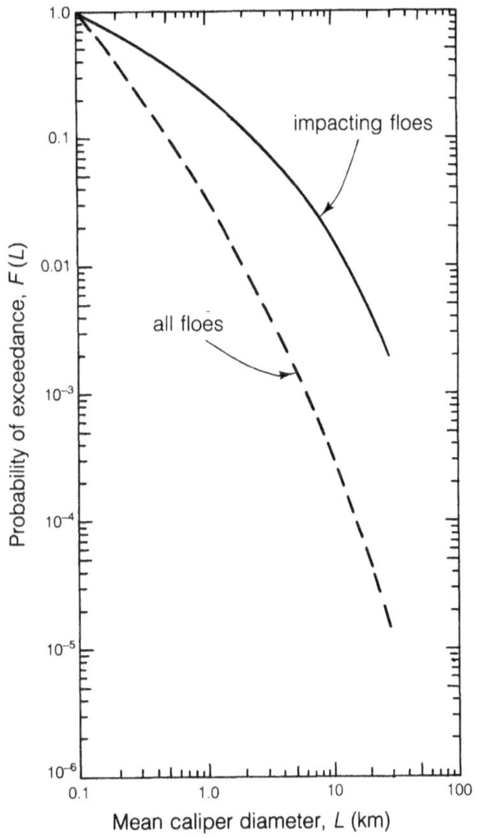

Fig. 7.22 Exceedance probability for diameter of an impacting floe (solid line) compared with exceedance probability for all floes (dashed line). The distribution is as in Fig. 3.13(b) for the Beaufort Sea.

(c) Similarly, we calculate the velocity distribution for floes, given that impact has occurred, by multiplying the velocity probability density function for all floes by the factor

$$U/\bar{U}$$

An example velocity distribution, based on Fig. 3.14(a), is shown in Fig. 7.23 (dashed line), and the modified exceedance function for impacting floes is shown by the solid line.

7.7.4 Design load recommendation

7.7.4.1 Method

A rigorous analysis of multi-year floe impact should really look in detail at the stress history of a large number of simulated impacts. The velocity distribution and diameter distribution of impacting floes can be derived from the foregoing analysis in Section 7.7.3, and these should ideally then be combined with a probability distribution for ice floe thickness, and inserted as input into an analysis of the form illustrated in Section 6.4.5. Carrying out a large number of such simulations would then yield a probability distribution for peak loads.

However, such a procedure is unwieldy and its details are not yet properly supported by experimental evidence. I propose instead a greatly simplified two-stage approach to modelling an impact event, again taking as its basis the concept of "stress at peak load during an event", which is the basis upon which the data set of Fig. 7.15 is plotted. The fundamental simplifying assumption we make is this: we assume that the peak load during an impact event occurs at, or close to, the stage of maximum penetration. Referring to Fig. 7.18, which shows a schematic impact history, this means that we are assuming that the highest load peak during the impact occurs towards the end of the process, just before the floe comes to rest. This may not always be true in reality: it seems to be broadly true of the impacts from Hans Island illustrated in Fig. 5.26, but not so true of the impact illustrated in Fig. 5.24. However, it is a reasonable assumption to make, and will actually lead, if anything, to slightly conservative estimates of load.

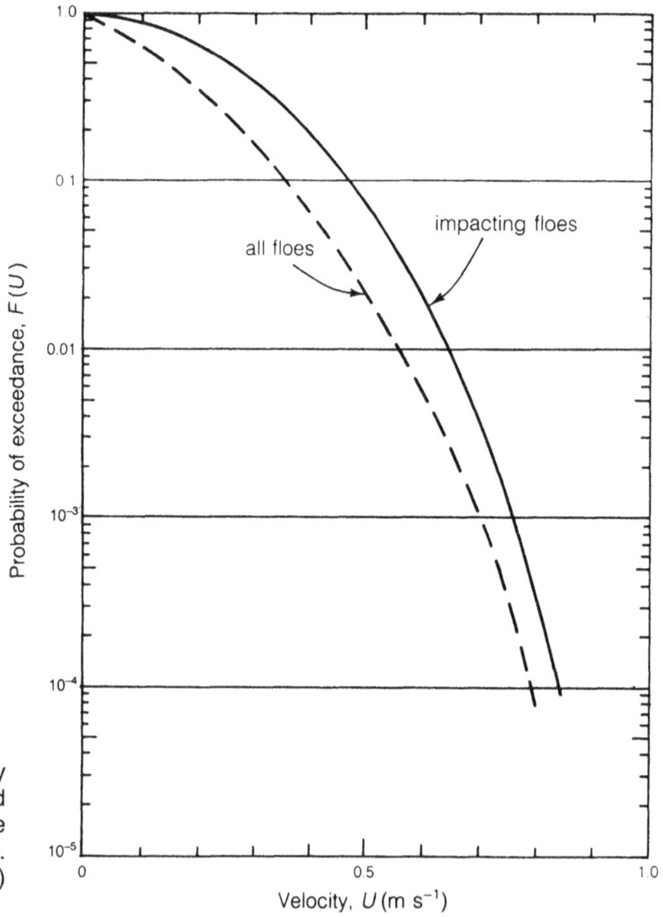

Fig. 7.23 Exceedance probability for velocity of an impacting floe (solid line) compared with exceedance probability for all floes (dashed line). The distribution is as in Fig. 3.14(a) for summer in the Beaufort Sea.

The proposed procedure is this:

(1) For a given floe, of velocity, diameter and thickness generated by probabilistic methods, calculate the maximum penetration it achieves, using the methodology of Section 7.7.2. The contact area A_i at maximum penetration is then given by equation (7.18).

(2) At this stage of maximum penetration, select one random value of normalised stress at peak load, σ_i^*, using the distribution specified in Fig. 7.16(c), and convert this to stress σ_i over the actual area A_i by the equation

$$\sigma_i = \sqrt{\frac{A_0}{A_i}}\,\sigma_i^*$$

The peak load P_i during this multi-year impact event is then simply

$$P_i = \sqrt{A_i A_0}\,\sigma_i^* \qquad (7.34)$$

Using the schematic picture shown in Fig. 7.18, this corresponds to allowing the impact to proceed to maximum penetration, and then selecting just one value of "stress at peak load during an event" to represent the maximum peak during the event.

7.7.4.2 Floe thickness

We still require a probability distribution for floe thickness h, since this is an essential ingredient of the expression for maximum contact area (equation (7.18)).

Floe thickness is not such a simple concept as it might seem. At least two different aspects of floe thickness should be involved when we consider an impact: firstly, to calculate the initial kinetic energy of a floe we need to know the areal average thickness $\bar{h}$ over the whole floe, since this helps determine the initial mass; secondly, to calculate the energy dissipation during the penetration process we need to know average and extreme thickness at the contact zone as the impact proceeds. These two aspects are quite different, but very hard to separate. Some data on ice thickness available from submarine sonar measurements have been shown in Fig. 3.7, but these represent statistics for a series of random spot measurements of ice thickness along a line. Naturally, these data show quite high variability, and the average thickness of a floe would not be expected to show such wide extremes since it is effectively the average of a large number of spot measurements. However, it is difficult to translate these data into terms of average ice floe thickness, since we would need additional information about the distance scale over which spot thicknesses effectively persist, and we would also need to relate average thickness statistics to floe diameter: the larger the diameter, the less variability we would expect in the floes' average thickness. An analysis of this type has been carried out by Marcellus and Morrison (1982) using data from *Sargo* and *Seadragon*. It involves considerable sophistication and complication, particularly in the definition of what, on the basis of an underwater profile, constitutes "a floe", and I will not pursue the matter here. The average thickness of a floe is, in any case, a relatively minor element in the calculation of initial kinetic energy of a floe. The kinetic energy of a floe of mean caliper diameter L, thickness h and velocity U is given approximately by:

$$\tfrac{1}{2}\rho_i(0.66\,L^2)hU^2$$

where ρ_i is the density of ice, and I have used equation (3.4). It is clear that diameter L and velocity U both contribute by their square to the total kinetic energy, while thickness h contributes only linearly. Variations of average thickness, as opposed to spot thickness, are thus relatively unimportant. Since the computation of statistics on average floe thickness is so difficult, I shall here work from the statistics on spot thicknesses as given by Fig. 3.7, and assume that they can be used to represent average floe thickness as well. This will lead to an over-estimate of extreme kinetic energies, and should ideally be refined.

As an example for the Beaufort Sea, we have already seen in Section 3.2.2 and Fig. 3.7 that the probability density for submerged ice draft, h_s, is approximately a negative exponential, of the form

$$f(h_s) = B\exp(-bh_s) \qquad (7.35)$$

with b determined as approximately 0.37 m^{-1} and $B = 0.41\ m^{-1}$ for the whole track taken through the Beaufort Sea. For the analysis of multi-year floe impact we need to modify this in two ways. First, we must express it in terms of ice thickness h instead of ice draft h_s, using the approximate relationship

$$h = 1.12h_s$$

Next we need to restrict the probability distribution to multi-year alone. The distribution as given refers to Beaufort Sea ice cover as a whole and includes first-year and second-year ice as well as multi-year ice. The simplest way to modify the probability distribution is to exclude from the population all ice of thickness less than a certain thickness h_0, where $h_0 \sim 2$ m, on the assumption that all such ice is first-year or second-year. If we apply this criterion to the distribution of equation (7.35), and renormalise it so that there is a probability of unity that ice thickness h lies in the range $h_0 < h < \infty$, we find

$$f(h) = B'e^{-b'h} \tag{7.36}$$

with $b' = 0.33\ \text{m}^{-1}$ and $B' = 0.64\ \text{m}^{-1}$. This analysis used $h_0 = 2$ m, and is a reasonable form to adopt for multi-year ice thickness. In terms of exceedance probability, the probability $F(h)$ that ice thickness exceeds h is given by the integral of equation (7.36), that is:

$$F(h) = 1.94e^{-0.33h} \tag{7.37}$$

This function is shown plotted in Fig. 7.24. It gives a median thickness (50% exceedance) of 4.1 m, and a thickness of 11.1 m at the 95% exceedance level.

It must be stressed again that this is not an exceedance function for average floe thickness. Strictly, $F(h)$ gives the probability that a random point in a field of multi-year ice has a thickness exceeding h. It tells us nothing about a point 1 km away, or, indeed, even 1 m away. I have already discussed briefly the shortcomings of using this

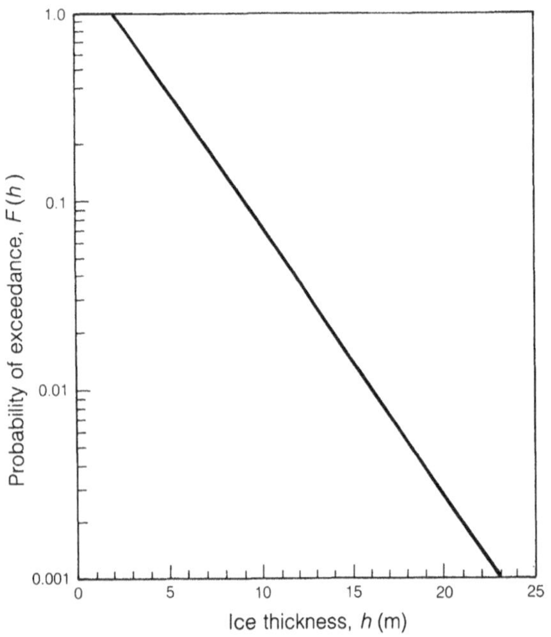

Fig. 7.24 Form assumed for the exceedance probability function for multi-year ice thickness, following equation (7.37).

distribution for the analysis of initial kinetic energy, but there are shortcomings also in its use for calculation of peak loads. To use equation (7.37) directly in the analysis of multi-year ice impact involves a further simplification and approximation. During an impact event we are not strictly just sampling one single point thickness on a multi-year floe. More realistically, we are sampling a finite track length X of multi-year ice, where X corresponds to the penetration distance during impact. During penetration we may encounter several keels; a single thickness value from the distribution of Fig. 7.24 can be used only if the penetration distance X is, in general, small compared with the distance scale over which multi-year ice thickness typically varies. This is a difficult statistic to define, since it depends so sensitively on the definition of a "keel". However, for practical purposes major keels occur with a frequency of the order about 5 per km (based on keels of draft exceeding 5 m as analysed by Wadhams and Horne, 1980). This is an average spacing of a few hundred metres. As long as the penetration distance during an impact event is considerably less than a few hundred metres, then it is approximately valid to assume that we are effectively taking just one sample thickness from the distribution of Fig. 7.24. In general, though not always, this is a reasonable assumption: in Table 7.9 we showed that a typical impact achieves up to about 10 m penetration, and that quite an extreme impact is required before penetration exceeding 50 m is achieved.

Finally, note that in adopting a generalised thickness distribution obtained from Fig. 3.7 we are effectively including all keels in the correct statistical fashion. This means that the analysis already caters automatically for multi-year ridge impacts and we do not have to make any separate provision for them.

7.7.4.3 An example

Let us, then, proceed through the design load calculation for a set of example environmental parameters using the methodology described. The parameters are intended to be broadly representative of summer in the Beaufort Sea, but of course they should not be adopted as formal site-specific design recommendations. We proceed as follows.

(a) First we calculate the impact frequency. The number of impacts expected per unit time is given by equation (7.27) as

$$n_0(\bar{L}+D)\bar{U}$$

and the area density of floes n_0 is given by

$$n_0 = c/\bar{\mathscr{A}}$$

where c is the areal concentration of multi-year ice and $\bar{\mathscr{A}}$ is the mean floe area. Now, let us assume an average concentration of 5% multi-year ice (see Fig. 3.12). We adopt the floe size distribution of Fig. 7.22, which yields a value

$$\bar{\mathscr{A}} = 0.23 \text{ km}^2$$

as the average area for all floes. The area density is then

$$n_0 = 2.2\times10^{-7} \text{ km}^{-2}$$

The mean floe diameter $\bar{L}$ is, again using Fig. 7.22,

$$\bar{L} = 0.34 \text{ km}$$

and the mean floe velocity is, from the distribution of Fig. 7.23, taken to be

$$\bar{U} = 0.18 \text{ m s}^{-1}$$

We take a structure diameter $D = 100$ m. Now, for the purposes of this calculation, we shall assume that the mobile summer season at the (fictional) site in question has a mean duration of $t = 57$ days. We find then that the expected number of impacts during this period is

$$n_0(\bar{L}+D)\bar{U}t = 85$$

This includes all impacts, of whatever size, and represents the annual number of summer impacts. Obviously if movement occurs during winter months as well a separate further analysis may be carried out.

(b) We now take the probability distributions for impacting floes (Figs 7.22, 7.23 and 7.24) and use the Monte Carlo method to generate a series of annual sets of 85 random permutations of diameter L, velocity U and thickness h. As before, we model the impact sequence as a regular process rather than a Poisson process.

(c) For each of these permutations we calculate a contact area at maximum penetration, using equation (7.18). For example, a floe of diameter 1 km, thickness 5 m, velocity 0.3 m s^{-1} achieves a theoretical maximum contact area of $A_i = 47$ m^2.

TABLE 7.10 Multi-year ice-impact design loads: example calculation for the summer season in the Beaufort Sea

Input data		
Structure diameter		100 m
Multi-year ice concentration		5%
Length of mobile season		57 days
Floe size		
mean diameter		0.34 km
mean area		0.23 km^2
distribution		as in Fig. 7.22
Floe velocity		
mean summer velocity		0.18 m s^{-1}
distribution		as in Fig. 7.23
Floe thickness		
median thickness		4.1 m
distribution		as in Fig. 7.24
Stress statistics		
	Mean, σ_m^*	*Standard deviation,* σ_s^*
Normalised indentation stress, σ^* (MPa) over $A_0 = 100$ m^2	0.92	0.45
Design loads		
25-year design load		530 MN
50-year design load		590 MN
100-year design load		640 MN

(d) At the point of maximum penetration we then use the probability distribution of Fig. 7.16 to select one value of normalised stress at peak load, σ_i^*. Let us suppose we select a random value of 1.0 MPa. The peak load P_i at maximum penetration is then

$$P_i = \sqrt{A_i A_0}\,\sigma_i^* = 69 \text{ MN}$$

(e) We now run this process many times and build up exceedance statistics for the annual peak load. I have carried out this calculation using the parameters summarised in Table 7.10, and the results are shown in Fig. 7.25. The expected load distribution during a single year is shown in Fig. 7.25(a), and the average peak load for all events is 42 MN. Note that although

85 events per year occur, more than 50% of them result in peak loads lower than 25 MN. The design loads may be read directly from the exceedance plot for peak annual loads, plotted in Fig. 7.25(b). For this example, the 25-year design load is 530 MN (annual exceedance probability of 4%), and the 100-year design load is 640 MN (annual exceedance probability of 1%). Again, the probability distribution for annual peak loads is approximately log-normal.

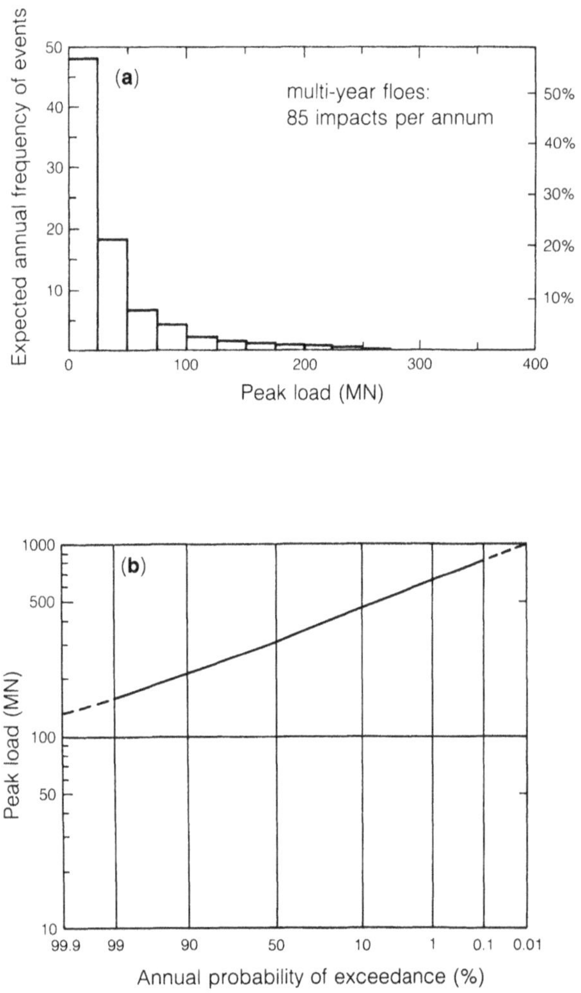

Fig. 7.25 (a) Histogram of expected multi-year ice impact loads generated during a year under the conditions of Table 7.10. (b) Annual exceedance probability for multi-year impact loads under the conditions described in Table 7.10.

7.8 DYNAMIC IMPACT: ICEBERGS

No full-scale measurements have ever been made of forces exerted during impact of an iceberg with a structure. However, the problem would seem not to be very substantially different from that of impact of a very thick multi-year floe with a structure: in both cases we expect progressive failure of ice over a steadily growing contact zone, and we expect the ice mass to come to rest when all its energy has been dissipated by the failure process. The principal differences between iceberg impact and ice floe impact are the following:

(a) An iceberg (if it is a blocky iceberg rather than a tabular iceberg) is generally free to rotate about all three axes during impact, while an ice floe is typically able to rotate significantly only in the horizontal plane, about the vertical axis.

(b) The contact zone of an iceberg with a structure typically grows progressively in two dimensions rather than just one – the contact width and the contact depth both grow as penetration proceeds.

(c) The failure modes of the ice may be rather different: in particular, full-thickness flexural failure is most unlikely during iceberg impact. We might also expect a higher degree of triaxial confinement during crushing of an impacting iceberg.

Because there are so few data on which to base an analysis, I propose to make a number of heroic assumptions in order to provide an illustration of the calculation process for iceberg impact. The quantitative conclusions should, of course, be treated with caution until full-scale experiments become available, but they provide a guide to the level of forces which might be expected.

First of all, to simplify the description of the interaction kinematics, we shall neglect all rotational components of motion. Nevel (1986) provides a good formal analysis of rotational motion during eccentric impacts. However, to apply it rigorously would involve complex numerical computations for a range of expected iceberg shapes and impact eccentricities, and this does not seem worthwhile at this stage. By neglecting rotational motion we are effectively over-estimating the amount of initial kinetic energy which is dissipated during the ice failure process, and so the calculation will provide, in this respect, an upper bound.

Secondly, let us assume that the iceberg is spherical, of radius R, and that it interacts with a flat vertical-faced structure of diameter D (Fig. 7.26). At a penetration distance x it is easily shown that the contact area A is simply given by

$$A = \pi(2Rx - x^2)$$

This can also be expressed in terms of θ, the half-angle subtended by the contact zone, as:

$$A = \pi R^2 \sin^2\theta \tag{7.38}$$

(As we shall see, this is an easier form for integration purposes.)

Thirdly, let us assume that the pressure–area curve concept holds for iceberg impact too, and that average pressures $\bar{\sigma}$ and peak pressures $\hat{\sigma}$ during impact are inversely proportional to the square root of contact area A. This is a bold assumption, and it must be admitted that it has little firm foundation. However, let us refer back to Fig. 5.38, where icebreaker indentation data are superimposed on simple edge-indentation data. It seems clear that icebreaker data, which may involve a significant degree of triaxial constraint, fall within the same bounds as the remainder of the data set, and certainly show a similar trend. I feel there is little doubt that iceberg indentation data, if and when they become available, will show a trend of decreasing failure pressure with increasing contact area. It is to be expected that this trend will follow approximately the inverse square-root dependence already assumed, though it may be that the numerical constants will be found to differ. Here, in the absence of evidence to the contrary, we shall assume the same constants as used above in Sections 7.6.2 and 7.7.4, which are based on the full-scale summary histograms of Fig. 7.16. Thus, we write the average stress $\bar{\sigma}(A)$ at contact area A as, by equation (7.14),

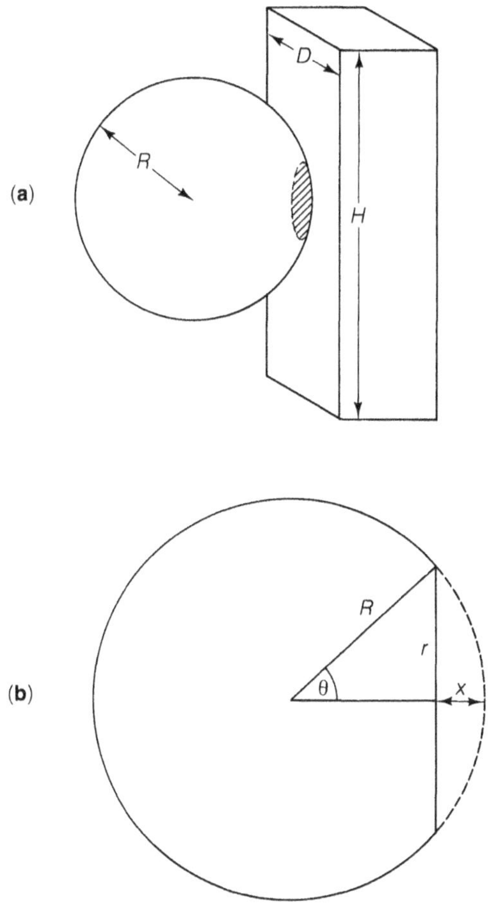

Fig. 7.26 Geometry assumed for impact of an iceberg. The iceberg is spherical, of radius R and the structure is flat-sided, of width D. (a) Perspective view. (b) Detail of geometry: contact zone of radius r at penetration x, with half-angle θ.

$$\bar{\sigma}(A) = \lambda_m \sigma_m^* \sqrt{A_0/A}$$

where $\lambda_m \sim 2/3$, $A_0 = 100\,\text{m}^2$ and $\sigma_m^* = 0.92\,\text{MPa}$.

We can now calculate degree of penetration during iceberg impact. Let the iceberg have mass m and initial velocity U_0. Its initial kinetic energy is thus $\frac{1}{2}mU_0^2$, and the iceberg comes to rest when this is all dissipated by the impact process. Equating initial kinetic energy to total work done, full penetration X is achieved when

$$\int_0^X A\bar{\sigma}(A)\,dx = \tfrac{1}{2}mU_0^2$$

From equation (7.14) we have, then:

$$\lambda_m \sigma_m^* \int_0^X \sqrt{AA_0}\,dx = \tfrac{1}{2}mU_0^2 \tag{7.39}$$

Now from the geometry of the contact zone (Fig. 7.26(b)) we can write

$$x = r\tan(\theta/2)$$

and

$$r = R\sin\theta$$

Unfortunately, use of these exact expressions leads to an intractable integral. Let us, however, make the reasonable assumption that under most circumstances $x \ll R$ – that is, penetration is small compared to iceberg radius, and the angle θ is small. As we shall see, this is normally true. Under this approximation we can write:

$$\tan(\theta/2) \sim \sin(\theta/2) \sim \tfrac{1}{2}\sin\theta$$

(This approximation is, incidentally, a surprisingly good one: even if θ is as much as 30°, the error of the approximation is only 7%.)

Writing, then,

$$x = \tfrac{1}{2}\sin\theta$$

we have:

$$x = \frac{R}{2}\sin^2\theta$$

Differentiating this gives:

$$dx = R\sin\theta\cos\theta\, d\theta$$

This enables us to use equation (7.38) to rewrite equation (7.39) in terms of θ as:

$$\lambda_m\sigma_m^* \int \sqrt{\pi A_0 R^2 \sin^2\theta}\, R\sin\theta\cos\theta\, d\theta = \tfrac{1}{2}mU_0^2$$

or

$$\lambda_m\sigma_m^*\sqrt{\pi A_0}\int R^2\sin^2\theta\cos\theta\, d\theta = \tfrac{1}{2}mU_0^2$$

This is an easy integral, yielding:

$$\tfrac{1}{3}\lambda_m\sigma_m^* R^2\sqrt{\pi A_0}\sin^3\theta = \tfrac{1}{2}mU_0^2$$

where θ is the half-angle of the contact area at full penetration. Substituting back into equation (7.38), the contact area at full penetration is then given by:

$$A = \left[\frac{3\pi R m U_0^2}{2\lambda_m\sigma_m^*\sqrt{A_0}}\right]^{2/3} \qquad (7.40)$$

We then adopt the same methodology as was used for multi-year impacts, and assume that the peak load occurs at or near this state of full penetration and that the normalised stress σ_i^* at peak load is drawn randomly from the distribution of Fig. 7.16. The peak force P_i is then given by

$$P_i = \sqrt{AA_0}\,\sigma_i^* \qquad (7.41)$$

An estimate of iceberg mass m in terms of its radius R can be given nominally by assuming the iceberg to be approximately spherical. We then have:

$$m \sim \tfrac{4}{3}\pi R^3\rho_i \qquad (7.42)$$

where ρ_i is the density of iceberg ice, which we shall here take to be an average of 0.88 Mg m^{-3}.

As illustration let us look first at peak loads calculated by the methodology of this section for two hypothetical impact cases. We shall then proceed to a full risk analysis for iceberg impact loads.

(1) *"Typical" iceberg impact.* Consider an iceberg of mass 1 million tonnes (10^9 kg) moving at velocity 0.25 m s^{-1}. This is a "medium"-sized iceberg as defined by Table 2.1 (from the International Ice Patrol) and corresponds to an iceberg of nominal radius about 65 m. Its velocity is at approximately the 50% exceedance level defined by Fig. 2.24. Using equation (7.40) we find that the contact area at full penetration is 213 m^2 (corresponding to a penetration distance X of only about 0.5 m for a perfect sphere). At this moment of full penetration the expected normalised peak stress is $\sigma_i^* = 0.92$ MPa, corresponding, by equation (7.41), to a peak load of 134 MN. At the two-standard-deviation level, the normalised peak stress expected from the distribution of Fig. 7.16 would be 1.82 MPa, corresponding to a peak load of 266 MN. This means that, during a long series of impacts by icebergs of mass 1 million tonnes moving at 0.25 m s^{-1}, approximately 1 in 40 of the impacts (2.5%) would be expected to give rise to peak loads in excess of 266 MN.

(2) *"Extreme" iceberg event.* Alternatively, consider a nominal "extreme" impact event (note, though, that I am not here ascribing the event a specific exceedance probability). Consider an iceberg of mass 10 million tonnes (10^{10} kg) moving at velocity 0.75 m s^{-1}. The iceberg has radius of approximately 140 m. From equation (7.40) its contact area at full penetration is 7136 m^2, corresponding to a penetration distance X of about 8 m. Note that this corresponds to a contact area of diameter about 95 m, so that even with this quite extreme hypothetical event, the iceberg would not be expected quite fully to envelop a structure of width 100 m. At the stage of full penetration, we would then expect a peak normalised stress of $\sigma_i^* = 0.92$ MPa and hence a peak load of 777 MN. During a series of similar impacts, the peak load at the 2.5% exceedance level would be 1537 MN.

Let us now calculate a typical impact frequency of icebergs with a structure. Now, Fig. 3.23 demonstrates graphically that iceberg flux is extremely variable and dependent on latitude. Most calculations of iceberg impact probabilities carried out in the literature have been performed using data appropriate to the Grand Banks of Newfoundland (e.g. Blenkarn and Knapp, 1969). However, at latitude 45°N, the Grand Banks are very much on the fringe of the Labrador Current "iceberg alley". Let us look at some representative statistics at higher latitudes.

From equation (7.27) the impact frequency in time period t is expected to be:

$$n_0(\overline{L}+D)\overline{U}t \qquad (7.43)$$

where n_0 is iceberg area density, $\overline{L}$ is mean iceberg diameter, D is structure width, and $\overline{U}$ is mean iceberg velocity. Representative values for these parameters at latitude about 55°–60°N are as follows:

n_0 – areal density of icebergs is, from Fig. 3.22, about 100 icebergs per 100×100 km square, or a density of approximately 10^{-8} m^{-2}. This, of course, shows strong seasonal and regional variations, which we shall ignore here.

$\overline{L}$ – precise statistics on iceberg diameter L and radius R have not yet been presented in this volume, and do not appear to be readily available in the public domain. However, consider the statistics given by Gustajtis (Table 3.1). Between latitudes 55° and 60° N, the statistics indicate that roughly 5% of icebergs are growlers, 53% are small, 47% are medium and 12% are large. In Fig. 7.27 (dashed line) I have plotted these statistics to indicate typical exceedance levels for radius and velocity. This has been carried out on the assumption that a large iceberg exceeds radius 60 m, a medium iceberg has radius in the range 30–60 m, a small iceberg has radius 7.5–30 m and that anything smaller is a bergy bit or a growler. In addition, I have assumed that we are concentrating on water depths in the range about 100–200 m: draft limitations then mean that an iceberg of radius greater than about 100 m cannot enter the area. The average radius, $\overline{R}$, of all icebergs defined by the distribution of Fig. 7.27 is 33 m and the average diameter $\overline{L}$ is $2\overline{R}$, or 66 m.

D – we shall assume a structure diameter of $D = 100$ m.

$\overline{U}$ – the velocity exceedance statistics of Fig. 2.24 define a mean iceberg velocity of 0.29 m s^{-1}. (These data are replotted, in a different form, in Fig. 7.28.)

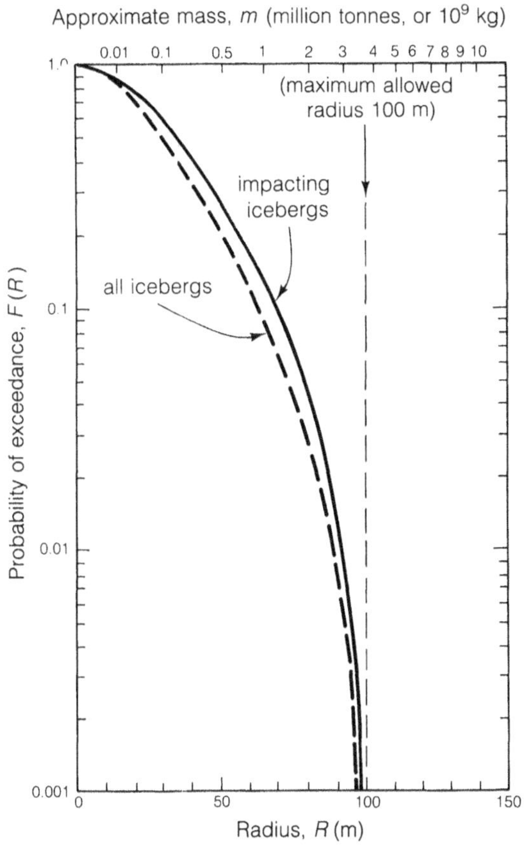

Fig. 7.27 Exceedance function for iceberg radius, representative of the east coast of Canada between latitude 55°N and 60°N: radius of all icebergs (dashed line), and radius of impacting icebergs (solid line). Structure diameter is assumed to be 100 m. Approximate iceberg mass is also indicated, on the basis of nominally spherical shape and average density of 0.88 Mg m^{-3}. The distribution is truncated at iceberg radius of 100 m, to model draft limitations in water depth 100 ~ 200 m.

We shall also assume that icebergs are a serious hazard only for about 9 months of the year: iceberg flux is typically at its peak during the April–June months, but may be almost negligible during winter months, if icebergs are locked in immobile landfast ice. Assuming, then an active period of $t = 275$ days per year, the average impact frequency during a year is, by equation (7.43):

$$n_0(\overline{L}+D)\overline{U}t = 11 \text{ impacts per year}$$

Thus, iceberg impact is by no means an extremely rare occurrence in medium to high latitudes off the east coast of Canada. It must be taken into account in just the same probabilistic fashion as we employed for multi-year floes in the Beaufort Sea (Section 7.7.4.3). Further south, however, in the region of the Grand Banks of Newfoundland, impact probabilities may typically be one or two orders of magnitude lower.

It is worth now going through a worked example to establish annual exceedance levels for peak

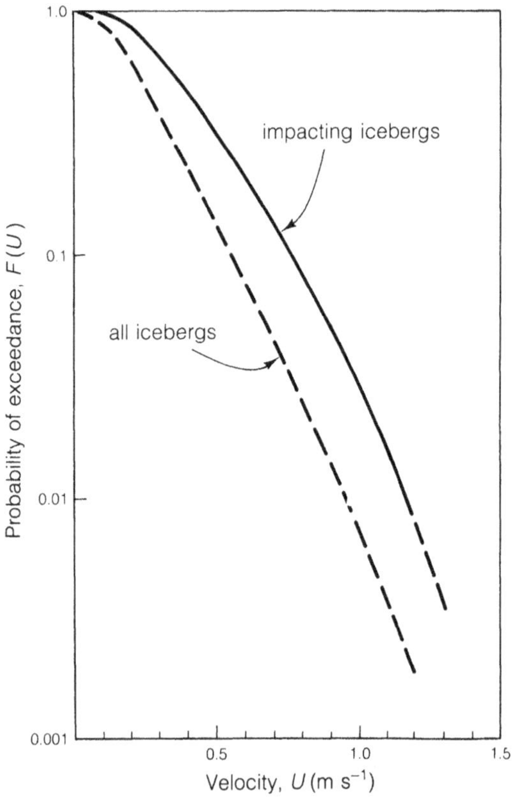

Fig. 7.28 Exceedance function for iceberg velocity (based on data of Figure 2.24). The dashed line shows the exceedance function for all icebergs; the solid line shows the exceedance function for impacting icebergs.

loads due to icebergs. Bear in mind, however, that the statistics used here are not precisely site-specific: the example is for purposes of illustrating the technique, and is broadly representative of the Labrador Coast.

First, we need to redefine our probability distributions for iceberg radius and velocity in accordance with Bayes' Theorem (Section 7.7.3.2). As with multi-year floes, there is of course a higher probability of larger, faster icebergs achieving impact.

For iceberg radius the probability density function for all icebergs needs to be multiplied throughout by the factor

$$(2R+D)/(2\bar{R}+D)$$

where $\bar{R} = 33$ m and $D = 100$ m. Note that we cannot here assume, as we did for multi-year floes, that the impacting feature is large compared with the structure – typically it is not, and the structure diameter is important. The revised exceedance function for *impacting* icebergs, as opposed to *all* icebergs, is shown in Fig. 7.27 (solid line).

The probability density function for iceberg velocity, similarly, needs to be modified, by multiplying by the factor

$$U/\bar{U}$$

where $\bar{U} = 0.29$ m s^{-1}. In Fig. 7.28 the same data as in Fig. 2.24 are shown (dashed line), plotted on a logarithmic scale. The modified exceedance function is shown by the solid line.

We are now almost in a position to run a Monte Carlo simulation of peak load during iceberg impact, using the exceedance functions described by Figs 7.27 and 7.28 and calculating peak loads by equations (7.40–7.42). We need, however, one further assumption: we need to define a limit to the contact area which may potentially be achieved if total penetration should occur. If the structure width is 100 m and it lies in water depth of 100–200 m it is reasonable to impose a limit of 15 000 m^2 to the contact area which can be achieved during impact. We shall impose this as a limit in the modelling process.

The results of running a simulation on this basis are tabulated in Table 7.11 and are shown in Fig. 7.29. The impact frequency is assumed to be 11 impacts per year. Under these conditions the 25-year design load is 620 MN and the 100-year design load is 770 MN. The probability distribution is approximately log-normal, but shows some tendency to flatten off at very high loads: this is due to truncation of the extremal iceberg distribution to conform with draft limitations.

This analysis is, of course, greatly simplified. I have made two rather sweeping assumptions:

(i) I have assumed that (normalised) indentation stress during iceberg impact can be described by the same statistics as we used for multi-year ice impact. There are no data publicly available to confirm this assumption. Stresses during iceberg impact might be higher on account of some extra degree of confinement being present (though as I have illustrated in Fig. 5.34(b) this is something of a myth: states of triaxial confinement during impact are actually rather rare). On the other hand, they may be lower on account of the generally higher porosity of iceberg ice, which is, after all, nothing more than highly compressed snow.

(ii) I have assumed the iceberg to be perfectly spherical. There are obvious reasons for doing

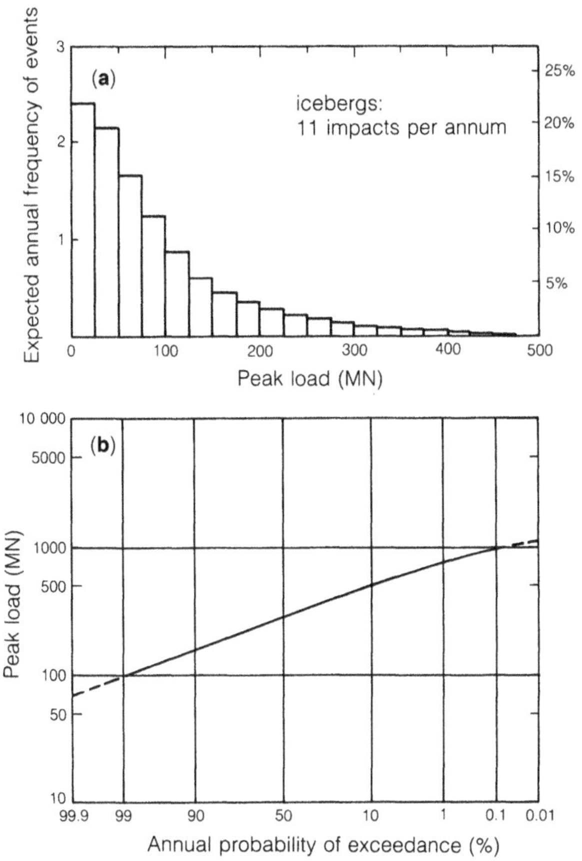

Fig. 7.29 (a) Histogram of expected iceberg loads experienced during a year, under the conditions described in Table 7.11. (b) Annual exceedance probability for iceberg impact loads.

TABLE 7.11 Iceberg impact design loads: example calculation representative of the east coast of Canada between 55° and 60° N

Input data		
Structure diameter		100 m
Iceberg concentration		10^{-8} m^{-2}
Length of season		275 days
Iceberg size mean radius distribution		 33 m as in Fig. 7.27
Iceberg velocity mean velocity distribution		 0.29 m s^{-1} as in Fig. 7.28
Average number of impacts per year		11
Iceberg shape		nominally spherical
Stress statistics		
	Mean, σ_m^*	*Standard deviation*, σ_s^*
Normalised indentation stress, (MPa) over $A_0 = 100$ m^2	0.92	0.45
Design loads		
25-year design load		620 MN
50-year design load		700 MN
100-year design load		770 MN

this – it makes analytical treatment of impact geometry possible, and enables a convenient relationship between iceberg mass and radius to be established (equation (7.42)). However, icebergs are far from spherical, and in reality we should expect impacts to differ significantly in two respects: firstly, the contact zone of an irregularly shaped iceberg should not grow so rapidly as that of a spherical iceberg, so we may be overestimating the contact area achieved; secondly, an estimate of mass based on horizontal radius at the waterline is almost invariably an overestimate – a stably floating iceberg is always less thick than it is wide. For both these reasons the assumption of spherical shape probably leads, if anything, to an overestimate of loads.

A similar analysis can of course be carried out, with appropriate statistics, for any other region prone to icebergs, such as East Greenland or the Barents Sea. Icebergs in the Beaufort Sea (in the form of ice islands) are very much rarer than off the east coast of Canada, and impact probabilities are close to negligible. Calculations of impact probability by Marcellus and Morrison (1982) indicate a return period of around 1000 years for a typical nearshore structure – an annual impact probability of only 10^{-3} per annum. In this case, and indeed for some areas offshore the east coast of Canada or northern Norway, it may be both prudent and permissible to make provision for detecting and removing icebergs by towing, if they should approach closely enough to place a structure at risk. In this case their impact loads may be ignored when specifying structural resistance, and effort needs to concentrate instead on ensuring the reliability of the detection and removal systems.

7.9 WINTER BLOCKING

Impact by large ice features with a structure has already been covered: if they are icebergs, their impact loads may be calculated by the methodology of Section 7.8; if they are major sea-ice features such as hummock fields, their statistics are tacitly already included in the cumulative probability distributions of Figs 7.22 and 7.24, and they are thus covered by the calculations of Section 7.7.

What happens, however, if such a feature becomes lodged against a structure after it has come to rest? Is the force it may then exert limited by driving forces (Section 7.5), or will sufficient driving forces always be available to fail the ice feature? The answer appears in general to be that driving forces *are* normally available, and that the concept of limited driving force can in most cases safely be ignored. This accords well with the observation that the Arctic is not, in fact, crowded with large invincible features which are incapable of being destroyed. As an example, consider a large, hypothetical, hummock field at rest against a structure of diameter 100 m. The hummock field has thickness 15 m, diameter 5 km and area 4×10^6 m^2. It is acted upon by wind, current and pack-ice of thickness about 2 m. Now the available driving forces would appear to be:

(a) *Wind*. Let us assume a wind of velocity 15 m s^{-1} to occur during the period that the feature is lodged against the structure. It exerts a drag force, by equation (7.6), of 2.2 MN.
(b) *Current*. Assume a current of 0.5 m s^{-1}. The current drag force is then, by Section 7.5.1.2, 4 MN.
(c) *Pack ice*. The surrounding pack-ice acts over an effective contact area of $A = 10^4$ m^2 across the diameter of the hummock field (Fig. 7.11). The peak load P expected is then

$$P = \sqrt{AA_0}\sigma_i^*$$

where σ_i^* has a value of $\sigma_m^* \pm \sigma_s^*$, or 0.92±0.45 MPa. The mean load expected is therefore 920 MN during a single event, with a probability of 2.5% of exceeding 1820 MN.

Obviously, the pack-ice driving force is the most important factor, since it vastly exceeds the force available from wind and current, which may therefore be ignored. However, consider now the force required to fail the feature itself against the structure. It has a gross contact area of $A = 1500$ m^2 undergoing failure, and thus requires an expected peak load P of

$$P = \sqrt{AA_0}\sigma_m^* = 360 \text{ MN}$$

or of 700 MN at the 2.5% exceedance level. It is clear that this force is in general theoretically available. Direct measurements of pack-ice driving forces which are now available (Croasdale *et al.*, in press) indicate, however, that typical pack-ice forces may in practice be rather less than the analysis of peak forces presented here suggests. Actual measurements of large scale stresses applied during a ridging event at the perimeter of a large multi-year floe have indicated a load intensity of about $3\text{–}5 \times 10^4$ N m^{-1} applied over a width of about 5 km (see Section 5.6.1). On this basis, a more typical driving force on the extreme floe we are considering here would lie in the range about 150–250 MN, and the interaction would indeed be governed by the limited driving force. However, it is important to appreciate that peak driving forces under more extreme conditions might very well substantially exceed these measured forces. Further measurements are certainly needed before we can confidently rely on limited driving forces as limits to the applied load on a structure. As always, there may very well be great difference between *typical* driving forces and *maximum possible* driving forces. I recommend therefore that driving force considerations may in general be ignored. It may be a useful check to make in any specific instance but, generally, it does not appear to enable us to reduce our estimates of peak loads exerted on structures.

7.10 LOCAL LOADS

Local load calculations are important to the design of structural skins and internal structural members. The meaning of "local" depends rather on the structure concerned (for instance, on what the spacing is between its internal members), but in general we mean loads over areas in the range 1–100 m^2.

It is not easy to make recommendations about local load design levels. There are two principal difficulties which complicate the issue.

7.10.1 Consequences

The consequence of local damage or failure is generally very much less severe than that of global failure, particularly for a fixed gravity structure which runs no risk of sinking if its local integrity is threatened. In many cases, local damage may have a consequence which is merely economic and which involves none of the more important considerations of loss of life or environmental damage. This means that the acceptable level of risk may be very much higher than that for global failure, and what is needed is an involved economic calculation in which two factors are weighed against each other. In simple terms, these two factors are:

(a) The *expected cost of damage:* that is, the probability of damage, p_1, during the lifetime of the structure, multiplied by the cost c_1 of the damage if it should occur. The cost of damage should, of course, include repair costs and all consequential losses such as lost production and deferred revenue.
(b) The *cost of further strengthening the structure* to such an extent that the risk of local failure is reduced to negligible proportions. Let this cost be c_2.

Then these two factors are weighed in the balance. If

$$p_1c_1 > c_2$$

then we are justified in strengthening the structure, since the cost of doing so is less than the expected loss if we fail to do so. On the other hand, if

$$c_2 > p_1c_1$$

it makes economic sense to risk local damage and simply repair it if it should occur.

7.10.2 Local "events"

The second problem is this: although we have fairly successfully defined peak stress statistics for global interaction "events", it is less easy to define what constitutes a local interaction "event". That is, if a structure undergoes, say, 1000 global interaction events during its lifetime, over contact areas of order 100–1000 m^2, how many local interaction events is it likely to undergo over an area of order, say, 10 m^2? Without knowing how many such local events are occurring it is, of course, impossible to quantify a probabilistic exceedance level.

Here I shall propose a simple pragmatic approach to the matter, but I have to admit that I do not know at this stage how to quantify the associated risk levels. It seems likely that local ice-design specifications will have to follow the traditional course of engineering design codes: that is, we set up an initial set of specifications, and in due course (year-by-year, or decade-by-decade) relax or strengthen the specification until the observed rate of failure strikes a satisfactory balance between cost of damage versus cost of avoiding it.

As a starting point, I show in Fig. 7.30(a) a family of design curve lines which may be found to provide guidance on local load design levels. The bold line represents a straight-line upper bound to publicly available observations to date, taken from Fig. 5.38 (plotted on logarithmic scales). The dashed lines represent a series of lines drawn at the one-, two- and three-standard-deviation levels of the "peak stress during an event", defined by equation (7.13).

Two acceptable approaches appear to be available:

(1) *Classical engineering approach.* Take the upper bound to measured pressures as a function of area, and multiply this by a "safety factor" of, say, 1.5. The safety factor represents prudence or ignorance, and simply reflects unease that future measured forces may exceed those measured to date. This is shown as the solid line in Fig. 7.30(b). The equation of this line is:

$$\sigma = 20.7A^{-0.43} \quad \text{(MPa)} \tag{7.44}$$

(2) *Confidence level approach.* Take a probabilistic level of deviation from the mean peak stresses measured; for instance, the three-standard-deviation level shown in Fig. 7.30(a). In some sense, this represents the stress level which, for an event occurring over a certain given area has a chance of about 0.14% of being exceeded (about 1 in 750). However, as I have already pointed out, we do not know to define the event population to which this refers, and unless we know how many events are expected to occur the comfort afforded by this probabilistic approach is purely illusory. The line obtained by this method is shown in Fig. 7.30(b), compared with the classical engineering approach explained above. The equation of this line is:

$$\sigma = 22.7A^{-1/2} \quad \text{(MPa)} \tag{7.45}$$

The two curves are really quite close, especially in the region of most interest, which is around the area of about 10 m^2 (where few data points

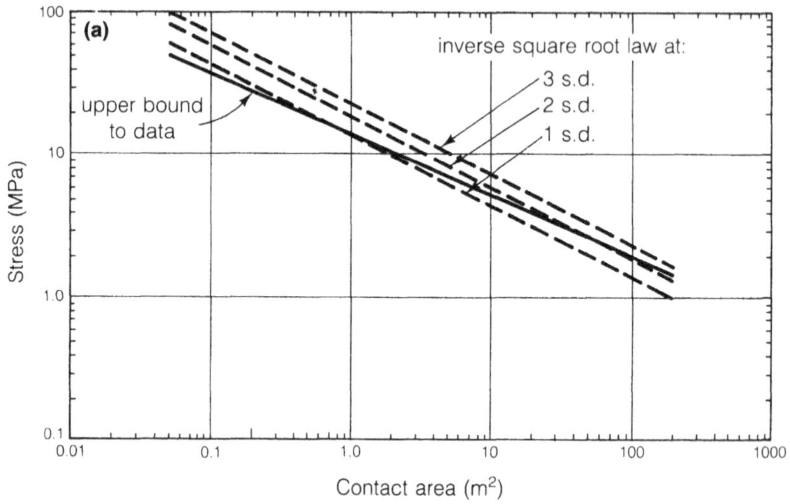

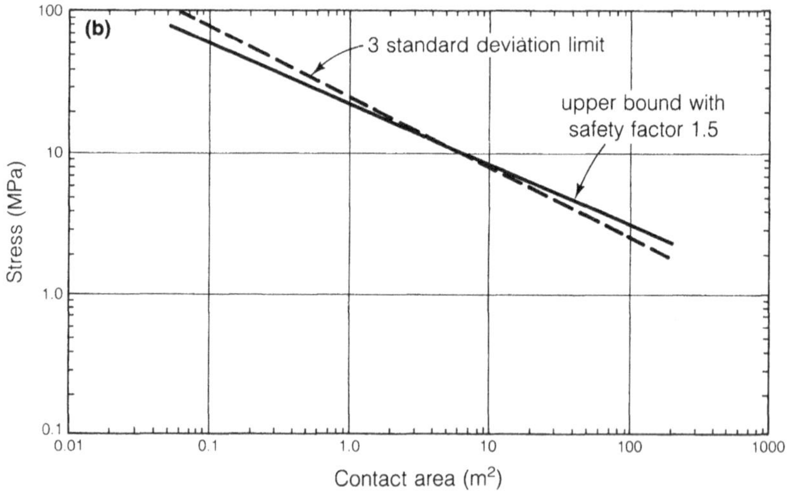

Fig. 7.30 Local load design curves: (a) Bold line representing straight-line upper bound to all publicly available data, and dashed lines showing one- two- and three-standard-deviation limits to inverse square root relationship of equation (7.13). (b) Alternative recommendations: solid line showing upper bound multiplied by safety factor of 1.5; or dashed line showing three-standard-deviation limit to equation (7.13).

exist). I would be inclined to adopt the solid line, the classical engineering approach, as a design curve, but both curves are bounds to all observed data and either is defensible.

7.11 EPILOGUE

I am conscious that this book contains some untied ends (and also some ends which have been rather hastily knotted). In this final chapter, I have tried to provide sample answers to the questions an engineer has to ask about ice design loads, but in all cases this has necessarily involved short-cuts and approximations. This is because the formalism does not yet exist to allow a rigorous scientific justification of any design load procedure.

The design load procedure I have recommended rests on two pillars. Primarily, it rests on straightforward statistical analysis of raw load data (the pressure–area curve); but it rests also, in a secondary fashion, on a large array of theoretical analysis (Chapter 6) which has not yet quite been fully developed. This theoretical analysis is used mainly as a qualitative back-up to our interpretation of the raw load data, and not as a predictive tool. Ideally, we should of course like to tackle the design load problem by means of deductions from well established physical laws; in practice we have to look directly at the full-scale data which are available, and use physical laws only as a means of understanding what may be causing the observed behaviour.

If progress is to be achieved towards more rigorous methodology then I recommend that work should concentrate on the following areas:

(1) a sounder theoretical understanding of non-simultaneous failure;
(2) analysis of imperfect contact conditions;
(3) more detailed measurement of statistics on natural irregularities in sea ice – thickness irregularities, crack density, etc.

No-one should underestimate the work which would be involved in order to enable rigorous theoretical predictions of ice loads to be made. In the meantime I feel it is more productive to concentrate on the continuing acquisition of good quality full-scale data.

Bibliography

Reports of the *Arctic Petroleum Operators Association* (APOA) may be obtained from:

Pallister Resource Management Ltd,
105, 4116 – 64th Avenue S.E.,
Calgary, Alberta, Canada T2C 2B3.

In general, these and most other references may in case of any difficulty be obtained through either:

Scott Polar Research Institute,
University of Cambridge
Lensfield Road, Cambridge CB2 1ER, UK.

or

US Army Cold Regions Research and
Engineering Laboratory (CRREL),
72 Lyme Road, Hanover, New Hampshire
03755-1290,USA.

The titles of proceedings of conferences of the *International Association for Hydraulic Research* (IAHR) and on *Port and Ocean Engineering under Arctic Conditions* (POAC) have been abbreviated as follows:

IAHR 70
Proceedings of the Symposium on Ice and its Action on Hydraulic Structures, 1970, Reykjavik, Iceland.
Available from: IAHR Secretariat, P.O. Box 177, Delft, The Netherlands.

IAHR 72
Proceedings of the Symposium on Ice and its Action on Hydraulic Structures, 1972, Leningrad, USSR.

IAHR 75
Proceedings of the International Symposium on Ice Problems, 1975, Hanover, New Hampshire, USA.
Available from: Mr Guenther Frankenstein, USA CRREL, 72 Lyme Road, Hanover, New Hampshire 03755-1290, USA.

IAHR 78
Proceedings of the IAHR Symposium on Ice Problems, 1978, Luleå, Sweden.
Available from: Town of Luleå, Information Office Fack, 5-951 85 Luleå, Sweden.

IAHR 81
Proceedings of the International Symposium on Ice, 1981, Quebec, Canada.
Available from: Dr Bernard Michel, Département de Génie Civil, Université Laval, Cité Universitaire, Québec, Canada G1K 7P4.

IAHR 84
Proceedings of the IAHR Ice Symposium, 1984, Hamburg, W. Germany.
Available from: Dr Joachim Schwarz, Ice Engineering Department, Hamburgische Schiffbau-Versuchsanstalt GmbH, P.O. Box 60 09 29, D-2000 Hamburg 60, W. Germany.

IAHR 86
Proceedings of the IAHR Symposium on Ice, 1986, Iowa City, USA.
Available from: Mr R. Ettema, Institute of Hydraulic Research, The University of Iowa, Iowa City, Iowa 52242, USA.

POAC 71
Proceedings of the 1st International Conference on Port and Ocean Engineering Under Arctic Conditions, 1971, Technical University of Norway, Trondheim.

POAC 73
Proceedings of the 2nd International Conference on Port and Ocean Engineering Under Arctic Conditions, 1973, University of Iceland, Reykjavik.

POAC 75
Proceedings of the 3rd International Conference on Port and Ocean Engineering Under Arctic Conditions, 1975, University of Alaska, Fairbanks.

POAC 77
Proceedings of the 4th International Conference on Port and Ocean Engineering Under Arctic Conditions, 1977, Memorial University of Newfoundland, St. John's.

POAC 79
Proceedings of the 5th International Conference on Port and Ocean Engineering Under Arctic Conditions, 1979, University of Trondheim, Norway.

POAC 81
Proceedings of the 6th International Conference on Port and Ocean Engineering Under Arctic Conditions, 1981, Laval University, Quebec.

POAC 83
Proceedings of the 7th International Conference on Port and Ocean Engineering Under Arctic Conditions, 1983, Technical Research Centre of Finland, Helsinki.

POAC 85
Proceedings of the 8th International Conference on Port and Ocean Engineering Under Arctic Conditions, 1985, Narssarssuaq, Greenland.

POAC 87
Proceedings of the 9th International Conference on Port and Ocean Engineering Under Arctic Conditions, 1987, Fairbanks, Alaska.

Aagaard, K. 1984. The Beaufort undercurrent. In *The Alaskan Beaufort Sea, ecosystems and environments*, ed. P. W. Barnes, D. M. Schell, and E. Reimnitz. Academic Press, pp. 47–71.

Afanasyev, Y. P., Dolgopolov, Y. V. and Shyeishtein, Z. I. 1971. Ice pressure on separate supporting structures in the sea. *Arkticheskii i Antarkticheskii Nauchno-Issledovatel'skii*, Trudy, Institut **300,** 61–80. (English translation: *CRREL Draft Translation No.* 346 (1972), Hanover, New Hampshire.)

Agerton, D. J. 1981. Large winter ice movements in the nearshore Alaskan Beaufort Sea. In *POAC 81*, Vol 2, pp. 599–608.

Ahlnäs, K. 1983. Sea-ice studies in Ross Sea, Antarctica, using NOAA-Satellite imagery. In *POAC 83*, Vol. 1, pp. 42–51.

Anderson, C. 1971. The flow of icebergs along the Canadian East Coast. In *Proceedings of the Canadian Seminar on Icebergs, 6–7 December 1971, Maritime Command Headquarters, Halifax, Nova Scotia.*

Anderson, K. G. (ed.) 1983. Tarsiut Island research programme 1982–1983. *APOA Project* No. 198.

Anderson, R. G., Gardner, L. R. T. and Hodgkins, W. R. 1963. Deformation of uniformly loaded beams obeying complex creep laws. *Journal of Mechanical Engineering Science* **5,** 238–244.

Ashby, M. F. and Duval, P. 1985. The creep of polycrystalline ice. *Cold Regions Science and Technology* **11,** 285–300.

Ashby, M. F. and Hallam, S. D. 1986. The failure of brittle solids containing small cracks under compressive stress-states. *Acta Metallurgica* **34**(3), 497–510.

Ashby, M. F., Palmer, A. C., Thouless, M., Goodman, D. J., Howard, M.W., Hallam, S. D., Murrell, S. A. F., Jones, N., Sanderson, T. J. O. and Ponter, A. R. S. 1986. Nonsimultaneous failure and ice loads on Arctic structures. In *Offshore Technology Conference 1986, Houston*, Paper No. OTC 5127, pp. 399–404.

Ashton, G. (ed.) 1986. *River and lake ice engineering*. Littleton, Colorado: Water Resources Publications.

Assur, A. 1960. Composition of sea ice and its tensile strength. *U.S. Snow, Ice and Permafrost Research Establishment, Research Report* No. 44.

Banke, E. G. and Smith, S. D. 1973. Wind stress on Arctic sea ice. *Journal of Geophysical Research* **78**(35), 7871–7882.

Barnes, P., Tabor, D. and Walker, J. C. F. 1971. The friction and creep of polycrystalline ice. *Proceedings of the Royal Society (London)*, Ser. A **324,** 127–155.

Baudais, D. J., Watts, J. S. and Masterton, D. M. 1976. A system for offshore drilling in the Arctic Islands. In *Offshore Technology Conference 1976, Dallas*, Paper No. OTC 2622, Vol. 3, pp. 31–43.

Bazant, Z. P. and Kim, J. -K. 1985. Fracture theory for non-homogeneous brittle materials with application to ice. In *Proceedings of the ASCE Arctic '85 Conference, San Francisco, 25–27 March 1985*, ed. F. L. Bennett and J. L. Machemehl, pp. 917–930.

Bergdahl, L. and Wernersson, L. 1977. Calculated and expected thermal ice pressures in five Swedish Lakes. *Department of Hydraulics, Chalmers University of Technology, Göteborg, Report Series* B: 7.

Bergdahl, L. and Wernersson, L. 1978. Probabilities of thermal ice pressures in five Swedish lakes. In *IAHR 78*, Vol. 1, pp. 349–362.

Blanchet, D. 1983. Interpretation of Tarsiut instrumentation, winter 81–82. Load results and analysis. *APOA Project* No. 197, Report No. 7(b).

Blenkarn, K. A. 1970. Measurement and analysis of ice forces on Cook Inlet structures. In *Offshore Technology Conference 1970, Houston*, Paper No. OTC 1261, pp. 365–378.

Blenkarn, K. A. and Knapp, A. E. 1969. Ice conditions on the Grand Banks. In *The Ice Seminar, CIM Special*, Vol. 10, pp. 61–72.

Bolin, B., Döös, B. R., Jäger, J. and Warrick, R. A. (ed.). In press. *The greenhouse effect, climatic change and ecosystems*. SCOPE Series, John Wiley.

Bourke, R. H. and Garrett, R. P. 1987. Sea ice thickness distribution in the Arctic Ocean. *Cold Regions Science and Technology* **13**, 259–280.

Bruce, J. and Harrington, G. 1982. Design aspects of a mobile Arctic caisson. In *Offshore Technology Conference 1982, Houston*, Paper No. OTC 4333.

Bruen F. J. and Vivatrat, V. 1984. Ice force prediction based on strain-rate field. In *Proceedings of the 3rd International Offshore Mechanics and Arctic Engineering Symposium, New Orleans, 12–17 February 1984*, Vol. 3, pp. 275–281.

Butkovich, T. R. 1959. On the mechanical properties of sea ice, Thule, Greenland, 1957. *US Snow, Ice and Permafrost Establishment, Research Report* No. 54.

Calladine, C. R. 1962. The effect of cross-section shape on the creep buckling behaviour of columns. *International Journal of Mechanical Sciences* **4**, 387–407.

Calladine, C. R. 1969. *Engineering plasticity*, first edition. Oxford: Clarendon Press.

Calladine, C. R. 1982 (unpubl.). *Creep buckling of floating ice sheet*. Personal communication to The British Petroleum Company plc, 5 August 1982.

Carslaw, H. S. and Jaeger, J. C. 1959. *Conduction of heat in solids*, second edition. Oxford: Clarendon Press.

Carter, D. and Michel, B. 1978. Lois et mécanismes de l'apparente fracture fragile de la glace de rivière et de lac. *Université Laval, Faculté des Sciences et de Génie. Département de Génie Civil, Section Mécanique des Glaces, Rapport* S-22.

Central Intelligence Agency. 1978. *Polar regions atlas*. Washington DC: CIA.

Chen, A. C. T. 1981. Ice pressure sensor inclusion factors. *Journal of Energy Resources Technology* **103**(1), 82–86.

Cherepanov, N. V. 1971. Spatial arrangement of sea ice crystal structure. *Problemy Arktiki i Antarktiki* **38,** 176–181.

Cole, D. M. 1986. Effect of grain size on the internal fracturing of polycrystalline ice. *CRREL Report* No. 86–5.

Collins, J. A. 1981. *Failure of materials in mechanical design*. John Wiley.

Colony, R. and Thorndike, A. S. 1985. Sea ice motion as a drunkard's walk. *Journal of Geophysical Research* **90,** 965–974.

Cooksley, S. D. 1984 (unpubl.). *Yield and fracture surfaces of brittle solids under multi-axial loading*. Ph.D. thesis, Cambridge University.

Cottrill, A. 1987. Untitled. *Offshore Engineer*, January 1987, pp. 42–48.

Cox, G. F. N. and Johnson, J. B. 1983. Stress measurements in ice. *CRREL Report* No. 83–23.

Cox, G. F. N. and Weeks, W. F. 1982. Equations for determining the gas and brine volumes of sea ice samples. *CRREL Report* No. 82–30.

Cox, G. F. N., Richter-Menge, J. A., Weeks, W. F., Mellor, M. and Bosworth, H. 1984. Mechanical properties of multi-year sea ice. Phase 1: test results. *CRREL Report* No. 84–9.

Cramér, H. 1955. *The elements of probability theory*. New York: John Wiley.

Croasdale, K. R. 1970. The 'Nutcracker' ice strength tests. *APOA Project* No. 1.

Croasdale, K. R. 1971. The 'Nutcracker' ice strength tests. *APOA Project* No. 9.

Croasdale, K. R. 1980. Ice forces on fixed, rigid structures. In *IAHR Working Group on Ice Forces on Structures, a State-of-the-Art Report*, ed. T. Carstens. *CRREL Special Report* No. 80–26, pp. 34–106.

Croasdale, K. R. 1984. The limited driving force approach to ice loads. In *Offshore Technology Conference 1984, Houston*, Paper No. OTC 4716, pp. 57–64.

Croasdale, K. R., Comfort, G., Frederking, R. M. W., Graham, B. W. and Lewis, E. L. (in press). *A pilot experiment to measure Arctic pack-ice driving forces*.

Croasdale, K. R., Morgenstern, N. R. and Nuttall, J. B. 1977. Indentation tests to investigate ice pressures on vertical piers. *Journal of Glaciology* **19**(81), 301–312.

Croasdale, K. R., Graham, B. W., Comfort, G. and Marcellus, R. W. 1986. Evaluation of ice pressure and strain sensors in a large test basin. In *IAHR 86*, Vol. 3, pp 213–225.

Danielewicz, B. W. and Cornett, S. 1984. Ice forces on Hans Island, 1983. *APOA Project* No. 202.

Danielewicz, B. W. and Metge, M. 1981. Ice forces on Hans Island, August, 1980. *APOA Project* No. 180.

Danielewicz, B. W. and Metge, M. 1982. Ice forces on Hans Island, 1981. *APOA Project* No. 181.

Dickins, D. F. and Wetzel, V. F. 1981. Multi-year pressure ridge study, Queen Elizabeth Islands. In *POAC 81*, Vol. 2, pp. 765–775.

Dieter, G. E. 1981. *Mechanical metallurgy*, second edition. McGraw-Hill.

Dunwoody, A. B. 1981. Hydrodynamic added mass derivation for multi-year ice. (In *Ice forces on Hans Island, August, 1980*, by B. W. Danielewicz and M. Metge, 1981, *APOA Project* No. 180, Appendix A).

Dunwoody, A. B. 1982 (unpubl.). *Energies dissipated on impact of extreme feature with a fixed offshore structure*. Dome Petroleum Ltd, January, 1982.

Dunwoody, A. B. 1983. The design ice island for impact against an offshore structure. In *Offshore Technology Conference 1983, Houston*, Paper No. OTC 4550, pp. 325–330.

Duval, P. and Le Gac, H. 1980. Does the permanent creep-rate of polycrystalline ice increase with crystal size? *Journal of Glaciology* **25**(91), pp. 151–157.

Edwards, R. Y., Tam, W. and Nawwar, A.M. (unpubl.). *Ship's hull as a test instrument.* Kanata: Arctec Canada Ltd.

Edwards, R. Y., Lewis, J. W., Wheaton, J. W. and Coburn, J. 1973. Full-scale and model tests of a Great Lakes icebreaker. *Transactions, Society of Naval Architects and Marine Engineers* **80,** 170–207.

El-Tahan, M. S. and El-Tahan, H. W. 1983. Forecast of iceberg ensemble drift. In *Offshore Technology Conference 1983, Houston,* Paper No. OTC 4460, pp. 151–158.

Engelbrektson, A. 1983. Ice force design of offshore structures in the light of experiences from the Baltic. In *Proceedings, Offshore Göteborg 83, Technology for the Arctic, Gothenburg, Sweden, 1–4 March 1983. Session (A3),* pp. 1–10.

Evans, A. G., Palmer, A. C., Goodman, D. J., Ashby, M. F., Hutchinson, J. W., Ponter, A. R. S. and Williams, G. J. 1984. Indentation spalling of edge-loaded ice sheets. In *IAHR 84,* Vol. 1, pp. 113–121.

Evers, K. U. and Wessels, E. 1986. Model test study of level ice forces on cylindrical multi-legged structures. In *Proceedings, Polartech '86 International Offshore and Navigation Conference and Exhibition, Helsinki, 27–30 October 1986,* Vol. 3, pp. 251–275.

Frankenstein, G. and Garner, R. 1967. Equations for determining the brine volume of sea ice from $-0.5°$ to $-22.9°C$. *Journal of Glaciology* **6**(48), 943–944.

Fransson, L. and Sandkvist, J. 1985. Brash ice shear properties – laboratory tests. In *POAC 85,* Vol. 1, pp. 75–87.

Frederking, R. M. W. 1977. Plane strain compressive strength of columnar-grained and granular-snow ice. *Journal of Glaciology* **18**(80), 505–516.

Frederking, R. M. W. 1980. Dynamic forces on an inclined structure. In *Proceedings of the IUTAM Symposium on the Physics and Mechanics of Ice, Copenhagen, August 1979.* Berlin: Springer-Verlag, pp. 104–116.

Frederking, R. M. W. and Gold, L. W. 1975. Experimental study of edge loading of ice plates. *Canadian Geotechnical Journal* **12,** 456–463.

Frederking, R. M. W., Sayed, M., Wessels, E., Child, A. J. and Bradford, D. 1984. Ice interaction with Adams Island, winter 1982–83. In *IAHR 84,* Vol. 3, pp. 187–201.

Frederking, R. M. W., Wessels, E., Maxwell, J. B., Prinsenberg, S. and Sayed, M. 1986a. Ice pressures and behaviour at Adams Island, winter 1983/84. *Canadian Journal of Civil Engineering* **13,** 140–149.

Frederking, R. M. W., Wessels, E., Maxwell, J. B., Prinsenberg, S. and Sayed, M. 1986b. Ice interaction with Adams Island, winter 1984/85. In *IAHR 86,* Vol. 3, pp. 127–143.

Freudenthal, A. M. and Geiringer, H. 1958. The mathematical theories of the inelastic continuum. In *Handbuch der Physik,* ed. S. Flügge, Springer-Verlag, Vol. 6, pp. 229–433.

Frost, H. J. and Ashby, M. F. 1982. *Deformation-mechanism maps.* Oxford: Pergamon Press.

Gaskill, H. S. and Rochester, J. 1982 (unpubl.). *Ocean currents derived from iceberg drift.* Department of Mathematics and Statistics. Memorial University of Newfoundland.

Geotechnical Resources Ltd 1986 (unpubl.). *Medium-scale multi-year ice indentation test program.* Calgary, Alberta: Geotechnical Resources Ltd.

Ghoneim, G. A. M. and Keinonen, A. J. 1983. Full-scale impact test of Canmar Kigoriak in thick ice. In *POAC 83,* Vol. 3, pp. 329–346.

Glen, J. W. 1955. The creep of polycrystalline ice. *Proceedings of the Royal Society (London),* Ser. A **228,** No. 1175, 519–538.

Glen, J. W. 1968. The effect of hydrogen disorder on dislocation movement and plastic deformation of ice. *Physics of Condensed Matter* **7,** 43–51.

Glen, I. F. and Blount, H. 1984. Measurement of ice impact pressures and loads onboard CCGS Louis St. Laurent. In *Proceedings of the 1984 ASME Symposium, New Orleans, 12–14 February 1984,* Vol. 3, pp. 246–252.

Glen, I. F. and Comfort, G. 1983. Ice impact pressure and load: investigation by laboratory experiments and ship trials. In *POAC 83,* Vol. 1, pp. 516–533.

Gold, L. W. 1972. The process of failure of columnar-grained ice. *Philosophical Magazine* **26,** 310–328.

Gold L. W. 1973. Activation energy for creep of columnar-grained ice. In *Physics and chemistry of ice,* ed. E. Whalley, S. J. Jones and L. W. Gold. Ottawa, Royal Society of Canada, pp. 362–364.

Gold, L. W. 1977. Engineering properties of freshwater ice. *Journal of Glaciology* **19**(81), 197–212.

Gold, L. W. and Krausz, A. S. 1971. Investigation of the mechanical properties of St. Lawrence River sea ice. *Canadian Geotechnical Journal* **8**(2), 163–169.

Goodman, A. M. and Goodall, I. W. 1978. Use of existing steel data in design for creep. In *Creep of engineering materials,* ed. C. D. Pomeroy. London: Mechanical Engineering Publications Ltd, pp. 127–151.

Goodman, D. J. 1979. Critical stress intensity factor (K_{IC}) measurements at high loading rates for polycrystalline ice. In *Proceedings of the IUTAM Symposium on the Physics and Mechanics of Ice, Copenhagen, August 1979.* Berlin: Springer-Verlag, pp. 129–146.

Goodman, D. J. 1980. Surface strain measuring instruments for use on sea ice. *Workshop on Sea Ice Field Measurements, St. John's, Newfoundland, C-CORE Report* No. 80–21, pp. 75–96.

Goodman, D. J., Frost, H. J. and Ashby, M. F. 1981. The

plasticity of polycrystalline ice. *Philosophical Magazine* **43**(3), 665–695.

Gow, A. J., Ackley, S. F., Weeks, W. F. and Govoni, J. W. 1982. Physical and structural characteristics of Antarctic sea ice. *Annals of Glaciology* **3**, 113–117.

Graham, B. W., Chabot, L.G. and Pilkington, G. R. 1983. Ice load sensors for offshore Arctic structures. In *POAC 83*, Vol. 4, pp. 547–562.

Gulf 1986 (unpubl.). *Proposed analysis of dynamic horizontal ice loading on an offshore structure*. Gulf Canada Resources Inc. Joint Industry Project Proposal.

Gumbel, E. J. 1958. *Statistics of extremes*. New York: Columbia University Press.

Gustajtis, K. A. 1979. Iceberg population distribution study in the Labrador Sea. *C-CORE Publication 79–8*, July 1979. St. John's, Newfoundland.

Hall, R. T. 1980. AIDJEX modeling group studies involving remote sensing data. In *Sea ice processes and models*, ed. R. S. Pritchard. University of Washington Press, pp. 151–162.

Hallam, S. D. 1986. The role of fracture in limiting ice forces. In *IAHR 86*, Vol. 2, pp. 287–319.

Hallam, S. D., Duval, P. and Ashby, M. F. 1986. A study of cracks in polycrystalline ice under uniaxial compression. In *Proceedings of the 7th Symposium on the Physics and Chemistry of Ice, Grenoble, 1–5 September 1986*. Published in *Journal de Physique*, Colloque C1, Supplément au No. 3, Tom. 48 (March 1987), pp. 303–311.

Hallam, S. D., Jones, N. and Howard, M. W. 1987. The effect of sub-surface irregularities on the strength of multi-year ice. In *Proceedings of the 6th International Symposium on Offshore Mechanics and Arctic Engineering (OMAE), Houston, 1–6 March 1987*, Vol. 4, pp. 235–238.

Hallam, S. D. and Nadreau, J. P. 1987 (in press). Failure maps for ice. In *POAC 87* (preprint).

Häusler, F. U. 1981. Multi-axial compressive strength tests on saline ice with brush-type loading platens. In *IAHR 81*, pp. 526–539.

Hawkes, I. and Mellor, M. 1972. Deformation and fracture of ice under uniaxial stress. *Journal of Glaciology* **11**(61), 103–131.

Hetenyi, M. 1946. *Beams on elastic foundation*. University of Michigan Press.

Hibler, W. D. 1980a Modelling a variable thickness ice cover. *Monthly Weather Review* **108**(12), 1943–1973.

Hibler, W. D. 1980b. Modeling pack ice as a viscous–plastic continuum: some preliminary results. In *Sea ice processes and models*, ed. R. S. Pritchard. University of Washington Press, pp. 163–176.

Hirayama, K., Schwarz, J. and Wu, H. C. 1974. An investigation of ice forces on vertical structures. *IIHR Report* No. 158. University of Iowa, Iowa City.

Hobbs, P. V. 1974. *Ice physics*. Oxford: Clarendon Press.

Hoikkanen, J. 1984. Measurements and analysis of ice pressure against a structure in level ice and in pressure ridges. In *IAHR 84*, Vol. 3, pp. 161–174.

Hotzel, I. S. and Miller, J. D. 1983. Icebergs: their physical dimensions and application of measured data. *Annals of Glaciology* **4**, 116–123.

Hutchinson, J. W. 1979. Recent developments in non-linear fracture mechanics. In *Proceedings of the 7th Canadian Congress of Applied Mechanics, Sherbrooke, 27 May–1 June, 1979*, pp. 24–36.

Inoue, M. and Koma, N. 1985. Field indentation tests on cylindrical structures. In *POAC 85*, Vol. 2, pp. 555–568.

Iyer, S. H. 1983. Size effects in ice and their influence on the structural design of offshore structures. In *POAC 83*, Vol. 3, pp. 414–432.

Jacka, T. H. 1984. The time and strain required for development of minimum strain-rates in ice. *Cold Regions Science and Technology* **8**, 261–268.

Jefferies, M. G., Stewart, H. R., Thomson, R. A. and Rogers, B. T. 1985. Molikpaq deployment at Tarsiut P-45. In *Proceedings of the ASCE Arctic '85 Conference, San Francisco, 25–27 March 1985*, ed. F. L. Bennett and J. L. Machemehl, pp. 1–27.

Jellinek, H. H. G. 1958. The influence of imperfections on the strength of ice. *Proceedings of the Physical Society (London)* **71**(5), 797–814.

Johnson, J. B. 1983. A surface integral method for determining ice loads on offshore structures from in situ measurements. *Annals of Glaciology* **4**, 124–128.

Johnson, J. B., Cox, G. F. N. and Tucker, W. B. 1985. Kadluk ice stress measurement programme. In *POAC 85*, Vol. 1, pp. 88–100.

Jones, S. J. 1982. The confined compressive strength of polycrystalline ice. *Journal of Glaciology* **28**(98), 171–177.

Jones, P.D., Wigley, T. M. L. and Wright, P. B. 1986. Global temperature variations between 1861 and 1984. *Nature* **322**, 430–434.

Jordaan, I. J. 1986. Numerical and finite element techniques in calculation of ice–structure interaction. In *IAHR 86*, Vol. 2, pp. 405–440.

Karas, J. H. W., Kelly, P. M. and Wigzell, M. 1984 (unpubl.). The operating season: Harrison Bay, Norton Sound, off Newfoundland and the Labrador Sea. *BP Petroleum Development Ltd., University of East Anglia Sea Ice Distribution Project, Report* No. CRU84–14.

Kelly, P. M., Cherry, B. S. G. and Lough, J. M. 1982 (unpubl.). Atlas of Northern Hemisphere ice conditions. *BP Petroleum Development Ltd., University of East Anglia Sea Ice Distribution Project, Report* No. CRU82–1.

Kendall, K. 1978. Complexities of compression failure. *Proceedings of the Royal Society (London)*, Ser. A **361**, 245–263.

Kerr, A. D. 1978. On the determination of horizontal forces a floating ice plate exerts on a structure. *CRREL Report* No. 78–15.

Kerr, A. D. 1980. On the buckling force of floating ice

plates. In *Proceedings of the IUTAM Symposium on the Physics and Mechanics of Ice, Copenhagen, August 1979*. Berlin: Springer-Verlag, pp. 163–178.

Kerr, A. D. and Palmer, W. T. 1970. The deformations and stresses in floating ice plates. *National Aeronautics and Space Administration, Report* No. S-70-3, NYU-AA-70-17.

Knott, J. F. 1973. *Fundamentals of fracture mechanics*. London: Butterworth.

Koch, L. 1945. The East Greenland ice. *Meddelelser om Grønland* **130**(3).

Korzhavin, K. N. 1962. Action of ice on engineering structures. *CRREL Draft Translation* No. 260.

Kovacs, A. 1979. Icebergs: an overview. *CRREL Special Report* No. 79–21.

Kovacs, A. 1983. Characteristics of multi-year pressure ridges, In *POAC 83*, Vol. 3, pp. 173–182.

Kovacs, A. and Mellor, M. 1974. Sea ice morphology and ice as a geologic agent in the southern Beaufort Sea. In *The coast and shelf of the Beaufort Sea*, ed. J. C. Reed and J. E. Slater. San Francisco: Arctic Institute of North America, pp. 113–161.

Kovacs, A., Sodhi, D. S. and Cox, G. F. N. 1982. Bering Strait sea ice and the Fairway Rock icefoot. *CRREL Report* No. 82–31.

Kry, P. R. 1977. Ice rubble fields in the vicinity of artificial islands. In *POAC 77*, Vol. 1, pp. 200–211.

Kry, P. R. 1979a High aspect ratio crushing tests. *APOA Project* No. 93.

Kry, P. R. 1979b. Stress distribution during continuous crushing of ice. *APOA Project* No. 148.

Kry, P. R. 1980a. Ice forces on wide structures. *Canadian Geotechnical Journal* **17**(1), 97–113.

Kry, P. R. 1980b. A statistical prediction of effective ice crushing stresses on wide structures. In *Proceedings of the ASCE Convention and Exposition, Portland, 14–18 April 1980, Preprint* No. 80–084.

Kujala, P. and Vuorio, J. 1985. On the statistical nature of the ice-induced pressures measured on board I. B. Sisu. In *POAC 85*, Vol. 2, pp. 823–837.

LaBelle, J. C., Wise, J. L., Voelker, R. P., Schulze, R. H. and Wohl, G. M. 1983. *Alaska marine atlas*. Arctic Environmental Information and Data Centre, University of Alaska.

Lake, R. A. and Lewis, E. L. 1970. Salt rejection by sea ice during growth. *Journal of Geophysical Research* **75**(3), 583–597.

Langhorne, P. J. 1983. Laboratory experiments on crystal orientation in NaCl ice. *Annals of Glaciology* **4**, 163–169.

Langleben, M. P. 1980. Water drag coefficient at AIDJEX, Station Caribou. In *Sea ice processes and models*, ed. R. S. Pritchard. University of Washington Press, pp. 464–471.

Lavrov, V. V. 1973. Scale effects as indications of icebreaking mechanism. In *Studies in ice physics and ice engineering*. Yakovlev.

Lawn, B. R., Evans, A. G. and Marshall, D. B. 1980. Elastic/plastic damage in ceramics: the median/radial crack system. *American Ceramics Society Bulletin* **63**(9–60), 574–581.

Lee, R. W. and Schulson, E. M. 1986. The strength and ductility of ice under tension. In *Proceedings of the 5th International Symposium on Offshore Mechanics and Arctic Engineering (OMAE) Tokyo, 13–18 April 1986*, Vol. 4, pp. 298–302.

Lee, J., Ralston, T. D. and Petrie, D. H. 1986. Full-thickness sea ice strength tests. In *IAHR 86*, Vol. 1, pp. 293–306.

Lewis, J. W. and Edwards, R. Y. 1970. Methods for predicting icebreaking and ice resistance characteristics of icebreakers. *Transactions, Society of Naval Architects and Marine Engineers* **78**, 213–249.

Lewis, E. L. and Weeks, W. F. 1971. Sea ice: some polar contrasts. In *Symposium on Antarctic ice and water masses, Tokyo, 19 September 1970*, ed. G. Deacon. Scientific Committee on Antarctic Research, pp. 23–34.

Lindsay, D. G. 1976. *Sea ice atlas of Arctic Canada, 1961–1968*. Ottawa: Department of Energy, Mines and Resources.

Lindsay, D. G. 1977. *Sea ice atlas of Arctic Canada, 1969–1974*. Ottawa: Department of Energy, Mines and Resources.

Lindsay, D. G. 1981. *Sea ice atlas of Arctic Canada, 1975–1978*. Ottawa: Department of Energy, Mines and Resources.

Lipsett, A. W. and Gerard, R. 1980. Field measurements of ice forces on bridge piers, 1973–1979. *Alberta Research Council Report* No. SWE 80–3, Edmonton.

Liu, H. W. and Miller K. J. 1979. Fracture toughness of freshwater ice. *Journal of Glaciology* **22**(86), 135–143.

Määttänen, M. 1981. Experiences with vibration isolated lighthouses. In *POAC 81*, Vol. 1, pp. 491–501.

Määttänen, M. 1983. Dynamic ice–structure interaction during continuous crushing. *CRREL Report* No. 83–5.

McBride, E. J. 1981 (unpubl.). *Ice stress sensor program*. Plano, Texas: ARCO Oil and Gas Company.

McGonigal, D. 1981. Ice stress panel results, Issungnak 1980. *APOA Project* No. 171–1, Vol. 5.

MacKenzie, A. C. 1968. On the use of a single uniaxial test to estimate deformation rates in some structures undergoing creep. *International Journal of Mechanical Sciences* **10**, 441–453.

McLaren, A. S., Wadhams, P. and Weintraub, R. 1984. The sea ice topography of M'Clure Strait in winter and summer of 1960 from submarine profiles. *Arctic* **37** (2), 110–120.

Mancini, C. V., Dowse, B. E. W. and Chevallier, J. -M. 1983. Caisson retained island for Canadian Beaufort sea – geotechnical design and construction considerations. In *Offshore Technology Conference 1983, Houston*, Paper No. OTC 4581, Vol. 4, pp. 17–22.

Mangor, K. and Zorn, R. 1983. Iceberg conditions offshore Greenland. *Iceberg Research* **4**, 4–20.

Marcellus, R. W. and Morrison, T. B. 1982. Ice design statistics for the Canadian Beaufort Sea. *Beaufort Sea Environmental Impact Statement (EIS) 1982, Report* No. RWI 15. Calgary: Pallister Resource Management Ltd.

Markham, W. E. 1981. *Ice atlas, Canadian Arctic Waterways.* Canada: Ministry of Supply and Services.

Marko, J. R., Fissel, D. B. and Birch, J. R. 1986. Physical approaches to iceberg severity prediction. *Environmental Studies Revolving Funds Report* No. 038. Ottawa.

Mellor, M. 1983. Mechanical behaviour of sea ice. *CRREL Monograph* No. 83–1.

Mellor, M. and Cole, D. M. 1982. Deformation and failure of ice under constant strain-rate. *Cold Regions Science and Technology* **5**, 201–219.

Metge, M. 1976. Ice conditions and ice defence at Netserk B-44 and Adgo P-25 during the winter of 1974–75. *APOA Project* No. 104–1.

Metge, M., Strilchuk, A. and Trofimenkoff, P. 1975. On recording stress in ice. In *IAHR 75*, pp. 459–468.

Metge, M., Pilkington, G. R., Strandberg, A. G. and Blanchet, D. 1983. A new sensor for measuring ice forces on structures: laboratory tests and field experience. In *POAC 83*, Vol. 4, pp. 790–801.

Michel, B. 1978. The strength of polycrystalline ice. *Canadian Journal of Civil Engineering* **5**, 285–300.

Michel, B. 1979. *Ice mechanics.* Québec: Les presses de l'Université Laval.

Michel, B. and Blanchet, D. 1983. Indentation of an S2 floating ice sheet in the brittle range. *Annals of Glaciology* **4**, 180–187.

Michel, B. and Paradis, M. 1976. Analyse statistique des lois de fluage secondaire de la glace de rivière et de lac. *Université Laval, Faculté des Sciences et de Génie. Département de Génie Civil, Section Mécanique des Glaces, Rapport* GCS-76-02.

Michel, B. and Toussaint, N. 1977. Mechanisms and theory of indentation of ice plates. *Journal of Glaciology* **19**(81), 285–300.

Miller, K. J. 1979. The application of fracture mechanics to ice problems. In *Proceedings of the IUTAM Symposium on the Physics and mechanics of Ice, Copenhagen, August 1979.* Berlin: Springer-Verlag, pp. 265–277.

Miller, T. W., McLatchie, A., Hedley, R. and Morris, G. 1974. Ice crushing tests. *APOA Project* No. 66.

Morgenstern, N. R. and Nuttall, J. B. 1971. The interpretation of ice strength from in-situ indentation tests. *APOA Project* No. 16.

Nakajima, H., Koma, N. and Inoue, M. 1981. The ice force acting on a cylindrical pile. In *POAC 81*, Vol. 1, pp. 517–525.

Nakawo, M. 1983. Measurements on air porosity of sea ice. *Annals of Glaciology* **4**, 204–208.

Nakawo, M. and Sinha, N. K. 1981. Growth rate and salinity profile of first-year ice in the high Arctic. *Journal of Glaciology* **27**(96), 315–330.

Nakawo, M. and Sinha, N. K. 1984. A note on brine layer spacing of first-year ice. *Atmosphere–Ocean* **22**(2), 193–206.

Naval Oceanography Command 1985. Sea ice climatic atlas: Volume I. Antarctic. *Publication Navair 50-1C-540*, Ashville, USA.

Nelson, R. 1975. Internal stress measurements in ice sheets using embedded load cells. In *POAC 75*, pp. 361–373.

Nelson, R. D. and Sackinger, W. M. 1976. Ice stress measurements at Adgo and Netserk Islands 1974–75. *APOA Project* No. 104–3.

Nemat-Nasser, S. and Horii, H. 1982. Compression induced non-planar crack extension with application to splitting, exfoliation and rockburst. *Journal of Geophysical Research* **87**(B8), 6805–6821.

Neshyba, S. 1980. On the size distribution of Antarctic icebergs. *Cold Regions Science and Technology* **1**, 241–248.

Nevel, D. E. 1972. Ultimate failure of floating ice sheet. In *IAHR 72*, Vol. 2, pp. 17–22.

Nevel, D. E. 1986 (in press). Iceberg impact forces. In *IAHR 86* (preprint).

Osterkamp, T. E. and Gosink, J. P. 1984. Observations and analyses of sediment-laden ice. In *The Alaskan Beaufort Sea, ecosystems and environments*, ed. P. W. Barnes, D. M. Schell, and E. Reimnitz. Academic Press, pp. 73–93.

Overgaard, S., Wadhams, P. and Leppäranta, M. 1983. Ice properties in the Greenland and Barents Sea during summer. *Journal of Glaciology* **29**(101), 142–164.

Palmer, A. C., Goodman, D. J., Ashby, M. F., Evans, A. G., Hutchinson, J. W. and Ponter, A. R. S. 1983. Fracture and its role in determining ice forces on offshore structures. *Annals of Glaciology* **4**, 216–221.

Parmerter, R. R. and Coon, M. D. 1972. Model of pressure ridge formation in sea ice. *Journal of Geophysical Research* **77**(33), 6565–6575.

Paterson, W. S. B. 1981. *The physics of glaciers*, second edition, Oxford: Pergamon Press.

Pease, C. H. 1980. Eastern Bering Sea ice processes. *Monthly Weather Review* **108**(12), 2015–2023.

Petrie, D. H. and Poplin, J. P. 1986. Comparison of small-scale and large-scale sea ice strengths. In *IAHR 86*, Vol. 1, pp. 265–277.

Peyton, H. R. 1966. Sea ice strength. *University of Alaska, Geophysical Institute, Report* No. UAG R-182.

Pilkington, G. R., Blanchet, D. and Metge, M. 1983. Full-scale measurements of ice forces on an artificial island. In *POAC 83*, Vol. 4, pp. 818–834.

Pilkington, G. R., Wright, B. D., Dixit, B. C., Woolner, K. S. and O'Dell, B. D. 1986. A review of Kulluk's performance after three years operations in the Beaufort Sea. In *IAHR 86*, Vol. 3, pp. 145–159.

Pomeroy, C. D. (ed.) 1978. *Creep of engineering materials.* London: Mechanical Engineering Publications Ltd.

Ponter, A. R. S. and Brown, P. R. 1978. The finite element solution of rapid cycling creep problems. *International Journal of Numerical Methods in Engineering* **12,** 1001–1024.

Ponter, A. R. S., Palmer, A. C., Goodman, D. J., Ashby, M. F., Evans, A. G. and Hutchinson, J. W. 1983. The force exerted by a moving ice sheet on an offshore structure. Part 1. The creep mode. *Cold Regions Science and Technology* **8,** 109–118.

Poplin, J. P., Ralston, T. D. and St. Lawrence, W. 1987. A thermal ice drill for profiling thick multi-year ice. *Cold Regions Science and Technology* **14,** 1–11.

Pritchard, R. S. 1977. The effect of strength on simulations of sea-ice dynamics. In *POAC 77*, pp. 494–505.

Pritchard, R. S. 1980. A simulation of nearshore winter ice dynamics in the Beaufort Sea. In *Sea ice processes and models*, ed. R. S. Pritchard. University of Washington Press, pp. 49–61.

Prodanovic, A. 1979. Two-dimensional limit analysis of ice failure described by a quadratic yield criterion. In *Recent advances in engineering science*, ed. R. L. Sierakowski. Gainsville: University of Florida Press, pp. 515–519.

Ralston, T. D. 1977a. Ice force design considerations for conical offshore structures. In *POAC 77*, pp. 741–752.

Ralston, T. D. 1977b. Yield and plastic deformation in ice crushing failure. *ICSI/AIDJEX Symposium on Sea Ice Processes and Models, Seattle, Washington.* (Also in *Sea ice processes and models*, ed. R. S. Pritchard, 1980, University of Washington press, pp. 234–245.)

Ralston, T. D. 1978. An analysis of ice sheet indentation. In *IAHR 78*, Part 1, pp. 13–31.

Ralston, T. D. 1979. Sea ice loads. In *Technical Seminar on Alaskan Beaufort Sea gravel island design.* Exxon Company, USA.

Reimnitz, E. and Kempema, E. W. 1984. Pack ice interaction with stamukhi shoal, Beaufort Sea, Alaska. In *The Alaskan Beaufort Sea, ecosystems and environments*, ed. P. W. Barnes, D. M. Schell, and E. Reimnitz. Academic Press, pp. 159–183.

Richter-Menge, J. A., Cox, G. F. N., Perron, N., Durrell, G. and Bosworth, H. W. 1986. Triaxial testing of first-year sea ice. *CRREL Report* No. 86–16.

Riedel, H. and Rice, J. R. 1980. Tensile cracks in creeping solids. In *Proceedings of the 12th Conference on Fracture Mechanics, American Society for Testing and Materials, Special Technical Publication* No. 7000, pp. 112–130.

Rogers, B. T., Hardy, M. D., Neth, V. W. and Metge, M. 1986. Performance monitoring of the Molikpaq while deployed at Tarsiut P-45. In *Proceedings of the 3rd Canadian Marine Geotechnical Conference, St. John's, Newfoundland, 11–13 June 1986.*

Rose, E. 1947. Thrust exerted by expanding ice sheets. *Transactions of the American Society of Civil Engineers* **112,** 871–885.

Rothrock, D. A. 1975. The energetics of the plastic deformation of pack ice by ridging. *Journal of Geophysical Research* **80**(33), 4514–4519.

Rothrock, D. A. and Thorndike, A. S. 1984. Measuring the floe size distribution. *Journal of Geophysical Research* **89**(C4), 6477–6486.

Sackinger, W. M., Serson, H. V., Jeffries, M. O., Shoemaker, H. D. and Miin-Huey Yan 1985. Ice island generation and trajectories north of Ellesmere Island, Canada. In *POAC 85*, Vol. 2, pp. 1009–1040.

Sackinger, W. M., Jeffries, M. O., Miin-Huey Yan and Tippens, H. 1986. Ice islands. *Ice*, No. 80, pp. 18.

Sadowsky, M. 1928. Zweidimensionale Probleme der Elastizitätstheorie. *Zeitschrift für Angewandte Mathematik und Mechanik* **8**(2), 107–121.

Saeki, H., Nomura, T. and Ozaki, A. 1978. Experimental study on the testing methods of strength and mechanical properties for sea ice. In *IAHR 78*, Part 1, pp. 135–150.

St. John, J. W. and Daley, C. 1984. Shipboard measurement of ice pressures in the Bering, Chukchi and Beaufort Seas. In *Proceedings of the 1984 ASME Symposium, New Orleans, 12–14 February 1984*, Vol. 3, pp. 260–266.

St. John, J. W., Daley, C. and Blount, H. 1984. Ice loads and ship response to ice. *Arctec Report* No. SR-1291, Arctec Inc., Columbia, and Arctec Canada Ltd., Kanata.

Sanderson, T. J. O. 1978. Thermal stresses near the surface of a glacier. *Journal of Glaciology* **20**(83), 257–283.

Sanderson, T. J. O. 1982 (unpubl.). *BP total ice load programme, Tarsiut 1983: how to use strain-rate data to calculate forces exerted on island.* Internal report, BP Petroleum Development Ltd.

Sanderson, T. J. O. 1984a Theoretical and measured ice forces on wide structures. In *IAHR 84*, Vol. 4, pp. 151–207.

Sanderson, T. J. O. 1984b. Thermal ice forces against isolated structures. In *IAHR 84*, Vol. 4, pp. 289–299.

Sanderson, T. J. O., Child, A. J., Duckworth, R. and Westermann, P. H. 1983. BP total ice load programme, Tarsiut Island, 1983. *APOA Project* No. 198.

Sanderson, T. J. O. and Child, A. J. 1986. Ice loads on offshore structures: the transition from creep to fracture. *Cold Regions Science and Technology* **12,** 157–161.

Sayed, M., Frederking, R. M. W. and Croasdale, K. R. 1986. Ice stress measurements in a rubble field surrounding a caisson-retained island. In *Ice technology*, ed. T. K. S. Murphy, J. J. Connor and C. A. Brebbia. Springer-Verlag.

Schulson, E. M., Lim, P. N. and Lee, R. W. 1984. A brittle to ductile transition in ice under tension. *Philosopical Magazine*, A **49**(3), 353–363.

Schwarz, J. and Weeks, W. F. 1977. Engineering properties of sea ice. *Journal of Glaciology* **19**(81), 499–531.

Semeniuk, A. 1977. Ice pressure measurements at

Arnak L-30 and Kannerk G-42. *APOA Project* No. 122–1.

Seng-Kiong, T. and Shyam Sunder, S. 1985. Constitutive modeling of sea ice with applications to indentation problems. *CSEOE Research Report* No. 3, Massachusetts Institute of Technology, Cambridge, Massachusetts.

Sih, G. C. M. 1973. *Handbook of stress intensity factors for research workers and engineers.* Institute of Fracture and Solid Mechanics, Leigh University, Bethlehem, Pennsylvania.

Sim, R. G. 1968 (unpubl.). *The creep of structures.* Ph.D. thesis, University of Cambridge.

Sinha, N. K. 1978a. Observations of basal dislocations in ice by etching and replication. *Journal of Glaciology* **21**(85), 385–395.

Sinha, N. K. 1978b. Rheology of columnar-grained ice. *Experimental Mechanics* **18**(12), 464–470.

Sinha, N. K. 1979. Grain boundary sliding in polycrystalline materials. *Philosophical Magazine*, A **40**(6), 825–842.

Sinha, N. K. 1982. Delayed elastic strain criterion for first cracks in ice. In *Proceedings of the Symposium on Deformation and Failure of Granular Materials, Delft, 31 August–3 September 1982. International Union of Theoretical and Applied Mechanics*, pp. 323–330. Rotterdam: Balkema.

Sinha N. K. 1983a. Creep model of ice for monotonically increasing stress. *Cold Regions Science and Technology* **8**, 25–33.

Sinha, N. K. 1983b. Field test on rate sensitivity of vertical strength and deformation of first-year columnar-grained ice. In *POAC 83*, Vol. 1, pp. 231–242.

Sjölind, S. -G. 1984. Viscoeleastic buckling of beams and plates on elastic foundation. In *IAHR 84*, Vol. 1, pp. 63–72.

Sjölind, S. -G. 1985. Viscoelastic buckling analysis of floating ice sheets. *Cold Regions Science and Technology* **11**, 241–246.

Slomski, S. and Vivatrat, V. 1983. Selection of design ice pressures and application to impact load prediction. In *POAC 83*, Vol. 2, pp. 909–919.

Smith, M. and Rigby, B. 1981. Distribution of polynyas in the Canadian Arctic. In *Polynyas in the Canadian Arctic*, ed. I. Stirling and H. Cleator. Canadian Wildlife Service, Occasional Paper No. 45.

Spedding, L. G. 1977. Ice island count Southern Beaufort Sea, 1976. *APOA Project* No. 99–3.

Spedding, L. G. 1983. A large landfast ice movement. In *POAC 83*, Vol. 3, pp. 203–213.

Stephens, M. A. 1974. EDF statistics for goodness of fit and some comparisons. *Journal of the American Statistical Association* **69**, 730–737.

Stirling, I. and Cleator, H. (ed.) 1981. Polynyas in the Canadian Arctic, *Canadian Wildlife Service, Occasional Paper* No. 45.

Strilchuk, A. R. 1977. Ice pressure measurements, Netserk F-40, 1975–76. *APOA Project* No. 105–1.

Stringer, W. J. 1986. Summertime floe size distribution. *Ice*, No. 80, 18–19.

Tabata, T. 1966. Studies of the mechanical properties of sea ice: X. In *Physics of Snow and Ice: International Conference on Low Temperature Science 1966, Hokkaido University*, ed. H. Oura, pp. 481–497.

Taylor, T. P. 1973. Ice crushing tests 1973. *APOA Project* No. 52.

Templeton, J. S. 1979. Measurement of sea ice pressures. In *POAC 79*, Vol. 1, pp. 73–87.

Templeton, J. S. 1981. Analysis for embedded ice pressure sensor. *Journal of Energy Resources Technology* **103**(1), 87–95.

Thouless, M. D., Evans, A. G. and Ashby, M. F. 1985 (unpubl.). *Edge spalling of brittle plates.* Engineering Department, University of Cambridge.

Timco, G. W. 1984. Ice forces on structures: physical modelling techniques. In *IAHR 84*, Vol. 4, pp. 117–150.

Timco, G. W. 1986a. Indentation and penetration of edge-loaded freshwater ice sheets in the brittle range. In *Proceedings of the 5th International Symposium on Offshore Mechanics and Arctic Engineering (OMAE) Tokyo, 13–18 April 1986*, Vol. 4, pp. 444–452.

Timco, G. W. 1986b. Ice forces on multi-legged structures. In *IAHR 86*, Vol. 2, pp. 321–337.

Timoshenko, S. and Goodier, J. N. 1951. *Theory of elasticity*, second edition. McGraw-Hill.

Tomin, M. J., Cheung, M., Jordaan, I. J. and Cormeau, A. 1986. Analysis of failure modes and damage processes of freshwater ice in indentation tests. In *Proceedings of the 5th International Symposium on Offshore Mechanics and Arctic Engineering (OMAE) Tokyo, 13–18 April 1986*, Vol. 4, pp. 453–460.

Tucker, W. B. and Hibler, W. D. 1981. Preliminary results of ice modeling in the East Greenland area. In *POAC 81*, Vol. 2, pp. 867–878.

Urabe, N. and Yoshitake, A. 1981. Strain-rate dependent fracture toughness (K_{IC}) of pure ice and sea ice. In *IAHR 81*, Vol. 2, pp. 410–420.

Urabe, N. 1980. Fracture toughness of sea ice. *Cold Regions Science and Technology* **3**, 29–37.

US Naval Oceanographic Office 1957. Oceanographic atlas of the polar seas, Part 1, Antarctic. *H.O. Publication* No. 705, Part 1.

US Navy Hydrographic Office 1958. Oceanographic atlas of the polar seas, Part 2, Arctic. *H.O. Publication* No. 705, Part 2.

Van Dusen, M. S. 1929. Thermal conductivity of nonmetallic solids. In *International critical tables of numerical data, physics, chemistry and technology*, ed. E. W. Washburn. New York: McGraw-Hill. Vol. 5, pp. 216–217.

Vaudrey, K. D. 1977. Ice engineering – study of related properties of floating sea-ice and summary of elastic and viscoelastic analyses. *Civil Engineering Laboratory, Port Huenemen, California 93043, Technical Report* No. R860.

Wadhams, P. 1980. Ice characteristics in the seasonal sea ice zone. *Cold Regions Science and Technology* **2**, 37–87.

Wadhams, P. 1983. Arctic sea ice morphology and its measurement. *Journal of the Society for Underwater Technology* **9**(2), 1–12.

Wadhams, R. and Horne, R. J. 1980. An analysis of ice profiles obtained by submarine sonar in the Beaufort Sea. *Journal of Glaciology* **25** (93), 401–424.

Wadhams, P., McLaren, A. S. and Weintraub, R. 1985. Ice thickness distribution in Davis Strait in February from submarine sonar profiles. *Journal of Geophysical Research* **90**(C1), 1069–1077.

Walsh, J.E. 1978. A data set on Northern Hemisphere sea ice extent, 1953–76. *World Data Centre A for Glaciology, University of Colorado, Boulder, Glaciological Data, Report* GD-2, pp. 49–51.

Wang, Y. S. 1979a. Sea ice properties. In *Technical Seminar on Alaskan Beaufort Sea gravel island design*. Exxon Company, USA.

Wang, Y. S. 1979b. Crystallographic studies and strength tests of field ice in the Alaskan Beaufort Sea. In *POAC 79*, pp. 651–655.

Wang, Q. 1983. A tentative view on ice load applied on jacket platforms in the Bo-hai Gulf. In *POAC 83*, Vol. 2, pp. 930–939.

Weeks, W. F. and Ackley, S .F. 1982. The growth, structure and properties of sea ice. *CRREL Monograph* No. 82–1.

Weeks, W. F. and Assur, A. 1967. The mechanical properties of sea ice. *CRREL Monograph* II-C3.

Weeks, W. F. and Gow, A. J. 1978. Preferred crystal orientations along the margins of the Arctic Ocean. *Journal of Geophysical Research* **84**(C10), 5105–5121.

Weeks, W. F. and Gow, A. J. 1979. Crystal alignments in the fast ice of Arctic Alaska. *Journal of Geophysical Research* **85**(C2), 1137–1146.

Weeks, W. F. and Mellor, M. 1984. Mechanical properties of ice in the Arctic seas. In *Arctic technology and policy. Proceedings of the 2nd Annual MIT Sea Grant College Program Lecture and Seminar Series*, ed. I. Dyer and C. Chryssostomidis. Washington, Hemisphere, pp. 235–259.

Weeks, W. F., Tucker, W. B., Frank, M. and Fungcharoen, S. 1980. Characterisation of surface roughness and floe geometry of sea ice over the continental shelves of the Beaufort and Chukchi Seas. In *Sea ice processes and models*, ed. R. S. Pritchard. University of Washington Press, pp. 300–312.

Weibull, W. 1951. A statistical function of wide applicability. *Journal of Applied Mechanics* **18**, 293–297.

Westermann, P. H. 1984 (unpubl.). Charts of sea ice properties in North American Arctic waters. *BP Research Centre, Sunbury-on-Thames, Technical Memorandum* No. 134100.

Wetzel, V. F. 1981. Statistical study of late winter ice thickness distribution in the Arctic Islands, 1979–80. *APOA Project* No. 174.

Wetzel, V. F., Atwater, R. K. and Huta, T. E. 1974. Arctic ice movement and environmental data stations. In *The coast and shelf of the Beaufort Sea*, ed. J. C. Reed and J. F. Sater. Arctic Institute of North America, pp. 269–284.

Wierzbicki, T. 1985. Spalling and buckling of ice sheets. In *Proceedings of the ASCE Arctic '85 Conference, San Francisco, 25–27 March 1985*, ed. F. L. Bennett and J. L. Machemehl, pp. 953–961.

Williams, R. 1986. Amoco probes Beaufort from spray ice island. *Oil and Gas Journal*, April 7, 1986, pp. 42–44.

World Meteorological Organisation 1970. WMO sea-ice nomenclature. *WMO/OMM/BMO* No. 259. Geneva, Switzerland.

Wright, B., Pilkington, G. R., Woolner, K. S. and Wright, W. H. 1986. Winter ice interactions with an offshore structure. In *IAHR 86*, Vol. 3, pp. 49–73.

Zabilansky, L., Nevel, D. E. and Haynes, F. D. 1975. Ice forces on model structures. *Canadian Journal of Civil Engineering* **2**(4), 400–407.

Zubov, N. N. 1945. *Arctic ice*. Izdatel'stvo Glavsermorputi, Moscow. English translation by US Navy Oceanographic Office/American Meteorological Society.

Index

activation energy 78–80
  columnar ice 79–80, 82
  granular ice, 79, 84
Adams Island 121
added mass 110, 123, 185–186
adfreeze 117, 140–142, 144, 205–206
Adgo Island (P-25) 114, 117, 211
AIDJEX 44–45
albedo 18
Amerk Island (O-09) 114, 122
amplification (*see*: dynamic amplification)
anchor ice 11
anisotropy 72, 82–83, 85–87, 158
Antarctica 32, 64–67, 69
  icebergs 66–67
  ice characteristics 10, 11, 66
  Ross Ice Shelf 24, 67
  Ross Sea 64
  sea-ice distribution 64–65
  Weddell Sea 11, 64, 66
  (*see also*: ice conditions)
Antarctic Convergence 66
AOGA (Alaska Oil and Gas Association) 104
APOA (Arctic Petroleum Operators Association) 104, 236
Arctic (general ice conditions) 32–65, 68
Arctic Islands (*see*: Canadian Arctic Islands)
Arctic Ocean 35–39
  (*see also*: ice conditions)
area density 219–220
Arrhenius Law 78, 81
Arnak Island (L-30) 119–120
artificial islands (*see*: islands)
aspect ratio 100–101, 142–144, 154–155
atlases
  Antarctic 69
  Arctic 68
Atlantic Drift 55
axysymmetric properties 9, 82
Azores 33

Baltic Sea 112–113
  (*see also*: ice conditions)
Barents Sea 13, 55–61
  (*see also*: ice conditions)
basal glide 6, 76
basal plane 6, 82
Bayes' Theorem 220–222
Beaufort Gyre 21, 29, 35–36, 61
Beaufort Sea 1, 8, 13, 20–21, 32–47
  Alaskan 39, 42
  Canadian 43, 45
  (*see also*: ice conditions)
bergy bit 25
Bering Sea 1, 50
  (*see also*: ice conditions)
Bering Straits 33, 50
break-up 39, 42
bridge piers 113
brine 9, 13–15, 80–84
  channels 15
  migration 14–15
  pockets 13–15
  volume 14
  effect on ice strength 80–84, 146–147
buckling 16, 98, 145–151
  (*see also*: elastic buckling; creep buckling)
bummocks 17–18

C* integral 90
  (*see also*: fracture mechanics)
caisson-retained islands 114–115
  (*see also*: islands)
caliper diameter 44
Canadian Arctic Islands 50–54
  (*see also*: ice conditions)
Canadian Beaufort Sea (*see*: Beaufort Sea; *see also*: ice conditions)
cataclastic failure 95, 99
  (*see also*: crushing)
c-axis 6, 9, 72, 82–83, 86, 88
  and sea currents 10–11
  orientation 10, 88
Chukchi Sea 21, 45, 47–50, 61–63
  (*see also*: ice conditions)
clearing forces 180–181
climate
  long-term temperature trends 67–68
columnar ice 9, 72, 79–83, 86–87
  axisymmetric 9, 82
  oriented 11, 72
  unoriented 11, 72
compliant structures 104, 106–107
compression ridges (*see*: ridges)

compressive strength 70–72, 77, 92–95, 164–167
  biaxial 87–88
  columnar ice
    horizontal 86, 79–83
    vertical 83, 87
  concept of 70–72
  granular ice 79, 82, 88
  multi-axial 96–97
  strain-rate dependence 77–88
  temperature dependence 78–80, 82
  triaxial 96–97, 144
  uniaxial 78, 86
concentration (*see*: ice concentration)
confining pressure 5, 96
constitutive laws 85–88
contact
  imperfect 159–161, 187, 235
  perfect 187
contact factor 160–161, 186–189
continuum behaviour 71, 72–88
  (*see also*: creep; elastic)
Cook Inlet 110–112, 140, 188
Coriolis effect 25, 66
crack
  circumferential 99–100
  cleavage 169–170
  density 93
  initial 88–90, 92, 162
  initiation (*see*: nucleation)
  length 93, 169
  nucleation 88–92
  orientation 93
  propagation 88, 90–92, 92–96
  radial 19, 98, 99, 146, 167–169
  spalling 99, 169–170
  tangential (see: circumferential)
  tensile 149
  (*see also*: fracture)
creep 74–88, 148–161, 205–208
  delayed elastic strain 73, 74, 77
  primary (*see*: delayed elastic strain)
  secondary 74–75, 77–88
  tertiary 75–76, 77
creep buckling 148–151
critical strain 75–76, 88–90, 92, 162–164
crushing 92–96, 99, 113, 125–126, 135, 164–167
crystallography 9, 28
crystal structure 5–7, 9, 28
current 10, 33
  and crystal orientation 10
  drag 25, 185, 200
  patterns 33–35
  velocity 200
  (*see also*: drag)

damage 233
deformation
  continuum 72–88
  delayed elastic 74
  elastic 73, 145–148
  plastic 153
  (*see also*: creep)
deformation mechanisms 76–77
degree-days (*see*: freezing degree-days)
delamination 169–170
delayed elastic strain 73, 74, 77
density
  air 199
  ice 7–8, 21
    pure ice 21
    sea-ice 13
  sea 7–8, 147
  snow 21
design
  code 194
  life 194, 213
design load recommendations
  first-year ice 210–215
  icebergs 226–232
  multi-year ice 222–226
diffusional flow 77
dislocation
  climb 76–77
  glide 76–77, 78
  pile-up 76–77
drag
  coefficients 199–200
  current 25, 185, 200
  wind 25, 185, 199–200
  (*see also*: driving force)
drift
  floe 33, 47, 51, 200–201, 217–218
  icebergs 25–27, 34, 55, 59
  ice islands 35
  ships 35–36
  terminal velocity 200–201
driving force 195, 198–205
  current 200
  limited 204
  pack-ice 143, 204–205
  thermal 201–204
  wind 199–200
ductile behaviour (*see*: creep)
ductile-brittle transition 71, 161–164, 205, 208–209
dynamic amplification 106–107, 111–112
  (*see also*: resonance; compliant structures)
dynamic impact 181–186, 195, 215–232

East Siberian Sea 61–63
  (*see also*: ice conditions)
eccentric
  impact 185–186
  loading 172–174
elastic
  buckling 16, 98, 145–148
  delayed elastic deformation (*see*: creep)
  modulus 73, 74, 146–148

Ellesmere Island 28–29, 51–54, 123
(*see also*: ice islands)
energy
kinetic 185–186, 216–218
stored resonant 106–107
EPR sensor (*see*: sensor)
extensometer 130–131

failure
concept of 70–71
flexural 170–176
non-simultaneous 176–180
(*see also*: creep; damage; fracture)
failure modes 98–100, 145–189
indentation 98–100
large-scale 126, 129, 135–136, 149, 177
uniaxial 72–77
failure probability 191–194
fast-ice (*see*: landfast ice)
finger-rafting 15–16
firn 25, 201
first cracks 88–90, 92, 162
first-year ice 9–16, 32, 42–43, 49–50, 51, 54, 55, 61, 66
(*see also*: ice types)
flaws
distribution 164–165
effect on strength 88–90, 92–95, 164–165
orientation 93
size 93, 164–167
flexible structures 110–113
flexural strength 170–176
flexure 170–176
floe
area 45, 225
distribution 43–46
mean caliper diameter 44, 225
size 44–45, 225
thickness 223–225
forces
on structures 104–144, 211–212 (*see also*: ice forces; load)
driving (*see*: driving forces)
foundation stiffness 145
fracture 88–103, 164–177, 209–210
circumferential 99–100
compressive 92–95
radial 19, 98
tangential (*see*: circumferential)
tensile 88–92
(*see also*: crack; failure; fracture mechanics)
fracture mechanics 88–95
linear 90–95
non-linear 90–91
fracture toughness 90–91
Fram Strait 36, 55, 60
Fram
voyage of 35–36

Franz Josef Land 21, 61–65
frazil ice 8, 10
(*see also*: ice types)
freezing degree-days 42
freezing point
sea water 7
friction
contact 174–176
internal crack 94

gas pockets 13, 80
composition 13
effect on density 13
glaciers 5, 21, 24, 51, 54
Glen's Law 74
(*see also*: creep)
grain-size
dependence of creep 74
nucleation of cracks 76
Grand Banks 1, 33, 55, 58
(*see also*: ice conditions, Newfoundland)
granular ice 72, 79, 82–84
grease ice 8
greenhouse effect 67
Greenland 21, 25, 27
East 55–61
Sea 55–61
West 54–55
growler 25
Gulf of Bothnia 112
Gulf Stream 33, 55
Gurnard 36, 38

Hans Island 123–130
1980 Project 124–127
1981 Project 127–130
1983 Project 123
Hexpack (*see*: sensor)
Hondo bridge 113
Hooke's Law 73
(*see also*: elastic)
hummocks 17, 18

icebergs 21, 24–27, 31–32, 226–232
distribution 34, 46–47, 50–51, 54, 58–59, 61, 65–67
rare sightings 33–34, 65
size 25, 55, 59
types 24–27
bergy bit 25
blocky iceberg 24
growler 25
dome iceberg 25
drydock iceberg 25

ice island 27–29, 31, 46–47
tabular iceberg 24
velocity 25–27, 228–230 (*see also*: drift)
(*see also*: ice conditions)
icebreakers 137–139, 140
ice cap 24
Antarctic 5, 21
Greenland 21
ice characteristics
drift velocity 46–47
floe-size 44–45, 225
irregularities 171–174
temperature profile 13, 207
thickness 36–39, 42, 45–46
ice classification 7–11, 72
ice clearing 180–181
ice concentration 30
ice conditions (regional) 30–69
Antarctica 64–67
Arctic Ocean 35–39
Baltic Sea 112
Barents Sea 55, 60–61, 65
Beaufort Sea 20, 39–47
Bering Sea 50
Canadian Arctic Islands 20, 50–54
Chukchi Sea 47–50
Cook Inlet 110–111
East Siberian Sea 61–65
Grand Banks 55–58
Greenland Sea 55–61
Iceland 67–68
Kara Sea 61
Labrador 54–55, 56–59
Laptev Sea 61–63
Newfoundland 54–55, 56–59
Novaya Zemlya 61
Severnaya Zemlya 61
Soviet Union 61–65
Spitsbergen (*see*: Svalbard)
Svalbard 55–61
ice distribution (*see*: ice conditions)
ice draft 36–39
icefoot 124
ice forces 104–144, 194–196
dynamic 181–186, 195, 215–232
global 195
local 195–196, 232–234
measurement of (*see also*: measurement techniques) 105–110
static 194, 205–215
ice formation
development 9–21
genesis 7–8
ice islands 27–29, 46–47, 51–54
(*see also*: icebergs)
ice loads (*see*: ice forces; *see also*: load)
ice mechanics
brittle 88–95
continuum 72–88
creep 74–88, 148–161
ductile-brittle transition 88, 92, 161–164
elastic 73–74, 145–148
fracture 88–95
scale effects 2, 164–167
ice properties
crystal structure 5–7, 9, 28
density 7–8, 13, 21
gas content 13
mechanical properties 70–103
molecular structure 6
porosity 13
salinity 7–8, 13
thermal diffusivity 202
thermal expansion 201–204
ice rind 8
ice run 113
ice shelf 24
ice thickness 36–39, 42, 45–46, 49–50, 54, 57, 61, 66, 223–224
ice types 5–29
anchor 11
columnar 9
fast (*see*: landfast)
first-year 9–16, 32
frazil 8
granular 8
grease 8
Ih 5
icebergs 24–29
ice islands 27–29, 46–47
infiltration 10–11
landfast 12
multi-year 18–21
old 18
pancake 8
rind 8
S-2 72
second-year 16–18
T-1 72
underwater 9
young 8
Immerk Island (B-48) 114
impact
dynamic 181–186, 195, 215–232
eccentric 185–186
probability 218–222
impact hammer 143–144
imperfect contact (*see*: contact; *see also*: non-simultaneous failure)
impurities
in sea ice 11
in ice crystals 7
inclusion factor 108–109
indentation 97–103, 151–161
full-scale tests 101–103, 104–144
laboratory tests 98–101
indentation factor 154–155, 157–159
infiltration ice 10–11

in situ
tests 101–103
measurement techniques 107–110
islands
artificial 113–122, 130–137
caisson-retained 114–115
gravel 114–115
sand 114–115
isotropy 72, 154
Issungnak Island (O-61) 114, 120

J-integral 90

Kadluk Island 114, 121–122
Kannerk Island (G-42) 114, 119–120
Kara Sea 10, 61–65
(*see also*: ice conditions)
keel 15, 21, 36
(*see also*: sail-to-keel ratio)
Korzhavin's equation 160, 186–189
Kulluk 138, 196

Labrador
Current 33
Sea 1, 33
(*see also*: ice conditions)
landfast ice 12, 39, 42–43, 57, 194, 199
(*see also*: ice types; ice conditions)
Laptev Sea 7
(*see also*: ice conditions)
lighthouses 112–113, 140
load
design 190–235
global 195
life 190
local 138–139, 195–196, 232–234
peak 138–144, 210–213
ratio of mean to peak 216
total (*see*: global)
types 194–196
loading-rate 162–164, 209
local loads (*see*: load)
lower bound theorem 154

MAC (Mobile Arctic Caisson) (*see*: Molikpaq)
M'Clure Strait 35, 54
mean caliper diameter (*see*: floe)
measurement techniques 105–110
accelerometers 110
extensometers 130–131
strainmeters 109–110
stress gauges 107–110
structural response 106–107, 130
Medof panel (*see*: sensors)
meltpools 18, 28
meso-scale models 140–143
modelling
physical 4, 198
numerical 181–186
meso-scale 140–143
Molikpaq 130–137
structure 130–131
at Tarsiut (P-45) 131–133
at Amauligak (I-65) 133–137
Monte Carlo simulation 214
multi-axial behaviour 96–97, 144
multi-legged structures 198
multi-year ice 18–21, 31, 35, 43–46, 47, 50, 51, 54, 55–59, 61, 66 (*see also*: ice types)

net section stress 80–81
Netserk Island
South (N-44) 117
North (F-40) 117–119, 163–164
Newfoundland 1, 32, 54–55
(*see also*: ice conditions)
Newton's Second Law of motion 110, 123
Newtonian flow 153
non-simultaneous failure 100, 176–189, 205
Norton's Law 74
(*see also*: creep)
Novaya Zemlya 61
(*see also*: ice conditions)
nucleation
process 70, 88–90, 92
stress 90, 92
(*see also*: cracks)
Nutcracker tests 140

operating season 39–42
orthotropic properties 82, 154

pack-ice 18–19, 35–36, 38, 143, 204–205
(*see also*: driving force; modelling)
pancake ice 8
(*see also*: ice types)
Pemridge bridge 113
penetration distance 216–218
pile-up 114
plane strain 97–98, 154–155
plane stress 97–98, 154–155
plasticity
theory 152–157
yield stress 153, 155–156

platelets 9
Poisson process 214, 225
Poisson's ratio 73
polymorphs of ice 5
polynyas 21, 50
porosity 13
pressure-area curve 3, 139–144, 205, 210, 233–234
pressure ridges 15–23
primary creep 74 (*see also*: delayed elastic strain)
probability density 190–195
probability distributions
  exponential 223
  extremal 213
  lognormal 213, 215
  normal 192–193, 212–213
probability
  of failure 191–194
  of impact 218–222
propagation of cracks 88
  in compression 92–95
  in tension 90–92
  radial 167–169

Queen Elizabeth Islands 1, 20–21, 50–54
  (*see also*: Canadian Arctic Islands; ice conditions)

rafting 11, 15–16
recommendations (*see*: design load recommendations)
reference stress method 157–159
resonance 111, 133
response (*see*: structural response)
ride-up 4, 114, 174–176
ridges 15–23
  compression 15
  first-year 15–16, 17
  multi-year 20–21, 22–23
  second-year 20
  shear 15, 16
  (*see also*: ice conditions)
risk
  analysis 190–194
  of failure 191–192, 194
Ross Ice Shelf 24, 67
Ross Sea 64
roughness
  surface 199
rubble pile 111, 114–115, 196–198

S-2 ice 72
safety factor 194, 233–234
sail 15
sail-to-keel ratio 20–23, 36–39
St Lawrence Island 21, 50
salinity 7, 8, 13, 18, 21
Sargo 46, 51
scaling 164–167
scale effects 2, 91, 139–144
sea ice 9–21
  (*see also*: ice conditions; ice types; ice mechanics)
Seadragon 46, 51
secondary creep (*see*: creep)
second-year ice 16–18
  (*see also*: ice types)
sediment-laden ice 11
sensors
  stress 105–109
    ARCO sensor 107
    BP sensor 107
    cylindrical 109, 121
    EPR sensor 105, 122
    Esso sensor 105, 118
    Hexpack 105
    Medof panel 105, 130–131
  strainmeters 109–110
  (*see also*: measurement techniques)
Severnaya Zemlya 7, 10, 21, 61–65
  (*see also*: ice conditions)
shape factor 186
ships 35–36, 104, 144
  (*see also*: icebreakers)
sikussak 57
sloping-sided structures 174–176, 196
snow 8, 10, 20, 21, 199, 201
Soviet Union 61–65
  (*see also*: ice conditions)
spalling 99, 169–170
Spitsbergen (*see*: Svalbard)
spray ice 114–115
SSDC (Single Steel Drilling Caisson) 18, 114, 116
stamukhi 39, 42
static loading 194–195, 205–215
strain
  critical (*see*: critical strain)
  delayed elastic (*see*: delayed elastic strain)
strain-rate
  constant 72
  minimum 74–75
  (*see also*: creep; ice mechanics)
stress
  deviator 96
  for failure 70
  hydrostatic 5, 96
  tensor 85
stress sensor (*see*: sensors)
structures 196–198
  caisson-retained 114–115
  compliant 110–113
  multi-legged 198
  narrow 110–113
  resonant 196
  wide 113–137
  (*see also*: islands)

structural response 111, 133, 196
Svalbard 21, 55–61
  (*see also*: ice conditions)

Tarsiut Island
  (N-44) 114, 120–121, 149
  (P-45) 114, 131–133
temperature profile 13
tensile strength 78, 88–92
  (*see also*: ice mechanics; strength)
tensor
  stress 85
  stress deviator 96
  second invariant 85
terminal velocity 200–201
tertiary creep 75–76, 77
testing methods 71–72
theoretical analysis 145–189
thermal
  conductivity 9
  diffusivity 202
  expansion 201–204
  stress 201, 204
thickness (*see*: ice thickness)
transition
  ductile-brittle 71, 161–164
Transpolar Drift 35
Tresca yield criterion 153–154
  (*see also*: yield)
triaxial strength (*see*: multi-axial behaviour)

underwater ice 9
uniaxial
  compression 71, 78–86
  tension 71–78
upper bound
  theorem 154
  to ice forces 206
Uviluk (P-66) 114, 149
USSR (*see*: Soviet Union)

velocity 25–27, 46–47, 221–222, 228–230
  (*see also*: drift)
von Mises yield criterion 153–154 (*see also*: yield)

Ward Hunt Ice Shelf 28–29
water
  fresh 5–7
  sea 7–8
Weddell Sea 11, 64, 66
  (*see also*: Antarctica)
Weibull statistics 166
wind drag (*see*: drag)
winter blocking 194–195, 204–205
Wrangel Island 48–49, 61

yield
  concept of 70–72, 152–153
  plastic 153
  stress 155
  Tresca criterion 153–154
  von Mises criterion 153–154
Ymer 57
young ice 8
  (*see also*: ice types)
Young's modulus 73, 74, 146–148 (*see also*: elastic)